NANOINDENTATION IN MATERIALS SCIENCE

Edited by **Jiří Němeček**

Nanoindentation in Materials Science
http://dx.doi.org/10.5772/2829
Edited by Jiří Němeček

Contributors

Qing Peng, Li Ma, Lyle Levine, Ron Dixson, Douglas Smith, David Bahr, Jaroslav Menčík, Jiří Němeček, L. Zhang, T. Ohmura, K. Tsuzaki, Bruno A. Latella, Michael V. Swain, Michel Ignat, Ksenia Shcherbakova, Akiko Hatakeyama, Yosuke Amemiya, Nobuo Shimamoto, Bruno B. Lopes, Rita C.C. Rangel, César A. Antonio, Steven F. Durrant, Nilson C. Cruz, Elidiane C. Rangel, Mamadou Diobet DIOP, Vasarla Nagendra Sekhar, Ondřej Jiroušek, Hongwei Zhao, Hu Huang, Zunqiang Fan, Zhaojun Yang, Zhichao Ma

Published by InTech

Janeza Trdine 9, 51000 Rijeka, Croatia

Copyright © 2012 InTech

Notice

Publishing Process Manager Oliver Kurelic
Technical Editor InTech Prepress, Novi Sad
Cover InTech Design team

Additional hard copies can be obtained from orders@intechopen.com

Nanoindentation in Materials Science, Edited by Jiří Němeček
p. cm.
ISBN 978-953-51-0802-3

For EU product safety concerns:
IN TECH d.o.o., Prolaz Marije Krucifikse Kozulić 3, 51000 Rijeka, Croatia,
info@intechopen.com or visit our website at intechopen.com.

Meet the editor

Dr. Jiří Němeček is a senior researcher who currently works as an associate professor at the Department of Mechanics, Faculty of Civil Engineering, Czech Technical University in Prague (CTU). He has been involved in micromechanical testing and modeling of structural materials by nanoindentation and AFM since the year 2000 when he defended his doctorate. He is a group leader at the Center of Nanotechnology in Civil Engineering found in 2010 at CTU. His main research interests include micro-level characterization of cementitious and other heterogeneous structural materials by means of combined experimental and numerical approach. He has organized or co-organized several nanotechnology related conferences (Nanotechnology in Construction, Local Mechanical Properties) where he also serves in scientific boards (Nanotechnology in Cement and Concrete). He has authored numerous scientific papers published in international journals and edited several conference proceedings.

Contents

Preface XI

Section 1 Testing Methodologies, Principles, Sources of Errors and Uncertainties 1

Chapter 1 **First-Principles Quantum Simulations 3**
Qing Peng

Chapter 2 **Effect of the Spherical Indenter Tip Assumption on the Initial Plastic Yield Stress 25**
Li Ma, Lyle Levine, Ron Dixson, Douglas Smith and David Bahr

Chapter 3 **Uncertainties and Errors in Nanoindentation 53**
Jaroslav Menčík

Section 2 Nanomechanical Performance of Composite Materials 87

Chapter 4 **Nanoindentation Based Analysis of Heterogeneous Structural Materials 89**
Jiří Němeček

Chapter 5 **Application of Nanoindentation Technique in Martensitic Structures 109**
L. Zhang, T. Ohmura and K. Tsuzaki

Section 3 Nanoindentation on Thin Layers and Films 131

Chapter 6 **Indentation and Fracture of Hybrid Sol-Gel Silica Films 133**
Bruno A. Latella, Michael V. Swain and Michel Ignat

Chapter 7 **Nanoindentation as a Tool to Clarify the Mechanism Causing Variable Stiffness of a Silane Layer on Diamond 161**
Ksenia Shcherbakova, Akiko Hatakeyama, Yosuke Amemiya and Nobuo Shimamoto

Chapter 8 **Mechanical and Tribological Properties of Plasma Deposited a-C:H:Si:O Films 179**
Bruno B. Lopes, Rita C.C. Rangel, César A. Antonio, Steven F. Durrant, Nilson C. Cruz, Elidiane C. Rangel

Section 4 Characterization of Microdevices and Integrated Circuits 203

Chapter 9 **Characterization of Microdevices by Nanoindentation 205**
Mamadou Diobet DIOP

Chapter 10 **Mechanical Characterization of Black Diamond (Low-k) Structures for 3D Integrated Circuit and Packaging Applications 229**
Vasarla Nagendra Sekhar

Section 5 Nanoindentation on Biological Materials 257

Chapter 11 **Nanoindentation of Human Trabecular Bone – Tissue Mechanical Properties Compared to Standard Engineering Test Methods 259**
Ondřej Jiroušek

Section 6 Design of New Nanoindentation Devices 285

Chapter 12 **Design, Analysis and Experiments of a Novel *in situ* SEM Indentation Device 287**
Hongwei Zhao, Hu Huang, Zunqiang Fan, Zhaojun Yang and Zhichao Ma

Preface

The book "Nanoindentation in Materials Science" is dedicated to the widespread audience ranging from students to scientists and industrial researchers to help them with orientation in a variety of up-to-date topics related to the field of nanoindentation. Nanoindentation, as a universal experimental technique for measuring mechanical properties in small volumes, has attracted strong attention between the material scientists. Nowadays, nanoindentation that was formerly dedicated to pure homogeneous materials has infiltrated to quite distant areas (e.g. textile composites, cement industry, electronics, biomechanics and others). Everyday, it helps researches to study intrinsic properties of low level material units. The variety of potential topics is reflected also in this book where multiple phenomena are discussed.

The book contains twelve chapters divided into six sections. Each chapter gives the reader basic overview of the solved problem and relevant information about the topic. The subject is elaborated in details and contains original author's ideas and new solutions.

The first section, which is devoted to testing methodologies, principles, sources of errors and uncertainties, contains a contribution about the first principles of quantum simulations, a chapter on theoretical aspects, basic deformation mechanisms and evaluation of results in spherical indentation and a valuable chapter about the critical aspects in data evaluation, limitations in measurements and extensive review of error sources.

The second section deals with nanomechanical properties of composite materials. Their heterogeneity is in focus and different approaches how to solve this problem is proposed in the contributions that find applications in structural materials (e.g. cement, gypsum and metals).

The third section contains contributions from the area of thin layers and films. Novel hybrid sol-gel silica films are investigated and a biomolecular layer (aminosilane) formed on diamond surfaces with a potential use in nanomachines is studied. The section is complemented by a contribution about plasma sprayed films.

The fourth section solves problems in characterization of microdevices (e.g. MEMS) and integrated circuits. The fifth section is devoted to nanoindentation applied in the study of human trabecular bone linking –up the micro and meso-material scales. Finally, the sixth section deals with the development of a novel in-situ device, its design, performance and calibration aspects.

Hopefully, the book will be inspiring source of information for the readers which will help them to improve their own research in their fields of study.

Jiří Němeček
Faculty of Civil Engineering,
Czech Technical University in Prague,
Czech Republic

Section 1

Testing Methodologies, Principles, Sources of Errors and Uncertainties

Chapter 1

First-Principles Quantum Simulations of Nanoindentation

Qing Peng

Additional information is available at the end of the chapter

http://dx.doi.org/10.5772/48190

1. Introduction

Modern nanotechnology can build advanced functional materials in molecular level in nanoscales. The computer modeling and simulations are very important in such materials design. The ability to perform quantum simulations of materials properties over length scales that are relevant to experiments represents a grand challenge in computational materials science. If one could treat multi-millions or billions of electrons *effectively* at micron scales, such first-principle quantum simulations could revolutionize materials research and pave the way to the computational design of advanced materials.

There are two principal reasons why quantum simulations at relevant experimental scales are important. First of all, it allows a direct comparison between theory and experiment. For example, the rapidly emerging field of nanotechnology demands realistic and accurate modeling of material systems at the nanoscale, including nano-particles, nano-wires, quantum dots, NEMS (Nano-Eletro-Mechanical Systems) and MEMS (Micro-Electro-Mechanical Systems). All these nano-systems could reach a length scale of microns and contain millions or billions of electrons, if not more. Secondly, quantum simulations at larger scales are essential even for extended bulk crystals where periodic boundary conditions may be used. This is due to the fact that a real bulk solid always contains lattice defects (or impurities) whose interactions are long range - dislocations being the prominent example. An insufficiently large periodic unit cell would lead to unrealistically high concentrations of defects and/or impurities, rendering the results of such simulations questionable.

The full knowledge of the mechanical properties of the nano-materials are very important for the advanced materials design and applications. Nanoindentation has now become a standard experimental technique for evaluating the mechanical properties of thin film materials and bulk materials in small volumes [25]. As it can measure nanometer penetration length scales, nanoindentation is an indispensable tool to assess elastic moduli and hardness of materials. It also can be used to derive strain-hardening exponents, fracture toughness, and viscoelastic properties of materials [6].

In this chapter, we will present our recent studies of nano-indentation with first-principles quantum simulations. We developed a multi-scale approach that is based *entirely* on density functional theory (DFT) [18, 19] and allows quantum simulations at micron scale and beyond. The method, termed Quasi-continuum Density Functional Theory (QCDFT) [28–31], combines the coarse graining idea of the quasi-continuum (QC) approach and the coupling strategy of the quantum mechanics/molecular mechanics (QM/MM) method, and allows quantum simulations at the micron scale and beyond.

It should be stated at the outset that QCDFT is *not* a brute-force electronic structure method, but rather a multiscale approach that can treat large systems - effectively up to billions of electrons. Therefore, some of the electronic degrees of freedom are reduced to continuum degrees of freedom in QCDFT. On the other hand, although QCDFT utilizes the idea of QM/MM coupling, it does not involve any classical/empirical potentials (or force fields) in the formulation - the energy calculation of QCDFT is entirely based on orbital-free DFT (OFDFT). This is an important feature and advantage of QCDFT, which qualifies it as a bona fide quantum simulation method.

The QCDFT method have been used to study the nanoindentation of pure Al [30], random distributed Mg impurities in Al thin film [31], and solid solution effects on dislocation nucleation during nanoindentation [29]. The method was also applied to study the fracture in Al recently [28]. Although these works had been published before individually, a small review of this method and its applications is necessary. Here I will give an introduction to QCDFT method in Section II, especially how QCDFT evaluate the energies and forces from *ab initio* first-principles calculations. The application of QCDFT in quantum mechanical simulations of nanoindentations will be illustrated in Section III, IV and V for pure Al [30], random distributed Mg impurities in Al thin film [31], and solid solution effects on dislocation nucleation during nanoindentation [29] respectively. And finally our conclusions and outlooks of quantum mechanical simulations of nanoindentations in Sec. VI.

2. QCDFT methodology

In QCDFT, the degrees of freedom of the system is reduced by replacing the full set of N atoms with a small subset of N_r "representative atoms" or *repatoms* ($N_r \ll N$) that approximate the total energy through appropriate weighting. This approach of reducing degrees of freedom is critical in the multiscale method and it is adopted directed from the Quasi-continuum method [32, 37, 40]. The energies of individual repatoms are computed in two different ways depending on the deformation in their immediate vicinity. Atoms experiencing large variations in the deformation gradient on an atomic scale are computed in the same way as in a standard atomistic method. In QC these atoms are called *nonlocal* atoms. In contrast, the energy of atoms experiencing a smooth deformation field on the atomic scale is computed based on the deformation gradient **G** in their vicinity as befitting a continuum model. These atoms are called *local* atoms because their energy is based only on the deformation gradient at the point where it is computed. In a classical system where the energy is calculated based on classical/empirical inter-atomic potentials, the total energy E_{tot} can be written

$$E_{\text{tot}}^{\text{QCDFT}} = \sum_{i=1}^{N^{\text{nl}}} E_i(\mathbf{R}) + \sum_{j=1}^{N^{\text{loc}}} n_j E_j^{\text{loc}}(\mathbf{G}). \tag{1}$$

The total energy has been divided into two parts: an atomistic region of N^{nl} nonlocal atoms and a continuum region of N^{loc} local atoms ($N^{\text{nl}} + N^{\text{loc}} = N_r$). The calculation in the nonlocal region is identical to that in atomistic methods with the energy of the atom depending on the coordinates **R** of the surrounding repatoms. Rather than depending on the positions of neighboring atoms, the energy of a local repatom depends on the deformation gradients **G** characterizing the finite strain around its position. Then the energies and forces on the local atoms and nonlocal atoms are treated differently, but entirely on *ab inito* first-principle calculations.

In the following, we outline the energy and force formulations for both local and nonlocal regions of QCDFT. In the local region, a finite element mesh is constructed with each repatom is on the vertex of several surrounding finite elements. The energy and force of each local repatom can be obtained from the strain energy density and the stress tensor of the finite elements that share the same repatom. More specifically, according to the Cauchy-Born rule, the deformation gradient **G** is uniform within a finite element, therefore the local energy density ε and the stress tensor for each finite element can be calculated as a perfect infinite crystal undergoing a uniform deformation specified by **G**. In other words, one could perform an OFDFT-based energy/stress calculation for an infinite crystal by using periodic boundary conditions with the primitive lattice vectors of the deformed crystal, $\mathbf{h}_i$ given by

$$\mathbf{h}_i = \mathbf{G}\,\mathbf{H}_i, \qquad\qquad i = 1, 2, 3. \tag{2}$$

Here $\mathbf{H}_i$ are the primitive lattice vectors of the undeformed crystal and the volume of the primitive unit cell is Ω_0. The details of the OFDFT calculation can be found in Sec. II C. Once the strain energy density $\varepsilon(\mathbf{G}_k)$ is determined, the energy contribution of the jth local repatom is given as

$$E_j^{\rm loc}(\{\mathbf{G}\}) = \sum_{k=1}^{M_j} w_k \varepsilon(\mathbf{G}_k)\Omega_0, \tag{3}$$

where M_j is the total number of finite elements represented by the jth repatom, and w_k is the weight assigned to the kth finite element. The force on the jth local repatom is defined as the gradient of the total energy with respect to its coordinate $\mathbf{R}_j^{\rm loc}$. In practice, the nodal force on each finite element is calculated from the stress tensor of the finite element by using the principle of virtual work [47]. The force on the repatom is then obtained by summing the nodal force contributions from each surrounding finite elements.

For the energy/force calculation in the nonlocal region, we resort to a novel QM/MM approach that was developed recently for metals [43, 44]. The coupling between the QM and MM regions is achieved quantum mechanically within an OFDFT formulation. Although the detailed implementation of the QM/MM approach is presented in Sec. II D, we wish to stress two important points here: (1) The original QC formulation assumes that the total energy can be written as a sum over individual atomic energies. This condition is not satisfied by quantum mechanical models. The energy of the nonlocal region is now a functional of total electron density, so instead of the expression in Eq. 1, the total energy of QCDFT should be expressed as:

$$E_{\rm tot}^{\rm QCDFT} = E^{\rm nl}[\rho^{\rm tot}] + \sum_{j=1}^{N^{\rm loc}} n_j E_j^{\rm loc}(\{\mathbf{G}\}). \tag{4}$$

Here ρ^{tot} is the total electron density in the nonlocal region as well as the coupling nonlocal/local region i.e., the buffer region in the following discussion. (2) The nonlocal energy, $E^{\rm nl}$ should be calculated with appropriate boundary conditions; that is to say, it should include the interaction energy between the nonlocal atoms and neighboring local atoms. In the original QC framework, this requirement is realized by including dummy atoms in the energy/force calculation of a given nonlocal repatom. These dummy atoms are in the local region and within the cut-off radius of the given nonlocal repatom. The dummy atoms are not independent degrees of freedom in the local region, but rather slaves to the local repatoms. In this way, the nonlocal calculation is carried out with the appropriate boundary conditions, and at the same time, the energy of the dummy atoms is still treated with the Cauchy-Born rule, consistent with their status. In the QCDFT approach, a buffer region including the dummy atoms and local repatoms that are adjacent to the nonlocal repatoms is selected as the "MM" region, and the nonlocal atoms constitute the QM region. The nonlocal atoms are treated by OFDFT, and the coupling between the "MM" and QM region is also formulated within OFDFT. Therefore the entire system is formulated with one energy functional, OFDFT. Note that "MM" here is actually a misnomer: the local atoms are treated by OFDFT with the Cauchy-Born rule as mentioned earlier, and we retain the designation "MM" solely to indicate the similarity to the earlier coupling scheme [43, 44].

The nonlocal region is modeled at the atomistic level with a QM/MM approach. In a typical QM/MM calculation, the system is partitioned into two separated domains: a QM region and an MM region. In QCDFT, the QM region refers to the nonlocal region and the MM region refers to the buffer region. The buffer region is introduced to provide the boundary conditions for the calculation of nonlocal energy and it contains both dummy atoms and local repatoms. The dummy atoms differ from the local and nonlocal repatoms in the following sense: (1) their positions are interpolated from the positions of local repatoms using finite element shape functions; (2) the energy and force on the dummy atoms does not need to be considered explicitly since they are not explicit degrees of the freedom in the QC formulation.

At present, there are two types of QM/MM coupling strategies: mechanical coupling and quantum coupling [7, 22, 24, 34, 36, 45]. The interaction energy between the QM and MM regions is formulated at the MM level for mechanical coupling, and at the QM level for quantum coupling. Although mechanical coupling is much simpler than quantum coupling, it has many drawbacks - the most important one being that the electronic interaction between the two regions is ignored. For example, with mechanical coupling, electrostatic, kinetic, and exchange-correlation interaction energies are not considered explicitly. As a consequence, the physics of the QM region is not accurately captured. Another problem with mechanical coupling is that the reliability and availability of empirical potentials for treating the coupling are severely limited. In contrast, quantum coupling should be more accurate as it accounts for all quantum mechanical interaction terms. Depending on the level of the quantum description, the extent of the electronic coupling varies from merely long-range electrostatic interaction to a full Coulomb interaction, including short-ranged exchange-correlations [7, 22].

In the present QCDFT method, we use an OFDFT-based quantum mechanical QM/MM coupling proposed by Zhang and Lu [43, 44] which considers the full Coulomb, kinetic energy, and exchange-correlation interactions. More specifically, both the energy of the nonlocal atoms and the interaction energy between the nonlocal atoms and the buffer atoms are calculated by OFDFT. To simply the notation, we denote the nonlocal region as region I, and the buffer region as region II. Typically, the buffer region consists of several atomic layers surrounding the nonlocal region. The nonlocal energy E^{nl} as defined in Eq. (4) can be expressed as:

$$E^{\mathrm{nl}}[\rho^{\mathrm{tot}}] = \min_{\rho^{\mathrm{I}}}\{E_{\mathrm{OF}}[\rho^{\mathrm{I}};\mathbf{R}^{\mathrm{I}}] + E^{\mathrm{int}}_{\mathrm{OF}}[\rho^{\mathrm{I}},\rho^{\mathrm{II}};\mathbf{R}^{\mathrm{I}},\mathbf{R}^{\mathrm{II}}]\}. \tag{5}$$

where $\mathbf{R}^{\mathrm{I}}$ and $\mathbf{R}^{\mathrm{II}}$ denote ionic coordinates in region I and II respectively. The OFDFT energy functional E_{OF} is obtained routinely [13–15, 21, 23, 30, 41]. The total charge density of the QM/MM system ρ^{tot} consists of two contributions: $\rho^{\mathrm{tot}} = \rho^{\mathrm{I}} + \rho^{\mathrm{II}}$ where ρ^{I} and ρ^{II} represent the charge density from region I and II respectively. While ρ^{I} is determined self-consistently by minimizing the total energy functional Eq. (5), ρ^{II} is defined as a superposition of atom-centered charge densities ρ^{at} via $\rho^{\mathrm{II}}(\mathbf{r}) = \sum_{i\in \mathrm{II}}\rho^{\mathrm{at}}(\mathbf{r}-\mathbf{R}_i)$. Note that ρ^{at} is spherically symmetric and can be constructed *a priori*. It is important to point out that ρ^{at} is *not* a charge density of an isolated atom, but rather an atom-centered charge density whose superposition gives rise to the correct bulk density of region II [43]. Therefore $\rho^{\mathrm{II}}(\mathbf{r})$ is fixed for a given ionic configuration of region II and it changes upon the motion of region II ions. In other words, the electronic degree of freedom in the formulation is ρ^{I} only and ρ^{II} is fixed during the electronic relaxation. The interaction energy is thus defined as following:

$$\begin{aligned} E^{\mathrm{int}}_{\mathrm{OF}}[\rho^{\mathrm{I}},\rho^{\mathrm{II}};\mathbf{R}^{\mathrm{I}},\mathbf{R}^{\mathrm{II}}] = {} & E_{\mathrm{OF}}[\rho^{\mathrm{tot}};\mathbf{R}^{\mathrm{tot}}] - E_{\mathrm{OF}}[\rho^{\mathrm{I}};\mathbf{R}^{\mathrm{I}}] \\ & - E_{\mathrm{OF}}[\rho^{\mathrm{II}};\mathbf{R}^{\mathrm{II}}], \end{aligned} \tag{6}$$

where $\mathbf{R}^{\mathrm{tot}} \equiv \mathbf{R}^{\mathrm{I}} \bigcup \mathbf{R}^{\mathrm{II}}$. The energy functional of Eq. (5) can be written as

$$E^{\mathrm{nl}}[\rho^{\mathrm{tot}}] = \min_{\rho^{\mathrm{I}}}\{E_{\mathrm{OF}}[\rho^{\mathrm{tot}};\mathbf{R}^{\mathrm{tot}}] - E_{\mathrm{OF}}[\rho^{\mathrm{II}};\mathbf{R}^{\mathrm{II}}]\}. \tag{7}$$

A basic ansatz of the present QM/MM formulation (Eq. (7)) is that ρ^{I} must be confined within a finite volume (Ω^{I}) that is necessarily *larger* than region I but much smaller than the entire QM/MM region. In addition, since some terms in the formulation Eq. (7) could be more efficiently computed in reciprocal space (discussed in the following), we also introduce a volume Ω^{B} over which the periodic boundary conditions are applied. The periodic box Ω^{B} should be larger than Ω^{I} so that ρ^{I} does not overlap with its periodic images [43]. Note that the QM/MM system is only a small fraction of the entire QCDFT system. To facilitate the introduction of the QCDFT method, we present a schematic diagram in Fig. (1) which demonstrates the typical partition of domains in a QCDFT calculation. This particular example is for a nanoindentation calculation of an Al thin film which is used to validate the QCDFT method (see Sec. III for details). The lower-right panel shows the entire system and the corresponding finite element mesh. The lower-left panel is a blow-up view of the entire system, which is further zoomed in as shown in the upper-left panel, focusing on the nonlocal region. The upper-right panel shows out-of-plane displacements of the nonlocal atoms, where the dislocations and the stacking faults are clearly visible. All lengths are given in Å. The blue and green circles represent the nonlocal and buffer atoms, respectively. The volumes Ω^{I} and Ω^{B} are represented by the black dash box and solid box in the upper-left panel, respectively. There is no constraint on ρ^{II}, which can extend to the entire QM/MM system. In addition to its computational efficiency as discussed in Sec. II B, OFDFT allows Eq. (7) to be evaluated over Ω^{I} rather than over the entire QM/MM system as Eq. (7) appears to suggest [3, 43]. This significant computational saving is due to the cancellation in evaluating the first and second term of Eq. (7), and it is rendered by the orbital-free nature of OFDFT and the localization of ρ^{I}.

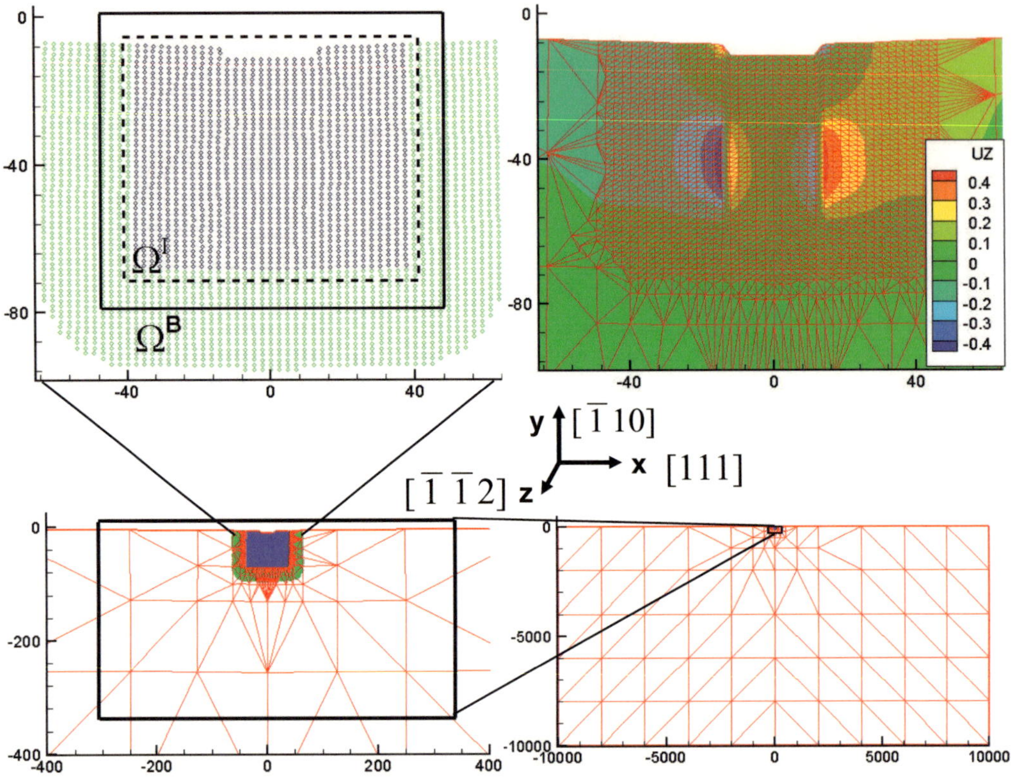

Figure 1. (Color online) The overview of the entire system and domain partition in QCDFT with nanoindentation as an example. The x, y and z axes are along [111],[$\bar{1}$10], and [$\bar{1}\bar{1}$2], respectively. Ω^I and Ω^B are 2.8 Å and 8 Å beyond the nonlocal region in $\pm x$ and $\pm y$ directions, respectively [43]. The colors indicate u_z, the out-of-plane displacement of atoms in the z-direction.

2.1. Ghost forces

In the QC method, the entire system is divided into two regions - local and nonlocal - and thus an artificial interface is introduced. The atoms in these two regions are modeled differently: in the local continuum region, the energy depends only on the deformation gradient, while in the nonlocal atomistic region, the energy depends on the position of the atoms. There is an inherent mismatch of the energy functional between the local and nonlocal regions. As a result, a well-defined energy functional for the entire QC model will lead to spurious forces near the interface, called "ghost forces" in the QC literature. Note that the ghost force only exists on local repatoms adjacent to the local/nonlocal (or QM/MM) interface. The principal reason for the ghost force is that we choose to focus on approximating the energy and not the force. One could opt to avoid the ghost force by formulating the force appropriately, but then one could no longer define an appropriate total energy of the system. There are two advantages of having a well-defined energy in atomistic simulations: (1) it is numerically more efficient to minimize energy, compared to the absolute value of a force; (2) one can potentially obey an energy conservation law in dynamical simulations.

In the QCDFT (or the QM/MM) approach, the ghost force is defined as the force difference between two distinctive formulations: (1) where the force is calculated by applying the

Cauchy-Born rule throughout the entire system; this corresponds to a "consistent" way of calculating force, thus no ghost force exists and (2) where the force is calculated based on the mixed local/nonlocal formulation aforementioned and hence the ghost force exists. In the first case, the force on a local repatom would be

$$\tilde{\mathbf{F}}(\mathbf{R}^{\mathrm{II}}) = -\frac{\partial E_{\mathrm{CB}}(\mathbf{R}^{\mathrm{tot}})}{\partial \mathbf{R}^{\mathrm{II}}}, \tag{8}$$

where $E_{\mathrm{CB}}(\mathbf{R}^{\mathrm{tot}})$ is the total energy of the system where the Cauchy-Born rule is used throughout. In the second case, the total energy of the QM/MM system can be written as:

$$E^{\mathrm{tot}}(\mathbf{R}^{\mathrm{tot}}) = E_{\mathrm{CB}}(\mathbf{R}^{\mathrm{II}}) + E_{\mathrm{QM}}(\mathbf{R}^{\mathrm{I}}) + E^{\mathrm{int}}_{\mathrm{QM}}(\mathbf{R}^{\mathrm{tot}}), \tag{9}$$

where $E_{\mathrm{CB}}(\mathbf{R}^{\mathrm{II}})$ is the local energy computed using the Cauchy-Born rule, $E_{\mathrm{QM}}(\mathbf{R}^{\mathrm{I}})$ is the nonlocal energy computed by a quantum mechanical approach, and $E^{\mathrm{int}}_{\mathrm{QM}}(\mathbf{R}^{\mathrm{tot}})$ is the quantum mechanical interaction energy. The force derived from this total energy functional is

$$\begin{aligned} \mathbf{F}(\mathbf{R}^{\mathrm{II}}) &= -\frac{\partial E_{\mathrm{CB}}(\mathbf{R}^{\mathrm{II}})}{\partial \mathbf{R}^{\mathrm{II}}} - \frac{\partial E^{\mathrm{int}}_{\mathrm{QM}}(\mathbf{R}^{\mathrm{tot}})}{\partial \mathbf{R}^{\mathrm{II}}} \\ &= -\frac{\partial E_{\mathrm{CB}}(\mathbf{R}^{\mathrm{tot}})}{\partial \mathbf{R}^{\mathrm{II}}} + \frac{\partial E^{\mathrm{int}}_{\mathrm{CB}}(\mathbf{R}^{\mathrm{tot}})}{\partial \mathbf{R}^{\mathrm{II}}} - \frac{\partial E^{\mathrm{int}}_{\mathrm{QM}}(\mathbf{R}^{\mathrm{tot}})}{\partial \mathbf{R}^{\mathrm{II}}} \\ &= \tilde{\mathbf{F}}(\mathbf{R}^{\mathrm{II}}) + \mathbf{F}^{\mathrm{ghost}}, \end{aligned} \tag{10}$$

where $E^{\mathrm{int}}_{\mathrm{CB}}(\mathbf{R}^{\mathrm{tot}}) = E_{\mathrm{CB}}(\mathbf{R}^{\mathrm{tot}}) - E_{\mathrm{CB}}(\mathbf{R}^{\mathrm{I}}) - E_{\mathrm{CB}}(\mathbf{R}^{\mathrm{II}})$ is the interaction energy calculated with the Cauchy-Born rule applied to the entire system, and $\mathbf{F}^{\mathrm{ghost}} = \frac{\partial\{E^{\mathrm{int}}_{\mathrm{CB}}(\mathbf{R}^{\mathrm{tot}}) - E^{\mathrm{int}}_{\mathrm{QM}}(\mathbf{R}^{\mathrm{tot}})\}}{\partial \mathbf{R}^{\mathrm{II}}}$ is the ghost force. Note that $E_{\mathrm{CB}}(\mathbf{R}^{\mathrm{I}})$ does not depend on $\mathbf{R}^{\mathrm{II}}$, hence its contribution to force is zero.

Having determined the ghost force in QCDFT, one can correct for it by adding a correction force $\mathbf{F}^{\mathrm{corr}} = -\mathbf{F}^{\mathrm{ghost}}$ on the local atoms ($\mathbf{F}^{\mathrm{corr}} + \mathbf{F}(\mathbf{R}^{\mathrm{II}})$) so that the resultant force is $\tilde{\mathbf{F}}(\mathbf{R}^{\mathrm{II}})$. The correction force is applied to the local repatoms as a dead load, computed each time the status of the representative atoms is updated, and remains fixed until the next update required due to the evolving state of deformation. In practice, the correction force is nonzero only for the local repatoms adjacent to the nonlocal region. Finally, the total energy expression has to be modified accordingly so that its gradient is consistent with the new formulation of force. This is achieved by incorporating the work done by the correction force into the original energy formulation:

$$\tilde{E}^{\mathrm{tot}} = E^{\mathrm{tot}} - \sum_{\alpha}^{N'_{\mathrm{rep}}} \mathbf{F}^{\mathrm{corr}}_{\alpha} \cdot \mathbf{u}_{\alpha}, \tag{11}$$

where N'_{rep} denotes the number of local repatoms whose correction force is nonzero, and $\mathbf{u}_{\alpha}$ is the displacement of the αth local repatom.

2.2. Parallelization of QCDFT

The present QCDFT code is parallelized based on the Message Passing Interface (MPI). The parallelization is achieved for both local and nonlocal calculations. The parallelization for the local part is trivial: since the energy/force calculation for each finite element is independent from others, one can divide the local calculations evenly onto each processor. The computational time is thus proportional to the ratio of the number of the local finite elements to the available processors. Parallelization of the nonlocal region is achieved through domain decomposition, since the calculations (e.g. charge density, energy and force computations) are all performed on real space grids, except the convolution terms in the kinetic energy. Grid points are evenly distributed to available processors, and results are obtained by summing up contributions from all grids. The calculation of the convolution terms is performed by parallelized FFT.

The speedup using parallelization in local part is linear. The nonlocal part is dependent on the nodes used. In our calculations, an over all of about 26 times speedup was achieved when the program runs on 32 nodes.

2.3. Model setup

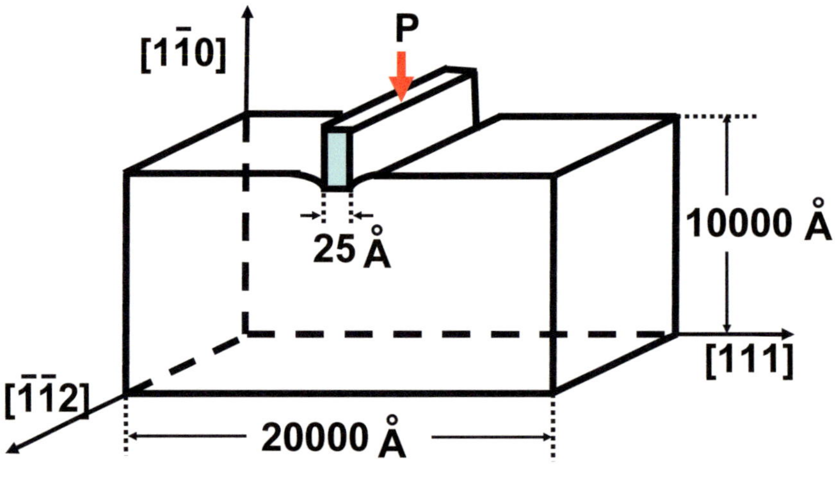

Figure 2. Schematic representation of the nanoindentation of Al thin film: geometry and orientation

Nanoindentation has now become a standard experimental technique for evaluating the mechanical properties of thin film materials and bulk materials in small volumes [25]. As it can measure nanometer penetration length scales, nanoindentation is an indispensable tool to assess elastic moduli and hardness of materials. It also can be used to derive strain-hardening exponents, fracture toughness, and viscoelastic properties of materials [6]. Moreover, nanoindentation also provides an opportunity to explore and better understand the elastic limit and incipient plasticity of crystalline solids [46]. For example, homogeneous nucleation of dislocations gives rise to the instability at the elastic limit of a perfect crystal. Exceeding the elastic limit can be manifested by a discontinuity in the load-displacement curve in a nanoindentation experiment [4, 8, 11, 26, 27, 35]. The onset of the discontinuity is an indication that the atomically localized deformation, such as dislocation nucleation occurs beneath the indenter. This correlation has been well established from both experimental and computational perspectives. For example, an *in situ* experiment by Gouldstone *et al.* using the Bragg-Nye bubble raft clearly demonstrated that homogeneous nucleation of dislocations corresponds to the discontinuity of the load-depth curve [12]. MD simulations have led to greater insight into the atomistic mechanism of nanoindentation [16, 48]. In particular, several QC simulations have been carried out for nanoindentation in Al thin films [13, 14, 17, 33, 39]. Tadmor *et al.* have used Embedded atom method based QC (EAM-QC) to study nanoindentation with a knife-like indenter with a pseudo-two-dimension (2D) model [39]. They observed the correspondence between the discontinuity in the load-displacement curve with the onset of plasticity. By using a much larger spherical indenter (700 nm), Knap *et al.* discovered that plasticity could occur without the corresponding discontinuity in the load-displacement curve [17]. However, when the indenter size was reduced (to 70 nm) the discontinuity reappeared. More recently, Hayes *et al.* have performed local OFDFT- and EAM-based QC calculations for nanoindentation of Al with a spherical indenter of 740 nm in radius [13, 14]. Using elastic stability criteria, they predicted the location of dislocation nucleation beneath the indenter, and obtained different results from EAM and OFDFT local QC calculations. Since many QC simulations have been carried out for nanoindentation of Al, it is not the purpose of the present paper (and we do not expect) to discover any new physics with QCDFT calculations. Instead, we use nanoindentation as an example to demonstrate the validity and the usefulness of the QCDFT method.

The present QCDFT approach is applied to nanoindentation of an Al thin film resting on a rigid substrate with a rigid knife-like indenter. The QC method is appropriate for the problem because it allows the modeling of system dimensions on the order of microns and thus minimizes the possibility of contaminating the results by the boundary conditions arising from small model sizes typically used in MD simulations. We chose this particular system for two reasons. First, there exists a good local pseudo-potential [9] and an excellent EAM potential [5] for Al. Secondly, results from conventional EAM-based QC simulations can be compared to the present calculations. An ideal validation of the method would require a full-blown OFDFT atomistic simulation for nanoindentation, which is not yet attainable. The second best approach would be a conventional QC simulation with an excellent EAM potential that compares well to OFDFT in terms of critical materials properties relevant to nanoindentation. Our reasoning is that the conventional QC method has been well established; thus as long as the EAM potential used is reliable, then the EAM-QC results should be reliable as well. In this paper, we have rescaled the "force-matching" EAM potential of Al [5] so that it matches precisely the OFDFT value of the lattice constant and bulk modulus of Al [3].

The crystallographic orientation of the system is displayed in Fig. (2). The size of the entire system is 2 μm $\times$ 1 μm $\times$ 4.9385 Å along the [111] (x direction), the $[\bar{1}10]$ (y direction), and the $[\bar{1}\bar{1}2]$ (z direction), respectively. The system is periodic in the z-dimension, has Dirichlet boundary conditions in the other two directions, and contains over 60 million Al atoms - a size that is well beyond the reach of any full-blown brute-force quantum calculation. The thickness of the thin film is selected to be comparable to the typical dislocation separation distance in well-annealed metals, which is of the order 1 μm. The unloaded system is a perfect single crystal similar to the experimental situation under the nanoindenter. The film is oriented so that the preferred slip system $\langle 110 \rangle$ $\{111\}$ is parallel to the indentation direction to facilitate dislocation nucleation. The indenter is a rigid flat punch of width 25 Å. We assume the perfect-stick boundary condition for the indenter so that the Al atoms in contact with it are not allowed to slip. The knife-like geometry of the indenter is dictated by the pseudo-2D nature of the QC model adopted. Three-dimensional QC models do exist and can be implemented in QCDFT [13, 14, 17]. We chose to work with the pseudo-2D model in this example for its simplicity. The prefix *pseudo* is meant to emphasize that although the analysis is carried out in a 2D coordinate system, out-of-plane displacements are allowed and all atomistic calculations are three-dimensional. Within this setting only dislocations with line directions perpendicular to the xy plane can be nucleated. The elastic moduli of C_{12}, C_{44}, C_{11} of Al are computed from three deformation modes, including hydrostatic, volume-conserving tetragonal and volume-conserving rhombohedral deformations. The shear modulus μ and Poisson's ratio ν are computed from the elastic moduli by a Voigt average: $\mu = (C_{11} - C_{12} + 3C_{44})/5$ and $\nu = \frac{C_{11}+4C_{12}-2C_{44}}{2(2C_{11}+3C_{12}+C_{44})}$. The values are listed in Table (1).

2.4. Loading procedure

The simulation is performed quasi-statically with a displacement control where the indentation depth (d) is increased by 0.2 Å at each loading step. We also tried a smaller loading step of 0.1 Å and obtained essentially the same results. Because OFDFT calculations are still much more expensive than EAM, we use EAM-based QC to relax the system for most of the loading steps. For load $d = 0$, the QCDFT calculation is performed to account for surface relaxations. From the resultant configuration, the depth of the indenter d is increased to 0.2 Å, again relaxed by QCDFT. After that, the calculations are done solely by EAM-QC except for the loading steps at $d = 1.8, 3.8, 9.2$ Å, when the corresponding EAM configurations are further relaxed by QCDFT. The onset of plasticity occurs at $d = 9.4$ Å. We increased the indenter depth of 0.2 Å from the relaxed QCDFT configuration at $d = 9.2$ Å, and then performed a QCDFT calculation to obtain the final structure at $d = 9.4$ Å. Such a simulation strategy is justified based on two considerations: (1) An earlier nanoindentation study of the same Al surface found that the onset of plasticity occurred at a smaller load with EAM-based local QC calculations comparing to OFDFT calculations [13]. The result was obtained by a local elastic stability analysis with EAM and OFDFT calculations of energetics and stress. The result suggests that we will not miss the onset of plasticity with the present loading procedure by performing EAM-QC relaxations preceding QCDFT. (2) Before the onset of plasticity, the load-displacement response is essentially linear with the slope determined by the elastic properties of the material. In other words, two QCDFT data points would be sufficient to obtain the correct linear part of the curve. Moreover, the fact that the EAM potential used in this study yields rather similar elastic constants to those from OFDFT suggests that the mixed EAM/OFDFT relaxation should not introduce large errors in the results.

2.5. Computation parameters

In Fig. (1), we present a schematic diagram illustrating the partition of domains for a QCDFT simulation of nanoindentation. The system shown in the diagram contains 1420 nonlocal repatoms, 736 local repatoms and 1539 finite elements, and is periodic along the z direction. The top surface is allowed to relax during the calculations while the other three surfaces of the sample are held fixed.

The parameters of the density-dependent kernel are chosen from reference [42], and Al ions are represented by the Goodwin-Needs-Heine local pseudo-potential [9]. The high kinetic energy cutoff for the plane wave basis of 1600 eV is used to ensure the convergence of the charge density. For the nonlocal calculation, the grid density for the volume Ω^I is 5 grid-points per Å. The Ω^I box goes beyond the nonlocal region by 8 Å in $\pm$x and $\pm$y directions so that ρ^I decays to zero at the boundary of Ω^I. The relaxation of all repatoms is performed by a conjugate gradient method until the maximum force on any repatom is less than 0.03 eV/Å.

At beginning of the simulation, the number of nonlocal repatoms is rather small, $\sim$ 80. As the load increases, the material deforms. When the variation of the deformation gradient between neighboring finite elements reaches 0.15, the mesh is refined, and the number of repatoms grows. The partitions, i.e. the size of nonlocal DFT region also grows. Close to the onset of plasticity, the number of nonlocal DFT atoms reaches 1420. The energy functional are the same for all the regions in these studies.

In order to validate the QCDFT method, we performed EAM-QC calculations of the nanoindentation with the same loading steps. We also calculated some relevant materials properties using the rescaled EAM and OFDFT method for bulk Al. The computational results along with experimental values extrapolated to T=0 K [5] are listed in Table (1). These numbers could shed light on the nanoindentation results from the QCDFT calculations.

	OFDFT	rescaled EAM	Experiment	(Unit)
elastic modulus C_{11}	117.17	97.13	118.0	GPa
elastic modulus C_{12}	41.36	51.16	62.4	GPa
elastic modulus C_{44}	29.76	30.23	32.5	GPa
Bulk modulus E	66.63	66.48	80.93	GPa
shear modulus μ	33.01	27.33	30.62	GPa
Poisson's ratio ν	0.287	0.319	0.332	
lattice constant a_0	4.032	4.032	4.032	Å
surface energy γ_{111}	0.867	0.72	1.14-1.2	J/m^2
stacking fault energy	0.10	0.10	0.12-0.14	J/m^2

Table 1. Elastic moduli, Poisson's ratio, lattice constant, (111) surface energy, and intrinsic stacking fault energy obtained by OFDFT and EAM calculations on bulk Al, and the corresponding experimental values extrapolated to T=0 K.

It is worth to note that the such quasi-2D simulations blocked the dislocations inclined in the third dimension. As a result, the twin formation may occur at a high loading rate and low temperature, compared with dislocation slip. Such quasi-2D limitations, however, could be validated in the systems where grain boundaries provides such geometrical confinement. For example, the deformation twinning has been observed in experimental studies of nonocrystalline Aluminum when the grain size is down to tens of nanometers[2, 20].

3. Quantum mechanical simulation of pure aluminum

The load-displacement curve is the typical observable for nanoindentation, and is widely used in both experiment and theory, often serving as a link between the two. In particular, it is conventional to identify the onset of plasticity with the first jump in the load-displacement curve during indentation [4, 10, 11, 13, 17, 33, 35, 39, 46]. In the present work, the loads are given in N/m, normalized by the length of the indenter in the out-of-plane direction.

Let us first discuss the QC results with the rescaled EAM potential. The load-displacement ($P - d$) curve shows a linear relation followed by a discrete drop at $d = 9.4$ Å, shown by the dashed line in Fig. (8). The drop corresponds to the homogeneous nucleation of dislocations beneath the indenter - the onset of plasticity. A pair of straight edge dislocations is nucleated at x=±13 Å, and y=-49 Å. In Fig. (9), we present the out-of-plane (or screw) displacement u_z of the nonlocal repatoms. The nonzero screw displacement of edge dislocations suggests that each dislocation is dissociated into two 1/6 <112> Shockley partials bound by a stacking fault with a width of about 14 Å. An earlier EAM-QC calculation [39] which has the same geometry as the present model but with a thinner sample (the thickness was ten times smaller than the present case), yields a separation distance of 13.5 Å. The activated slip planes are those {111} planes that are adjacent to the side surfaces of the indenter. The linear relation in the $P - d$ curve is due to: (1) the elastic response of the material before the onset of plasticity and (2) the particular choice of the rectangular indenter; a spherical indenter would have given rise to a parabolic $P - d$ curve [11, 13]. The slope for the linear part of the curve is 20.8 GPa, which is less than the shear modulus and C_{44}. The critical load, P_{cr} for homogeneous dislocation nucleation is 18.4 N/m, corresponding to a hardness of 7.3 GPa. Earlier EAM-QC calculations predicted the hardness to be 9.8 GPa [39]. The drop in applied load due to the nucleation of dislocations is $\Delta P = 3.4$ N/m. The value of ΔP from the previous EAM-QC calculation is 10 N/m [39], which is three times of the present result. Using the same sample size as in [39], we found $\Delta P = 9.02$ N/m, which is very close to the value reported in [39]. Thus, the discrepancy of ΔP is mainly due to the different sample sizes used in the two calculations indicating the importance of simulations at length scales relevant to experiments.

For QCDFT calculations, the load-displacement curve shows a linear relation up to a depth of 9.2 Å, followed by a drop at $d = 9.4$ Å, shown by the solid line in Fig. (8). The slope of initial linear part of the load-displacement curve is 23.9 GPa, rather close to the corresponding EAM value. The maximum load in linear region is $P_{cr} = 21.4$ N/m, corresponding to a hardness of 8.6 GPa. The fact that OFDFT predicts a larger P_{cr} than EAM is consistent with the results of Hayes *et al.* using local QC simulations for the same Al surface [13]. A pair of edge dislocations is nucleated at x=±13 Å, and y=-50 Å. The partial separation distance is about 19 Å, larger than the corresponding EAM value. The drop in the applied load due to dislocations nucleation is 7.8 N/m, which is more than twice of the corresponding EAM value. The large difference in ΔP between QCDFT and EAM-QC is interesting. It may suggest that although OFDFT and EAM produce rather similar results before the onset of plasticity, they differ significantly in describing certain aspects of defect properties. In particular, although both methods predict almost the same location for dislocation nucleation, they yield sizeable differences in partial dislocation width and ΔP. This result justifies the use of more accurate quantum simulations such as KS-DFT for nonlocal region where defects are present. Overall, we find that QCDFT gives very reasonable results comparing to the conventional EAM-QC. Although more validations are underway, we are optimistic that the QCDFT method is indeed reliable and offers a new route for quantum simulation of materials at large length scales.

4. Random distributed magnesium impurities in aluminum thin film

The effect of Mg impurities on the ideal strength and incipient plasticity of the Al thin film. In the calculations, five Mg impurities are introduced randomly below the indenter, as schematically shown in Fig. (5). The results of the randomly distributed Mg impurities are referred as *random*, distinguishing from the results of the pure system, referred as *pure*. At

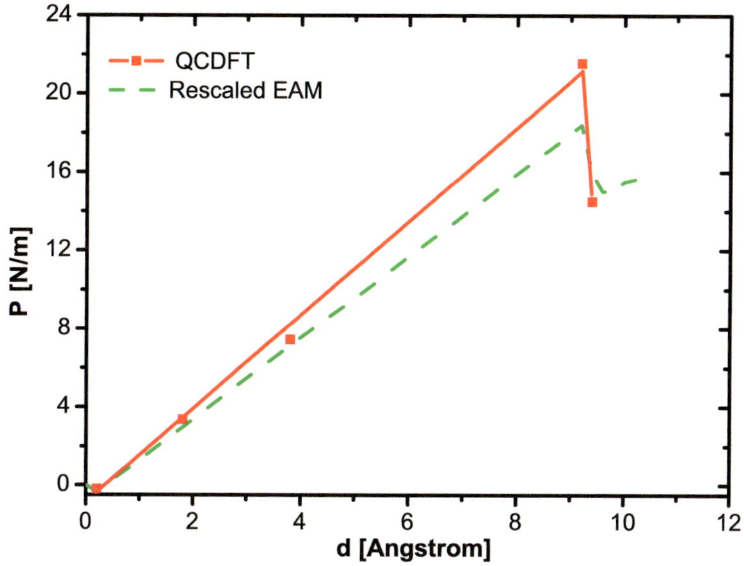

Figure 3. Load-displacement curve for nanoindentation of an Al thin film with a rigid rectangular indenter: with QCDFT (red solid line) and rescaled EAM-QC (green dashed line). The red squares are actual QCDFT data points and the solid line is the best fit to the data points. All EAM-QC data points are on the dashed line.

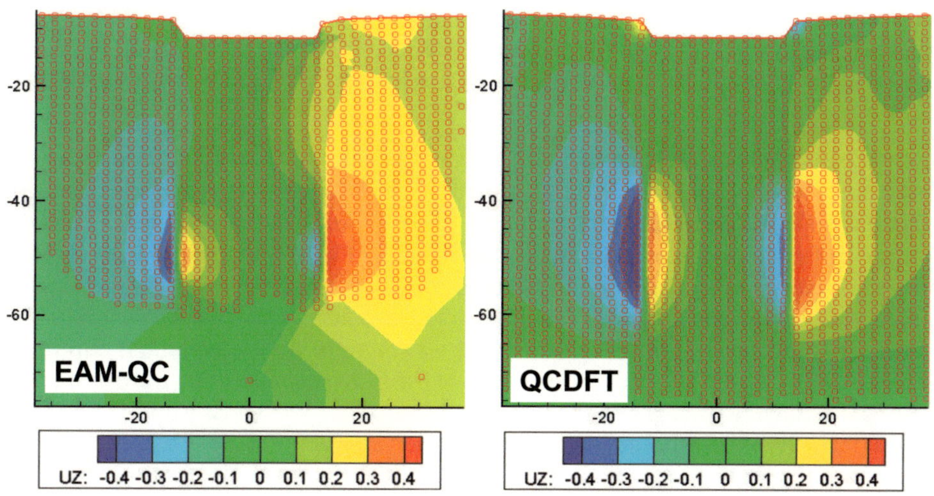

Figure 4. The out-of-plane displacement u_z obtained from the rescaled EAM-QC (left) and QCDFT (right) calculations. The circles represent the repatoms and the displacement ranges from -0.4 (blue) to 0.4 (red) Å.

$d = 3.0, 6.0, 7.5$ Å, the *random* results are obtained after full relaxations of the *pure* Al system. The QCDFT loading is carried out after $d = 7.5$ Å starting from the full relaxed configuration of a previous loading step, until the onset of the plasticity occurs at $d = 8.1$ Å.

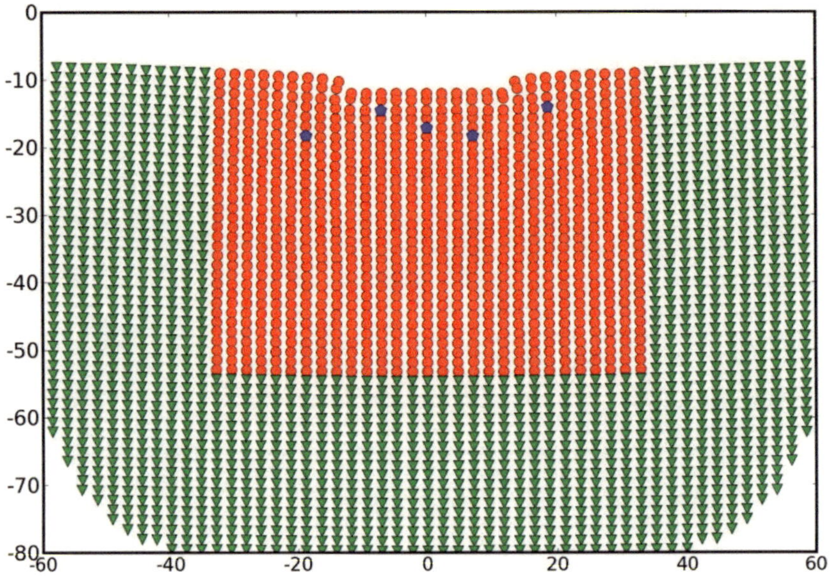

Figure 5. Schematic diagram of the randomly distributed Mg impurities in the Al thin film. The red spheres and blue pentagons represent nonlocal Al and Mg atoms, respectively. The green triangle represents Al buffer atoms. The dimensions are given in Å .

The load-displacement curve is the typical observable for nanoindentation, and is widely used in both experiment and theory, often serving as a link between the two. In particular, it is conventional to identify the onset of incipient plasticity with the first drop in the load-displacement curve during indentation [4, 10, 11, 13, 17, 30, 33, 35, 39, 46]. In the present work, the load is given in N/m, normalized by the length of the indenter in the out-of-plane direction.

For pure Al, the load-displacement ($P - d$) curve shows a linear relation followed by a drop at $d = 8.2$ Å, shown by the dashed line in Fig. (8). The drop corresponds to the homogeneous nucleation of dislocations beneath the indenter - the onset of plasticity. A pair of straight edge dislocations are nucleated at x=±13 Å, and y=-50 Å. In Fig. (9), we present the out-of-plane (or screw) displacement u_z of the nonlocal repatoms. The non-zero screw displacement of the edge dislocations suggests that each dislocation is dissociated into two 1/6 <112> Shockley partials bound by a stacking fault with a width of about 19 Å. The activated slip planes are those {111} planes that are adjacent to the edges of the indenter. The slope for the linear part of the curve is 27.1 GPa, which is less than the shear modulus μ=33.0 GPa and $C_{44} = 29.8$ GPa. The critical load, P_{cr} for the homogeneous dislocation nucleation is 18.4 N/m, corresponding to a hardness of 7.3 GPa (the critical load normalized by the area of the indenter), which is 0.22 μ . The drop in applied load due to the nucleation of dislocations is $\Delta P = 6.8$ N/m, agreeing with the load drop estimated by the elastic model[39] which is $\Delta P = 7.7$ N/m.

For randomly distributed impurities in the Al thin film, the load-displacement curve shows a linear relation up to a depth of 8.0 Å, followed by a drop at $d = 8.1$ Å, as shown by the

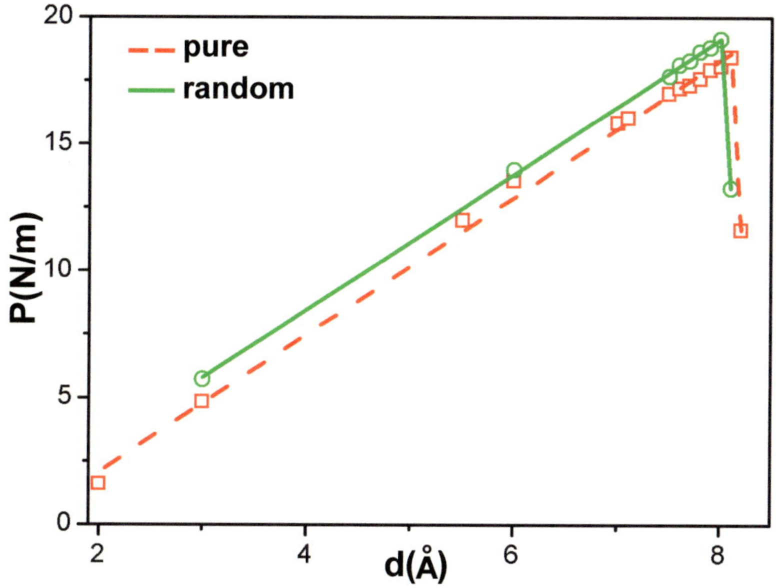

Figure 6. Load-displacement plot for the nanoindentation of the Al thin film with a rigid rectangular indenter: pure Al (red squares) and randomly distributed Mg impurity system (green circles). The corresponding lines are the best fit to the data points.

solid line in Fig. (8). The slope of initial linear part of the load-displacement curve is 26.7 GPa, rather close to the corresponding pure Al value. The maximum load in linear region is $P_{cr}^{im} = 19.2$ N/m, corresponding to a hardness of 7.6 GPa, which is 0.3 GPa greater than the pure Al system. A pair of Shockley partial dislocations is nucleated at x=-13 Å, y=-25 Å and x=13 Å, y=-22 Å respectively as shown in the right panel of Fig. (9). The drop in the applied load due to the dislocation nucleation is 5.9 N/m. The estimated load drop by the elastic model is $\Delta P = 7.6$ N/m. The smaller drop of the load for the random case than the elastic model is probably due to the presence of the Mg impurities, which is not accounted for in the elastic model [39]. The fact that the critical load and the hardness of the Al-Mg alloy are greater than that of the pure Al system demonstrates that the Mg impurities are responsible for the solid solution strengthening of the Al thin film. The presence of Mg impurities also hinders the formation of full edge dislocations and as a result, only partial dislocations are nucleated and they are pinned near the surface as shown in Fig. (9).

Finally we point out the possibility that the emitted dislocations may be somewhat constrained by the local/nonlocal interface from going further into the bulk. Because the critical stress to move an edge dislocation in Al is vanishingly small ($10^{-5}\mu$) comparing to that to nucleate a dislocation ($10^{-1}\mu$), a small numerical error in stress could easily lead to a large difference in the equilibrium dislocation position. The four-order-of-magnitude disparity poses a significant challenge to all atomistic simulations in predicting dislocation nucleation site, QCDFT method included. One can only hope to obtain a reliable critical load for the incipient plasticity, rather than for the equilibrium position of dislocations. The same problem has been observed and discussed by others [38]. However, despite the problem, the dramatic difference observed in the two panels of Fig. 7 unambiguously demonstrates the strengthening effect of Mg impurities. Therefore the conclusion is still valid.

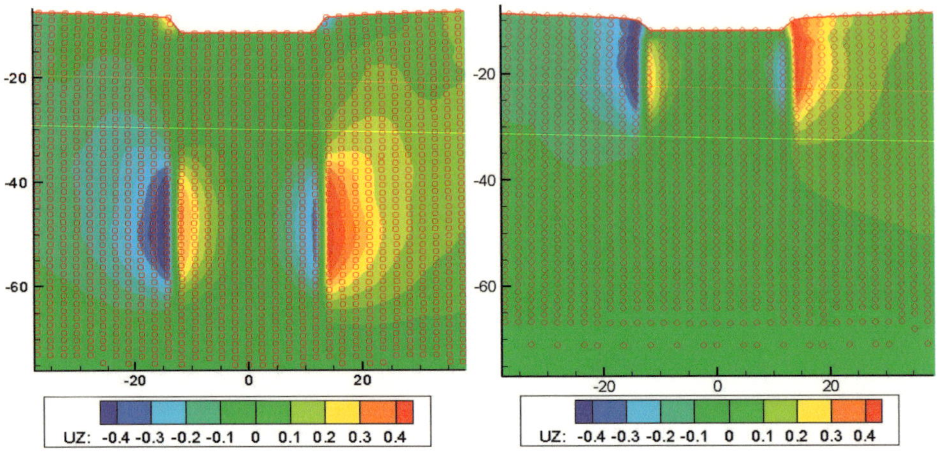

Figure 7. The out-of-plane displacement u_z obtained from the pure (left) and with Mg impurities (right) QCDFT calculations. The circles represent the repatoms and the displacement ranges from -0.4 (blue) to 0.4 (red) Å.

5. Solid solution effects on dislocation nucleation during nanoindentation

The load-displacement curve is the typical observable for nanoindentation, and is widely used in both experiment and theory, often serving as a link between the two. In particular, it is conventional to identify the onset of incipient plasticity with the first drop in the load-displacement curve during indentation [4, 10, 11, 13, 17, 30, 33, 35, 39, 46]. In the present work, the load is given in N/m, normalized by the length of the indenter in the out-of-plane direction.

For pure Al, the load-displacement ($P - d$) curve shows a linear relation initially, followed by a drop at $d = 8.2$ Åin Fig. (8). The drop corresponds to the homogeneous nucleation of dislocations beneath the indenter - the onset of plasticity. A pair of edge dislocations are nucleated at x=±13 Å, and y=-45 Å. In Fig. (9), we present the out-of-plane (or screw) displacement u_z of the nonlocal repatoms. The non-zero screw displacement of the edge dislocations suggests that each dislocation is dissociated into two 1/6 <112> Shockley partials bound by a stacking fault with a width of about 20 Å. The activated slip planes are {111} type and adjacent to the edges of the indenter. The slope for the linear part of the curve is 27.1 GPa, which is greater than the shear modulus μ=24.6 GPa but less than $C_{44} = 30.1$ GPa. The critical load, P_{cr} for the homogeneous dislocation nucleation is 18.4 N/m, corresponding to a hardness of 7.2 GPa. The drop in the applied load due to the nucleation of dislocations is $\Delta P = 6.8$ N/m, similar to the value estimated by an elasticity model[39], which is $\Delta P = 7.7$ N/m.

For randomly distributed impurities, the load-displacement curve shows a linear relation up to a depth of 8.0 Å, followed by a drop at $d = 8.1$ Åin Fig. (8). The slope of the initial linear part of the load-displacement curve is 26.7 GPa, rather close to the corresponding pure Al value. The maximum load in linear region is $P_{cr}^{im} = 19.2$ N/m, corresponding to a hardness of 7.5 GPa, which is 0.3 GPa or 4% greater than that of the pure Al. A pair of Shockley partial dislocations is nucleated at x=-13 Å, y=-40 Åand x=13 Å, y=-38 Årespectively as shown in Fig. (9). The fact that the hardness of the Al-Mg alloy is greater than that of the pure Al is an indication of solid solution strengthening. However, the magnitude of the strengthening

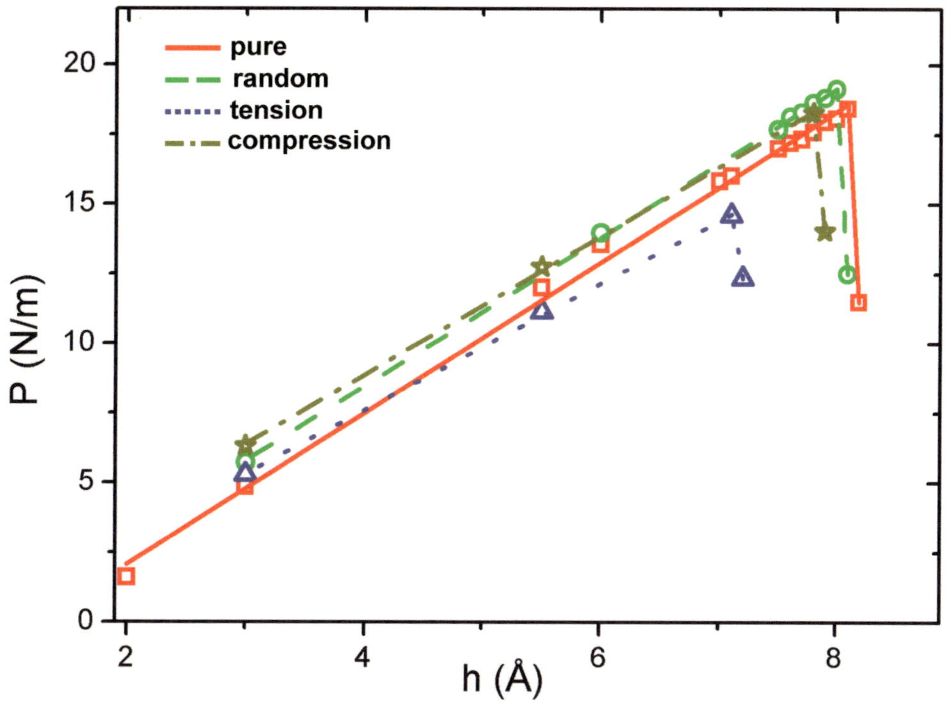

Figure 8. Load-displacement plot for the nanoindentation of the Al thin film with a rigid rectangular indenter: *pure* (solid line), *random* (dashed line), *tension* (dotted line), *compression* (dash-dotted line) obtained from QCDFT calculations. The lines are the best fit to the corresponding simulation data points.

is insignificant, and is on the order of the changes in the shear modulus. This finding is consistent to an experimental study on Cu-Ni solid solution alloys [1] the nanoindentation measurements demonstrated the effects of solute impurities on the formation of dislocations in a previously dislocation-free region to be minimal. Moreover, the experimental study suggested that overall dislocation nucleation is strongly related to shear modulus in this system. It is clear from our results that the origin of the strengthening is not due to the propagation of dislocations, but rather the nucleation of the dislocations. The presence of randomly distributed Mg impurities hinders the nucleation of dislocations. In fact, only the leading partial dislocations are nucleated trailing by stacking faults as shown in Fig. (9)b. This result is in contrast to the pure system where full dislocations are nucleated at a larger distance below the surface. The drop in the applied load due to incipient plasticity is 5.9 N/m, less than that of the pure system (6.8 N/m). The smaller drop of the load in the random case is due to the fact that the partial dislocations are nucleated instead of full dislocations as in the pure system. We expect that the critical load/hardness corresponding to the nucleation of full dislocations in this case will be higher because of work hardening.

The reason why partial dislocations are nucleated as opposed to full dislocations in the presence of random impurities can be understood from the following energetic consideration. The total energy of the system can be approximated by the dislocation elastic energy and the stacking fault energy. The former is given by $\frac{\mu}{2\pi(1-\nu)}b^2$ where μ, ν and b are the shear modulus, Poisson's ratio and Burgers vector, respectively. The latter energy can be expressed as $\gamma \times w$ where γ is the ISF energy and w is the width of the stacking fault. Because μ is

	pure	*random*	*compression*	*tension*
μ [GPa]	24.55	25.08	25.07	25.07
γ_{isf} [J/m^2]	0.12	0.11	0.11	0.11
b	2.85	1.65	1.65	1.65
w [Å]	20,20	32,30	28	7
E_{tot} [eV]	3.94	3.03	1.37	0.67

Table 2. The shear modulus μ, intrinsic stacking fault energy γ_{isf}, Burgers vector, stacking fault width and the approximate total energy for the dislocations, corresponding to the *pure*, *random*, *compression*, and *tension* cases.

increased in the presence of the impurities, the system could lower its energy by reducing b, i.e., dissociation into partials. Of course, it is energetic favorable only if the γ and w are not too large. The fact the γ value is slightly reduced in the presence of Mg impurities helps the dissociation. Using the quantities tabulated in Table II, we find that the total energy of the random impurities system (with two partial dislocations) is 0.91 eV lower than that of the pure system (two full dislocations). In Table II, the Burgers vectors for full and partial dislocation are determined to be 2.85 Å and 1.65 Å respectively. The width of the stacking fault (w) has two entries for the pure and random cases since there are two dislocations nucleated.

For Mg impurities below the slip plane as in the *tension* case, the load-displacement curve shows a linear relation up to a depth of 7.1 Å, followed by a drop at $d = 7.2$ Å(dotted line in Fig. (8). The maximum load in linear region is $P_{cr}^{im} = 17.8$ N/m, corresponding to a hardness of 6.5 GPa, which is 0.7 GPa or 10 % smaller than the pure Al system. A single Shockley partial dislocation is nucleated at x=-13 Å , y=-14 Å as shown in Fig. (9). The drop in the applied load due to the dislocation nucleation is 2.3 N/m.

Similarly, for Mg impurities above the slip plane as in the *compression* case, the load-displacement curve is linear up to a depth of 7.8 Å, followed by a drop at $d = 7.9$ Å, as shown by the dot-dashed line in Fig. (8). The maximum load in linear region is $P_{cr}^{im} = 18.3$ N/m, corresponding to a hardness of 7.1 GPa, which is 0.1 GPa smaller than the pure Al system. A single Shockley partial dislocation is nucleated at x=-13 Å , y=-35 Å in Fig. (9). The drop in the applied load due to the dislocation nucleation is 4.2 N/m.

From the above results, we conclude that the linear distribution of Mg impurities can actually soften the material and render dislocation nucleation easier than the pure system. In other words, the solid solution strengthening effect depends sensitively on the local configuration of the impurities. In the three cases (of the same Mg concentration) studied here, the hardness of the alloys varies and the impurities can either increase or decrease the hardness, depending on their configuration. It should be emphasized that the change in hardness is associated with dislocation nucleation, not with dislocation propagation. Although it is well-known that dislocation propagation (or dislocation-impurity interaction) depends sensitively on the impurity configuration, it is less recognized that the impurities configuration is important for dislocation nucleation.

The reason that the hardness in the tension case is less than that in the compression case can be understood from the atomic size consideration. The atomic or ionic radius of Al is 0.54 Å, which is less than that of Mg (0.86 Å). In the compression case, since smaller Al atoms are replaced by larger Mg atoms in a compression region, the substitution increases the compressive stress, and makes it more difficult to form edge dislocations, thus a higher hardness. On the other hand, the replacement of smaller Al atoms by larger Mg atoms in a tension region reduces the tension and makes it easier to form the dislocations, hence a lower hardness. Because the impurities are located on one side of the indenter - the symmetry is broken, dislocation is also nucleated at one side of the thin film. Because only one partial dislocation is nucleated in the compression and tension cases, the energy of the two cases is much smaller as shown Table II.

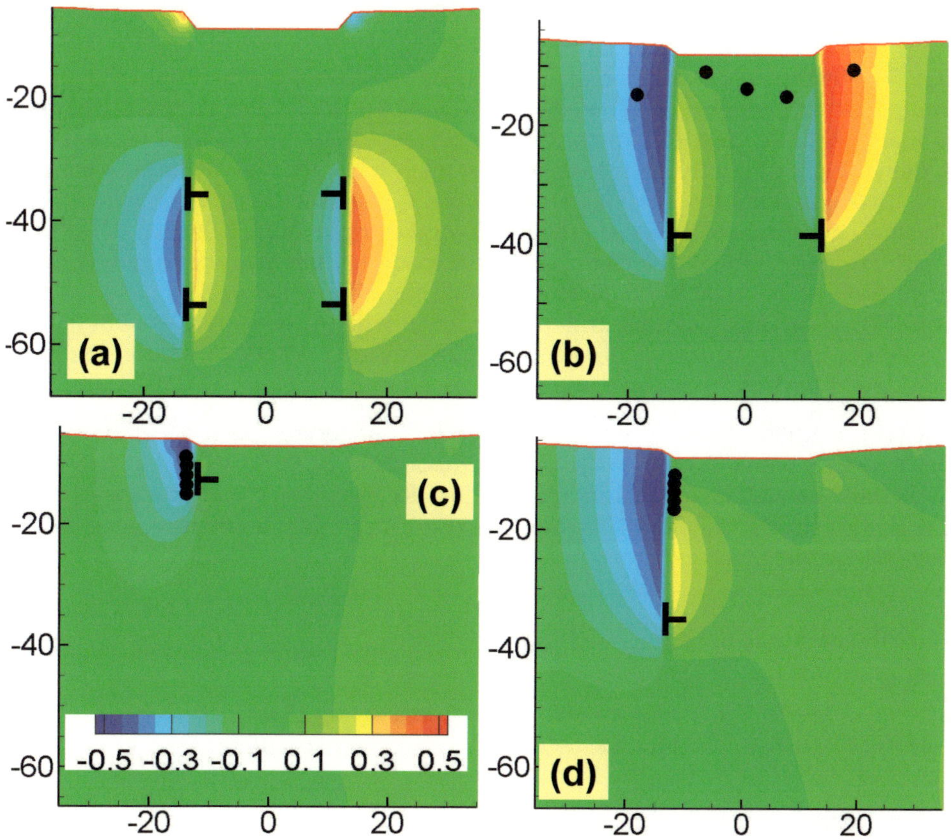

Figure 9. The out-of-plane displacement u_z corresponding to (a) *pure*, (b) *random*, (c) *tension*, (d) *compression* cases obtained from QCDFT calculations. The the displacement ranges from -0.5 (blue) to 0.5 (red) Å. The black dots indicate the position of the Mg impurities. All distances are given in Å.

6. Conclusions

We presented a concurrent multiscale method that makes it possible to simulate multi-million atoms based on density functional theory. The method - QCDFT - is formulated within the framework of the QC method, with OFDFT as its sole input, i.e., there is only one underlying energy functional (OFDFT) involved. Full-blown OFDFT and OFDFT-based elasticity theory are the two limiting cases corresponding to a fully nonlocal or a fully local version of QCDFT. The QC ghost force at the local-nonlocal interface is corrected by a dead load approximation. The QCDFT method is applied for a nanoindentation study of an Al thin film. The QCDFT results are validated by comparing against conventional QC with a OFDFT-refined EAM potential. The results suggest that QCDFT is an excellent method for quantum simulation of materials properties at length scales relevant to experiments.

For the study of the nanoindentation of an Al thin film in the presence and absence of randomly distributed Mg impurities, the Mg impurities are found to strengthen the hardness of Al and hinder the dislocation nucleation. The results suggest that QCDFT is a promising method for quantum simulation of materials properties at length scales relevant to experiments.

We also find that the solid solution effect depends sensitively on the local configuration of the impurities. Although a random distribution of the impurities increases the hardness of the material, linear distributions of the impurities actually lower the hardness. In both cases, the effects are entirely due to dislocation nucleation; the solid solution strengthening owing to dislocation motion is not considered here. Consistent to the experimental results on Ni/Cu alloys, the extent of the solid solution strengthening is found to be insignificant - in the same order of magnitude of the change in shear modulus. On the other hand, the incipient plasticity is observed to be quite different among the different cases. In the pure material, two full dislocations are nucleated under the indenter with the opposite sign. For the random distribution of the impurities, two partial dislocations are nucleated instead. For the linear distributions of the impurities, only one partial dislocation is nucleated.

The QCDFT method could be used for other FCC materials, such as gold and copper. Besides materials with FCC lattice, QCDFT method could be directly used to study the BCC (body-centered-cubic) materials, iron and its alloys for example. It will be also applicable for complex lattice structrues, such as hexagonal close-packed (hcp) structures, magnesium and its alloys for example.

Author details

Qing Peng
Department of Mechanical, Aerospace and Nuclear Engineering, Rensselaer Polytechnic Institute, Troy, NY 12180, USA

7. References

[1] Bahr, D. & Vasquez, G. [2005]. Effect of solid solution impurities on dislocation nucleation during nanoindentation, *J. Mater. Res.* 20(8): 1947–1951.

[2] Chen, M., Ma, E., Hemker, K. J., Sheng, H., Wang, Y. & Cheng, X. [2003]. Deformation twinning in nanocrystalline aluminum, *Science* 300(5623): 1275.

[3] Choly, N., Lu, G., E, W. & Kaxiras, E. [2005]. Multiscale simulations in simple metals: A density-functional-based methodology, *Phys. Rev. B* 71(9): 094101.

[4] Corcoran, S. G., Colton, R. J., Lilleodden, E. T. & Gerberich, W. W. [1997]. Anomalous plastic deformation at surfaces: Nanoindentation of gold single crystals, *Phys. Rev. B* 55(24): R16057–R16060.

[5] Ercolessi, F. & Adams, J. B. [1994]. Interatomic potentials from 1st-principles calculations - the force-matching method, *Europhys. Lett.* 26(8): 583–588.

[6] Fischer-Cripps, A. C. [2004]. *Nanoindentation*, Vol. 2nd ed., Springer-Verlag, New York. pp 1.

[7] Gao, J. & Truhlar, D. G. [2002]. Quantum mechanical methods for enzyme kinetics, *Annu. Rev. Phys. Chem.* 53: 467.

[8] Gerberich, W., Nelson, J., Lilleodden, E., Anderson, P. & Wyrobek, J. [1996]. Indentation induced dislocation nucleation: The initial yield point, *Acta Materialia* 44: 3585–3598.
[9] Goodwin, L., Needs, R. J. & Heine, V. [1990]. A pseudopotential total energy study of impurity-promoted intergranular embrittlement, *J. Phys. Condens. Matter* 2: 351.
[10] Gouldstone, A., Chollacoop, N., Dao, M., Li, J., Minor, A. M. & Shen, Y.-L. [2007]. Indentation across size scales and disciplines: Recent developments in experimentation and modeling, *Acta Mater.* 55(12): 4015 – 4039.
[11] Gouldstone, A., Koh, H. J., Zeng, K. Y., Giannakopoulos, A. E. & Suresh, S. [2000]. Discrete and continuous deformation during nanoindentation of thin films, *Acta Mater.* 48(9): 2277 – 2295.
[12] Gouldstone, A., Van Vliet, K. J. & Suresh, S. [2001]. Nanoindentation - Simulation of defect nucleation in a crystal, *NATURE* 411(6838): 656.
[13] Hayes, R. L., Fago, M., Ortiz, M. & Carter, E. A. [2005]. Prediction of dislocation nucleation during nanoindentation by the orbital-free density functional theory local quasi-continuum method, *Multiscale Mod. Sim.* 4: 359.
[14] Hayes, R. L., Ho, G., Ortiz, M. & Carter, E. A. [2006]. Prediction of dislocation nucleation during nanoindentation of Al3Mg by the orbital-free density functional theory local quasicontinuum method, *Philos. Mag.* 86: 2343.
[15] Hung, L. & Carter, E. A. [2009]. Accurate simulations of metals at the mesoscale: Explicit treatment of 1 million atoms with quantum mechanics, *Chemical Physics Letters* 475(4-6): 163 – 170.
[16] Kelchner, C. L., Plimpton, S. J. & Hamilton, J. C. [1998]. Dislocation nucleation and defect structure during surface indentation, *PHYSICAL REVIEW B* 58(17): 11085–11088.
[17] Knap, J. & Ortiz, M. [2003]. Effect of indenter-radius size on au(001) nanoindentation, *Phys. Rev. Lett.* 90(22): 226102.
[18] Kohn, W. [1995]. Density functional theory for systems of very many atoms, *Int. J. Quantum Chem.* 56: 229.
[19] Kohn, W. & Sham, L. J. [1965]. Self-consistent equations including exchange and correlation effects, *Phys. Rev.* 140(4A): A1133–A1138.
[20] Liao, X. Z., Zhou, F., Lavernia, E. J., He, D. W. & Zhu, Y. T. [2003]. Deformation twins in nanocrystalline Al, *Appl. Phys. Lett.* 83(24): 5062.
[21] Lignères, V. & Carter, E. [2005]. *Handbook of Materials Modeling*, Springer,Dordrecht, The Netherlands, pp. 127–148.
[22] Lin, H. & Truhlar, D. G. [2007]. QM/MM: What have we learned, where are we, and where do we go from here?, *Theor. Chem. Acc.* 117: 185.
[23] Lu, G., Peng, Q., Zhang, X., Hung, L. & Carter, E. A. [2008]. Quantum simulation of materials at micron scales and beyond, *Oberwolfach Reports* 5(2): 1117.
[24] Mordasini, T. & Thiel, W. [1998]. Combined quantum mechanical and molecular mechanical approaches, *Chimia* 52: 288.
[25] NIX, W. D. [1989]. Mechanical-properties of thin-films, *METALLURGICAL TRANSACTIONS A-PHYSICAL METALLURGY AND MATERIALS SCIENCE* 20(11): 2217.
[26] Oliver, W. & Pharr, G. [1992]. An improved technique for determining hardness and elastic modulus using load and displacement sensing indentation experiments, *J. Mater. Res.* 7(6): 1564–1583.
[27] Page, T. F., Oliver, W. C. & McHargue, C. J. [1992]. The deformation-behavior of ceramic crystals subjected to very low load (nano)indentations, *J. Mater. Res.* 7(2): 450–473.

[28] Peng, Q. & Lu, G. [2011]. A comparative study of fracture in Al: Quantum mechanical vs. empirical atomistic description, *JOURNAL OF THE MECHANICS AND PHYSICS OF SOLIDS* 59(4): 775–786.
[29] Peng, Q., Zhang, X., Huang, C., Carter, E. A. & Lu, G. [2010]. Quantum mechanical study of solid solution effects on dislocation nucleation during nanoindentation, *MODELLING AND SIMULATION IN MATERIALS SCIENCE AND ENGINEERING* 18(7): 075003.
[30] Peng, Q., Zhang, X., Hung, L., Carter, E. A. & Lu, G. [2008]. Quantum simulation of materials at micron scales and beyond, *Phys. Rev. B* 78: 054118.
[31] Peng, Q., Zhang, X. & Lu, G. [2010]. Quantum mechanical simulations of nanoindentation of Al thin film, *COMPUTATIONAL MATERIALS SCIENCE* 47(3): 769.
[32] Shenoy, V. B., Miller, R., Tadmor, E. B., Rodney, D., Phillips, R. & Ortiz, M. [1999]. An adaptive finite element approach to atomic-scale mechanics - the quasicontinuum method, *J. Mech. Phys. Solids* 47: 611.
[33] Shenoy, V. B., Phillips, R. & Tadmor, E. B. [2000]. Nucleation of dislocations beneath a plane strain indenter, *J. Mech. Phys. Solids* 48(4): 649 – 673.
[34] Sherwood, P. [2000]. *Modern Methods and Algorithms of Quantum Chemistry*, Vol. 3, NIC, Princeton, p. 285.
[35] Suresh, S., Nieh, T. G. & Choi, B. W. [1999]. Nano-indentation of copper thin films on silicon substrates, *Scripta Materialia* 41(9): 951 – 957.
[36] T. K. Woo, P. M. Margl, L. D. L. C. & Ziegler, T. [n.d.]. *ACS Symp. Ser.* .
[37] Tadmor, E. B. & Miller, R. E. [2005]. *Handbook of Materials Modeling*, Vol. 1, Kluwer Academic Publishers, chapter The Theory and Implementation of the Quasicontinuum Method.
[38] Tadmor, E. B., Miller, R. & Phillips, R. [1999]. Nanoindentation and incipient plasticity, *J. Mater. Res.* 14: 2249.
[39] Tadmor, E. B., Miller, R., R., P. & Ortiz, M. [1999]. Nanoindentation and incipient plasticity, *J. Mater. Res.* 14: 2233.
[40] Tadmor, E. B., Ortiz, M. & Phillips, R. [1996]. Quasicontinuum analysis of defects in solids, *Philos. Mag. A* 73: 1529.
[41] Wang, Y. A. & Carter, E. A. [2000]. *Theoretical Methods in Condensed Phase Chemistry*, Dordrecht, Kluwer, chapter 5.
[42] Wang, Y. A., Govind, N. & Carter, E. A. [1999]. Orbital-free kinetic-energy density functionals with a density-dependent kernel, *Phys. Rev. B* 60: 16350.
[43] Zhang, X. & Lu, G. [2007]. Quantum mechanics/molecular mechanics methodology for metals based on orbital-free density functional theory, *Phys. Rev. B* 76: 245111.
[44] Zhang, X., Peng, Q. & Lu, G. [2010]. Self-consistent embedding quantum mechanics/molecular mechanics method with applications to metals, *Phys. Rev. B* 82: 134120.
[45] Zhao, Y., Wang, C., Peng, Q. & Lu, G. [2010]. Error analysis and applications of a general QM/MM approach, *COMPUTATIONAL MATERIALS SCIENCE* 50(2): 714–719.
[46] Zhu, T., Li, J., Vliet, K. J. V., Ogata, S., Yip, S. & Suresh, S. [2004]. Predictive modeling of nanoindentation-induced homogeneous dislocation nucleation in copper, *J. Mech. Phys. Solids* 52(3): 691 – 724.
[47] Zienkiewicz, O. C. & Taylor, R. L. [2000]. *the Finite Element Method*, Vol. 1, Oxford, Butterworth-Heinemann. pp 23.
[48] Zimmerman, J. A., Kelchner, C. L., Klein, P. A., Hamilton, J. C. & Foiles, S. M. [2001]. Surface step effects on nanoindentation, *PHYSICAL REVIEW LETTERS* 87(16): art. no.–165507.

Chapter 2

Effect of the Spherical Indenter Tip Assumption on the Initial Plastic Yield Stress

Li Ma, Lyle Levine, Ron Dixson, Douglas Smith and David Bahr

Additional information is available at the end of the chapter

http://dx.doi.org/10.5772/48106

1. Introduction

Nanoindentation is widely used to explore the mechanical properties of small volumes of materials. For crystalline materials, there is a growing experimental and theoretical interest in pop-in events, which are sudden displacement-burst excursions during load-controlled nanoindentation of relatively dislocation-free metals. The first pop-in event is often identified as the initiation of dislocation nucleation, and thus the transition from purely elastic to elastic/plastic deformation. The maximum shear stress at this first pop-in event, or the onset of plastic yielding, is generally found to be close to the theoretical strength of the material and is frequently estimated from Hertzian elastic contact theory. However, an irregular indenter tip shape will significantly change the stress distribution in magnitude and location, and therefore the maximum shear stress, from a Hertzian estimation. The aim of this chapter is to state the challenges and limitations for extracting the initial plastic yield stress from nanoindentation with the spherical indenter tip assumption. We assess possible errors and pitfalls of the Hertzian estimation of initial plastic yield at the nanoscale.

2. Background

Instrumented nanoindentation has been widely used to probe small scale mechanical properties such as elastic modulus and hardness over a wide range of materials and applications (Doerner & Nix, 1986; Fisher-Cripps, 2002; Oliver & Pharr, 1992, 2004). The response of a material to nanoindentation is usually shown by plotting the indentation load, P, as a function of the indenter penetration depth, h.

2.1. Nanoindentation of crystalline materials

For crystalline materials, nanoindentation can be used to study defect nucleation and propagation events, which are detected by discontinuities in the load-depth relationship.

Generally, there are three types of discontinuities as illustrated in Fig.1. First are "pop-in" events (as shown in Fig. 1a), which are sudden displacement excursions into the target materials during load-controlled nanoindentation of relatively dislocation-free metals. Pop-ins were first observed and associated with dislocation nucleation, or the sudden onset of plasticity, by Gane and Bowde in 1968 (Gane & Bowden, 1968) using fine stylus indentation of metal crystals. Pop-ins may also be associated with crack nucleation and propagation (Morris et al., 2004; Jungk et al., 2006), phase transformations (Page et al., 1992), and mechanically induced twinning (Bradby et al., 2002; Misra et al., 2010).

The second type of discontinuity is a "pop-out" event (as shown in Fig. 1b), which is a discontinuous decrease in the indentation displacement, usually during unloading. Pop-outs may also be ascribed to dislocation motion (Cross et al., 2006) and phase transformations (Juliano et al., 2004; Ruffell et al., 2007; Haq et al., 2007; Lee & Fong, 2008). A lower unloading rate or a higher maximum indentation load promotes the occurrence of a pop-out (Chang & Zhang, 2009).

The third type of load-depth discontinuity is the "load drop" found during a displacement-controlled experiment (Kiely & Houston, 1998; Warren et al., 2004), as shown in Fig. 1c. A molecular dynamics study showed that load drops are associated with local rearrangements of atoms (Szlufarska et al., 2007).

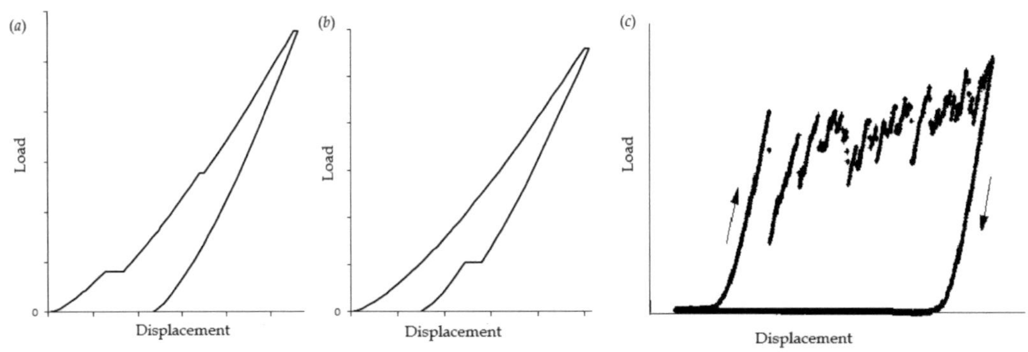

Figure 1. Schematic diagram of nanoindentation load displacement curve illustrating the (a) pop-in, (b) pop-out, and (c) load-drop behaviors.

2.2. Dislocation nucleation stress

Quantitative study of these nanoindentation phenomena requires reasonable estimates of the stresses that drove the particular event. Here, attention is focused on yield in metallic crystals. It is believed that the first pop-in event is most frequently the result of the initiation of dislocation nucleation, and thus the transition from purely elastic to elastic/plastic deformation (Gouldstone et al., 2000; Kelchner et al., 1998, 2009; Suresh et al., 1999; Li et al., 2002; Lorenz et al., 2003; Minor et al., 2004; Manson et al., 2006; Nix et al., 2007). The load-displacement curve before the pop-in occurs is often fully reversible, and is usually interpreted using the Hertzian contact theory (Johnson, 1999),

$$P = \frac{4}{3}\overline{E}_R R^{1/2} h^{3/2}, \tag{1}$$

where R is the indenter tip radius and

$$\overline{E}_R = \left[\left(1-\nu_I^2\right)/E_I + \left(1-\nu_S^2\right)/E_S\right]^{-1}$$

is the contact elastic modulus between the indenter (I) and specimen (S). In this case, the deformation is purely elastic prior to the first pop-in; if the indenter tip is unloaded before the first pop-in, atomic force microscopy (AFM) images show no indent on the specimen surface, whereas, if unloading occurs after the pop-in, a residual indent is observed (Chiu & Ngan, 2002; Schuh & Lund, 2004).

Nanoindentation pop-in tests can be a powerful tool for studying homogeneous and heterogeneous dislocation nucleation. When a pop-in event is caused by the sudden onset of crystal plasticity, whether through dislocation source activation (Bradby & Williams, 2004; Schuh et al., 2005) or homogeneous dislocation nucleation (Bahr et al., 1998; Chiu & Ngan, 2002), the maximum stress at the yield event is generally interpreted as the maximum shear stress in the body (Minor et al., 2006). This maximum shear stress, τ_{MAX}, at the first pop-in load, P_{CRIT}, is generally estimated from elastic contact theory (Johnson, 1999) as

$$\tau_{MAX} = 0.31\left[\frac{6\overline{E}_R^2}{\pi^3 R^2}\right]^{\frac{1}{3}} P_{CRIT}^{\frac{1}{3}} \tag{2}$$

For a variety of materials, when the first pop-in occurs, the maximum shear stress in the specimen is in the range of G/30 to G/5, where G is the shear modulus; this stress is very close to the theoretical strength calculated by the *ab initio* method (Van Vliet et al., 2003; Ogata et al., 2004).

A recent study using molecular dynamics simulations found that the stress components other than the resolved shear stress also affect the dislocation nucleation process (Tschopp & McDowell, 2005; Tschopp et al., 2007). Based on an anisotropic elasticity analysis, Li et al. (2011) derived in closed form the stress fields under Hertzian contact theory and computed the indentation Schmid factor as a ratio of the maximum resolved shear stress to the maximum contact pressure.

2.3. Indenter tip shape

It must be emphasized that equations (1) and (2) are restricted to spherical indentation and cannot be applied to arbitrary geometries. Thus, the radius of the indenter probe is an essential component for estimating dislocation nucleation shear stress inferred from spherical indentation responses. Also, access to the nanometer length scales needed to find a dislocation-free region in a metallic crystal may require very small radii. Here, experimentalists can take advantage of the imperfect manufacture of nominally sharp geometries, such as the three-sided pyramidal Berkovich indenter. The most common view is that the manufacturing process produces an approximately spherical cap on the apex of the indenter tip. The radius of the assumed spherical tip is generally obtained from the tip manufacturer, AFM, scanning electron microscopy (SEM), or by a Hertzian fit to the elastic load-displacement data (Chiu & Ngan, 2002; Constantinides et al., 2007; Gerberich et al., 1996; Gouldstone et al., 2000).

2.4. Challenge of extracting dislocation nucleation stress from nanoindentation

It is known that real pyramid indenter tips may have irregular shapes, especially at the nanometer-scale where the first pop-in event occurs. Previous finite element analysis (FEA) simulations and combined experimental and FEA studies have shown that even a highly irregular probe (which cannot simply be decomposed into "sphere" and "cone") will produce an elastic load-displacement relationship that could be perceived as having been from a spherical contact (Ma & Levine, 2007; Ma et al., 2009, 2012). Unsurprisingly, the simulations showed that the irregular shape generated shear stresses in the body that were significantly different, both in magnitude and location, from those produced by a true spherical probe. Using the common Hertzian spherical approximation to interpret experimental data can lead to a substantial underestimation of the maximum shear stress in the body at the initiation of plasticity. An assessment of the potential errors in experimental estimates of nucleation stresses is critical, especially in materials that exhibit the elastic-plastic transition at small indentation depth. We need to accurately measure the three dimensional shape of the true indenter, prepare a sample that has both low dislocation density and a smooth surface, and conduct nanoindentation experiments with accurate load and displacement measurements. In addition, several groups have reported that the rate at which the indenter tip penetrates the specimen can have a significant effect on the plastic deformation mechanisms in materials as diverse as Si (Jang et al., 2005), single-crystal Ni_3Al (Wang et al., 2003) and single-crystal Al_2O_3 (Mao et al., 2011). In the rest of this section, we will describe some of the difficulties involved in these measurements and some of the directions we, and others, are pursuing to overcome them.

2.4.1. *Direct measurement of three dimensional shape of true indenter*

There has been interest in direct measurement of the indenter geometry for at least two decades. At the larger micrometer to millimeter scale of Rockwell hardness indenters, for example, NIST has played a leading role in the drive toward indenter standardization (Song et al., 1997). Direct metrology of indenter geometry, using well-calibrated stylus profilers, is the most effective method at these size scales, and it is able to provide uncertainties low enough to support the uncertainty goals in Rockwell hardness measurements themselves.

For instrumented nanoindentation, however, the smaller sizes greatly increase the challenges to direct metrology of the indenter geometry. In the 1980s, Doerner and Nix (Doerner & Nix, 1986) measured the geometry of a Vickers indenter with transmission electron microscopy (TEM) by using a process for making carbon replicas of indents on soft surfaces.

In the early 90s, the work of Oliver and Pharr (Oliver & Pharr, 1992) led to the now prevalent method of inferring tip area from indentations on a sample of known modulus (McElhaney et al., 1998). By the late 90s, interest in the direct metrology of indenters using scanned probe microscopes was growing.

In 1998, researchers at PTB (Hasche et al., 1998) used scanned probe microscopy to characterize the geometry of Vickers microhardness indenters. Subsequently, they expanded their efforts to include Berkovich indenters (Herrmann et al., 2001).

During the following decade, work in this area significantly increased. Aldrich-Smith and coworkers (Aldrich-Smith et al., 2005) at the National Physical Laboratory (NPL) of the UK used a traceably-calibrated AFM to measure the tip area function of indenters, and compared these with the area function derived from indentation.

VanLandingham and coworkers (VanLandingham et al., 2005) undertook a detailed investigation of indenter tip shape as measured by AFM compared with that inferred by indentation on fused silica. They also developed a heuristic model of the indenter geometry as a spherical cap in place of an ideal apex on the Berkovich pyramid. For a vertical calibration range of less than 1 μm, they concluded that measurement of tip area by AFM typically performed better than the indentation method.

In 2007, McMinis and coworkers (McMinis et al., 2007) used the heuristic two-term polynomial model to fit the tip area function measured by AFM. The motivation for their model also was the assumption of an approximately spherical cap. However, in contrast with VanLandingham and coworkers, McMinis, et al. (2007) focussed on much smaller indentation depths – and found excellent performance of their approach relative to indentation on fused silica – particularly for depths less than 7 nm.

More recently, Munoz-Paniagua and coworkers (Munoz-Paniagua et al., 2010) effectively used a hyperbolic model to fit the geometry of worn Berkovich indenters measured by AFM. One of their illuminating observations was that the use of the indentation-derived shape area function could result in errors of the same scale as the assumption of an ideal Berkovich geometry.

Measurement of nanoindenter tip geometry by AFM remains an area of active investigation, and the next decade may well be more prolific than the last in terms of new insights and refinement of methods.

2.4.2. *Sample preparation*

It has been clearly demonstrated that irregularities in the tip shape can produce substantial changes in the perceived internal shear stresses. It is likely that irregularities in the sample topography can produce similar problems, and quantitative assessments of this affect are planned. In the meantime, it is important to minimize these problems by using samples that are as flat as possible. Unfortunately, mechanical polishing (even using colloidal silica) of face-centered cubic (FCC) metals can introduce substantial numbers of dislocations into the near-surface region of the sample which can adversely affect the measurement. The following procedure has been effective at largely eliminating these problems for Ni specimens, but quantitative assessments still need to be done. The measurements described in the next section used tungsten specimens, which are less prone to damage during polishing.

We start with a {111}-oriented, chemi-mechanically polished, metal single crystal that is acquired from a commercial vendor. Optical microscopy generally shows no scratches and an apparently perfect mirror finish. Further characterization using optical interferometry and AFM have demonstrated that these samples have a surface roughness, *Rq*, of about 2.5 nm and exhibit numerous small scratches that are about 20 nm deep. To relieve any residual stresses and decrease the dislocation density of the sample, the Ni single crystal is slowly heated to 850 °C in ultra-high-vacuum and annealed at this temperature for about 140 h before slow cooling back to room temperature.

Remarkably, optical microscopy of the annealed sample shows numerous long, straight ridges that resemble reversed polishing scratches. The ridge orientations are random and do not correspond to intersections between the {111} slip planes and the {111} surface. The most likely explanation is that early polishing stages introduced substantial compressive residual stresses that are "locked in" by dislocation structures. The annealing removes these dislocations and the stress relief is accompanied by projection of the previously compressed material above the sample surface. These ridges are removed by electropolishing the sample surface for 90 s at a constant current density of 1 A/cm^2 using a electropolishing solution of 37 % phosphoric acid, 56 % glycerol, and 7 % water, by volume. This final step produces an optically perfect surface, although measurements show residual broad features with maximum heights of around 20 nm to 30 nm. These features can be avoided by using AFM to locate flat regions suitable for nanoindentation.

Future work will use sub-micrometer, depth-resolved diffraction (Levine et al., 2006; Levine et al., 2011) at sector 34 of the Advanced Photon Source at Argonne National Laboratory to characterize the local dislocation density, residual elastic strain tensor, and crystallographic orientation throughout the volume beneath the touchdown point, both before and after nanoindentation. Such characterization can both verify the suitability of the target location and provide a quantitative measure of the microstructural changes and residual stresses that result from a nanoindentation measurement.

2.4.3. Nanoindentation experiment

In addition to having detailed knowledge of indenter tip shape and initial sample conditions, accurate experimental determination of properties such as initial plastic yield stress and elastic moduli also requires that the instrumented nanoindentation instrument produces accurate load and displacement data. As will be shown below, the measurement ranges of interest for studying the onset of plastic yield are typically below 1 mN for applied force (load) and below 20 nm for the displacement of the indenter tip into the specimen surface. If the accuracy in mechanical properties calculated from force and displacement data is desired at the level of a few percent, then both force and displacement data must be accurate to approximately 1 % in these ranges; this is a challenging task.

In the calibration of force, the two greatest challenges are obtaining accurate reference forces and transferring those reference forces to the nanoindentation instrument. Recently, however, techniques have been developed to realize forces down to 10 nN that are traceable to the International System of Units (SI) through the use of electrostatic forces, and those forces can be transferred traceably to nanoindentation instruments with the same level of accuracy using transfer force cells (Pratt et al., 2005) that can be placed directly in the instrument, in place of a specimen.

The problem of measuring indentation depth poses a different set of problems. In principle, length metrology with the necessary level of accuracy is easily available through the use of laser interferometer systems (see, for example, Smith et al., 2009). However, the indenter penetration depth, *h*, is necessarily defined as the depth to which the indenter penetrates the original plane of the specimen surface. This, however, is not the displacement that many commercial nanoindentation instruments measure; very often, they measure the motion of the indenter tip and the shaft to which it is mounted relative to housing (often referred to as a "load head") that contains the indenter shaft. In this arrangement, the compliance of the load frame – the mechanical path between the load head and the specimen – becomes a direct source of error in determining *h* from the measured displacement (Oliver & Pharr, 1992). Future work will include the design, construction and use of a nanoindention instrument capable of non-contact sensing of the location of the specimen surface, allowing a more direct measurement of penetration depth. This instrument will provide an SI-traceable path for force calibration as well.

3. Combined experimental and modeling method

In this section, we introduce our combined experimental and finite element analysis (FEA) method to study the effect of spherical indenter tip assumption on the initial plastic yield stress of <100>-oriented single crystal tungsten from nanoindentation.

3.1. Indenter tip shape measurement

The near-apex shape of two real diamond Berkovich indenters, one lightly and another heavily used, were measured using an Asylum Research MFP-3D[1] (Santa Barbara, CA) AFM in intermittent contact mode. This instrument uses a two-dimensional flexure stage under closed-loop control with position measurement by linear variable differential transformers (LVDTs). Motion of the stage is therefore decoupled from the vertical motion of the scanner, eliminating cross-coupling artifacts found in tube-type scanners. The vertical motion is also under closed-loop control with displacement measured by LVDT.

The Berkovich probes were cleaned using carbon-dioxide "snow" (Morris 2009). Silicon cantilevers with tip radii of 5 nm to 10 nm and approximately 300 kHz bending resonance

[1] Certain commercial equipment, instruments, software, or materials are identified in this paper to foster understanding. Such identification does not imply recommendation or endorsement by the National Institute of Standards and Technology, nor does it imply that the materials or equipment identified are necessarily the best available for the purpose.

frequencies were used to image the indenters. The AFM images were taken at a resolution of 256 x 256 pixels in a 1.25 μm × 1.25 μm region. A sharp silicon-spike tip characterizer (TGT-1, NT-MDT, Moscow, Russia) was scanned before and after imaging to verify that there were no double-tip artifacts or other deformities. Four hours were allowed before the final shape measurement to allow the temperature to stabilize within ± 0.1 °C. Drift-rate characterization experiments utilizing image correlation software showed that this reduced thermal expansion/contraction drift to less than 0.5 nm min^{-1}.

3.2. Nanoindentation experiment

The nanoindentation experiments were previously described in Li, et al. (2012). Cylindrical tungsten crystals were prepared by first grinding to SiC 600 grit. Grinding damage at the surface was then removed by electropolishing for several minutes at 20 V in a solution of 50 % glycerol (by volume) and 50 % water with 1.25×10^{-3} mol m^{-3} NaOH (Vander Voort, 1984; Zbib & Bahr, 2007). Electropolished tungsten at room temperature has a negligible (sub-nanometer thick) oxide present (Bahr et al., 1998).

The indentations on the electropolished tungsten were performed with a Hysitron (Minneapolis, MN) Tribo-Scope attachment to a Park (Santa Clara, CA) Autoprobe CP scanning probe microscope in ambient laboratory conditions. For the lightly used indenter, the loading schedule was as follows: loading to 100 μN, a partial unloading to 80 μN, then final loading to 700 μN at rates of 20 μN s^{-1}. The indenter was held at the peak load for 5 s, then unloaded to 140 μN at 60 μN s^{-1}. The purpose of the initial load-unload sequence was to inspect for plastic deformation prior to the first pop-in event. If the load-displacement data were different prior to the first pop-in event, then the elastic analysis was not used.

3.3. Finite element modeling

The commercial FEA package, Abaqus (Abaqus, 2011), was used for modeling the nanoindentation of a single-crystal tungsten sample. To accurately simulate nanoindentation in the elastic regime, the two AFM-measured indenter shapes were directly input into FEA models (Ma & Levine, 2007; Ma et al., 2009, 2012). First, high-frequency measurement noise in the AFM data was removed through a combination of median and averaging filters. No change in the orientation of the image was made (as would occur if the image were plane-fitted and leveled). After filtering, the discrete AFM data were interpolated to a 2.44 nm spacing using bi-cubic interpolation. Next, since W is a very stiff material, the AFM-measured indenter tip shapes were meshed with three dimensional (3D) deformable elements with a flat rigid plate at the top. A 2D elastic-deformation parameter study was carried out to determine the distance from the rigid plate to the 3D deformable indenter tip that was needed to avoid introducing artifacts into the nanoindentation simulations. In addition, complementary simulations of indentation by parabolic probes took advantage of the full symmetry of a sphere and four-fold symmetry of {100} W, so only one quarter of a sphere and specimen were modeled to reduce the computational burden. In all cases, contact between the indenter and specimen was assumed to be frictionless.

For the details of the FEA specimen mesh for both indenters, please see (Ma et al., 2009, 2012). The complete tungsten specimen and diamond indenter mesh was created using eight-node reduced-integration linear brick elements. In the contact region, the FEA mesh is high-density. The diamond element length is 2.44 nm. For the tungsten specimen, each element has a length of 1.266 nm (4 unit cells across) with nodes aligned to tungsten atom positions. This will facilitate future atomistic simulations that use the under-load state as a boundary condition (Wagner et al., 2008). Constitutive behavior was linear elastic, and fully anisotropic for the tungsten and isotropic for the diamond. Expressed using the crystallographic body-centered cubic (BCC) basis vectors, each edge of the tungsten mesh elements is <100>. The stiffnesses of tungsten are C_{11} = 522 GPa, C_{12} = 204 GPa, and C_{44} = 161 GPa (Lide, 2007); For the diamond indenter, the Young's modulus is selected from the published indentation modulus of {100} diamond (1126 GPa) and Poisson's ratio is 0.07 (Vlassak et al., 2003).

4. Results and discussion

4.1. Indenter tip shape

Figures 2(a) and 2(b) are the AFM rendering images of the lightly- and heavily-used Berkovich indenters, respectively. Figures 2(c) and 2(d) are the corresponding contour plots of the near-apex region with 2 nm contour spacing after interpolation and filtering, as described in Sec. 3.3. Note that the contour plots for both indenters are plotted to the same scale. It can be seen that the lightly-used indenter is much sharper while the heavily-used indenter is flattened and blunt in the near-apex region, as expected. From the data shown in Figs. 2(c) and 2(d), the radius of curvature at the apex may be estimated, although not in a unique way. It is difficult to assign a unique "radius of curvature" to a point on a general 3D object (Bei et al., 2005; Constantinides et al., 2007; Ma et al., 2009; Tyulyukovskiy & Huber, 2007). Constantinides, et al. (2007) approached this problem when characterizing conospherical diamond probes by fitting osculating paraboloids to sub-areas within the AFM images, and assigning a local radius to the pixel located at the vertex of the paraboloid. Tyulyukovskiy & Huber (2007) characterized AFM-measured shapes of spherically machined diamond probes by fitting the projected-area function to an equation that described a sphere with increasing imperfection closest to the apex. In the current study, the radius of curvature was estimated by fitting parabolas to several evenly spaced profiles through the apex (Ma et al., 2009). The mean radius estimated by this method is 121 nm ± 13 nm for the lightly-used indenter tip and 1150 nm ± 14 nm for the heavily-used indenter tip. All measurement uncertainties in this chapter are one standard deviation.

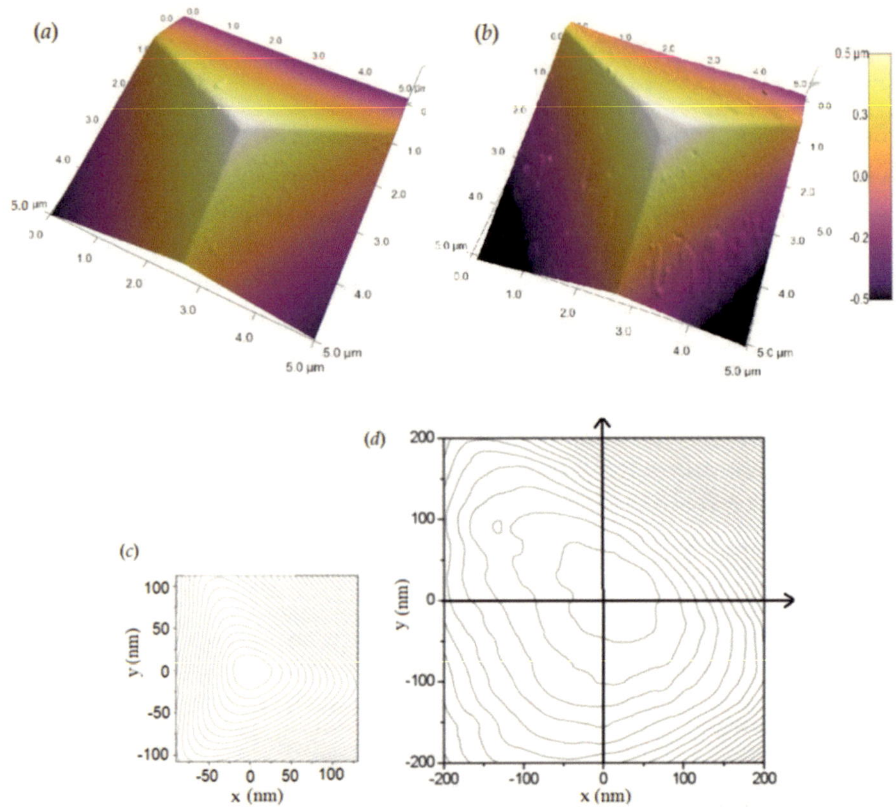

Figure 2. Three-dimensional AFM rendering image of the lightly (a) and heavily (b) used indenters, and contour plot of the surface of the lightly (c) and heavily (d) used indenter shape in the near-apex region with 2 nm contour spacing after interpolation and filtering.

4.2. Nanoindentation experiments

The AFM-measured sharp and blunt Berkovich probes were used to perform load-controlled nanoindentation experiments on the same <100>-oriented single-crystal tungsten specimen. For both indenters, ten indentation tests were performed as described in Sec. 3.2. All indentations exhibited a significant first pop-in event, but only six indentations using the sharp indenter and five indentations using the blunt indenter showed no plastic deformation during the initial unloading-reloading sequence, as described above. These are the only indentation tests that were analyzed further. Figure 3 shows load versus displacement data for two representative indentation experiments on tungsten, one using the sharp indenter and the other using the blunt indenter. In both sets of data, the nucleation event is seen clearly at the critical load, P_{CRIT}. The indenter radius was estimated by a least-squares fit of the Hertzian relationship to data preceding P_{CRIT},

$$P = \frac{4}{3}\bar{E}_{R} R^{1/2} \left(h - h_{OFF}\right)^{3/2}, \tag{3}$$

where an additional offset displacement, h_{OFF}, is used to adjust for uncertainty in the point of first contact. A reduced elastic indentation modulus of $\overline{E}_R = 317$ GPa, found from the published indentation modulus of {100} diamond (Vlassak et al., 2003) and the stiffness coefficients of tungsten listed in Section 3.3 (for $\overline{E}_W = 442$ GPa), was used to estimate R. Offset displacements found from fitting were less than 1 nm for both indentation experiments. Averaging over all of the accepted indentation tests, the mean radii are 163 nm ± 13 nm for the sharp indenter and 1160 nm ± 22 nm for the blunt indenter. This radius for the sharp indenter is significantly larger (≈ 35 %) than that inferred from the AFM measurement, while those for the blunt indenter are in good agreement.

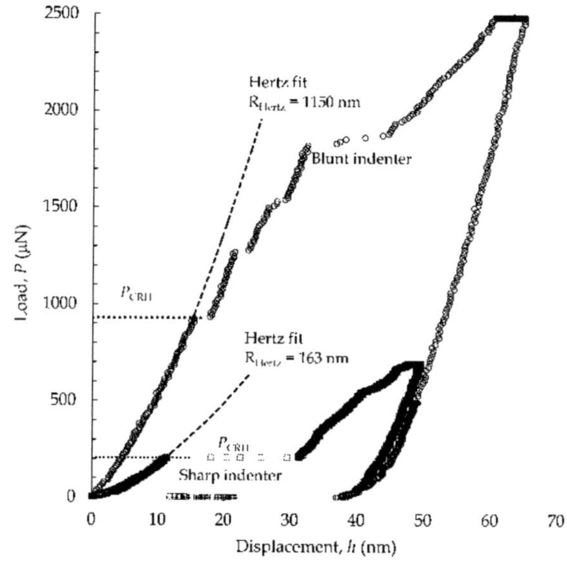

Figure 3. Load as a function of displacement for two representative indentation experiments on tungsten, one using the sharp indenter and the other using the blunt indenter. The critical load for plastic deformation, P_{CRIT}, is shown on each. A Hertzian fit to the initial elastic loading portion is drawn. The statistical uncertainty in P_{CRIT} is described in the text.

From the average indentation load at the first pop-in event (P_{CRIT} = 233 μN ± 90 μN for the sharp indenter and P_{CRIT} = 851 μN ± 131 μN for the blunt indenter) and the radius of the probe as inferred from fits to Eq. 3, the maximum shear stress, τ_{MAX}, may be estimated from Eq. 2, which assumes an isotropic material with Poisson's ratio equal to 0.3. The test-by-test average τ_{MAX}, estimated at the displacement excursion, was τ_{MAX} = 16.5 GPa ± 1.6 GPa for the sharp indenter and 7.1 GPa ± 0.4 GPa for the blunt indenter. The τ_{MAX} for the sharp indenter agrees with our earlier nucleation shear stress estimates on W using a rigid indenter simulation with a scaled modulus (Ma et al., 2009; Zbib & Bahr, 2007; Bahr et al., 1998). The τ_{MAX} obtained using the blunt indenter is much smaller, only about 43 % of that from the sharp indenter. This difference in τ_{MAX} is consistent with nanoindentation size effects reported by other researchers (Morris et al., 2004; Shim et al., 2008). In those studies, the maximum shear stress under small spherical indenters at the first pop-in was found to be very high, on the order of the theoretical strength; for larger spheres, the maximum stress decreased with increasing indenter radius.

The most likely explanation for this observed behavior is that the number of dislocations in the stressed volume increases with indenter radius. Thus, for sufficiently small indenters, the material in the highly stressed zone underneath the indenter is likely to be dislocation-free, requiring dislocation nucleation at the theoretical stress. As the indenter radius increases, there is an increasing likelihood that a nearby dislocation assists in the onset of plasticity. When a sufficiently large volume of material is probed, plasticity tends to be initiated by the motion of preexisting dislocations rather than by the nucleation of new dislocations.

Mason et al. (2006) and Schuh and Lund (2004) have suggested that pop-in events should have a rate-dependence wherein higher loading rates will generally correspond to higher pop-in loads. To examine this we have carried out a series of indentations into a model BCC system [Fe-3%Si by weight single crystal in the (100) orientation] using a Berkovich indenter similar to that shown in the remainder of this study. The indentations were made in the same Hysitron Triboindenter system, and regions of the sample were examined using scanning probe imaging prior to indentation. Between 50 and 100 indentations were made at loading rates of 5, 50, and 100 mN/s, and all samples that exhibited solely elastic loading prior to yield were selected. The cumulative probability of yield, as described by Schuh and Lund (2006), was used to determine the effect of loading rate on yield behavior. As shown in Fig. 4, there is no statistically significant effect of the load (and therefore stress) at yield due to indentation rate over the range tested. The experimentally observed results are similar to the results by Rajulapati et al. (2010), Wang et al. (2011), and Vadalakonda et al. (2006) in that the onset of plasticity is not significantly impacted by the loading rate for Ta (bcc), Mo (bcc), and Ni (fcc) single crystals, and polycrystalline W, Fe, and Ni. This suggests that the rate dependence observed by Mason et al. (2006) and Schuh and Lund (2004) in Pt may imply there are multiple processes that control the onset of plastic deformation, and that loading rate may impact some, but not all, of these processes.

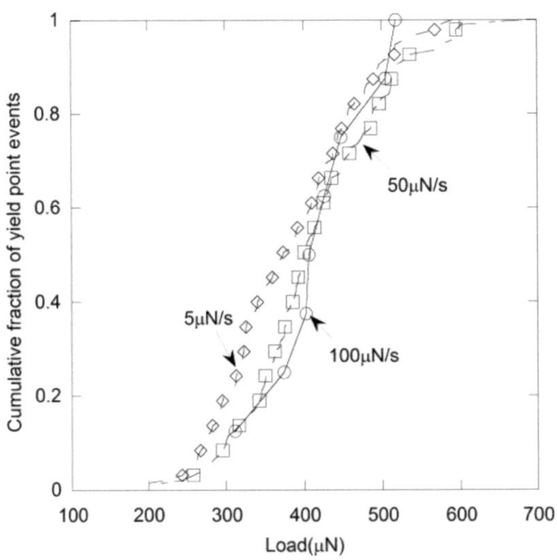

Figure 4. Rate dependence on yield behavior in Fe-3%Si sample. No effect of loading rate is noted in this material for these rates, the cumulative fraction of yield points as a function of load at which the yield point occurs is indistinguishable over these loading rates.

4.3. Comparison of FEA results to experiment

In general, directly modeling an indentation experiment using the FEA should produce a more accurate estimate of the indentation stresses than the simple Hertz approximation. However, it should be understood that there are computational difficulties with 3D FEA which prohibit a completely faithful virtual representation of the experiment and the computational times must be kept manageable. Generally, the indenter is kept rigid to avoid a full interior meshing and consequent increase in simulation size. This is not a problem when simulating the indentation of materials that are much more compliant than diamond; however, the stiffness of tungsten is an appreciable fraction of that for diamond. Nevertheless, our earlier work on a sharp indenter showed that a simulation using a rigid indenter matched the experimental results extremely well when these results were scaled using the "effective modulus" (Ma et al., 2009).

Since tungsten is very stiff, the high loads encountered during indentation can produce appreciable elastic deformation of the indenter tip, and it is necessary to quantify these changes and determine their effect on the nanoindentation experiment. Fortunately, the advent of highly-parallel computer architectures has made it possible to complete large 3D FEA models within manageable computational time. In this research, both AFM-measured indenters were input into FEA models with 3D deformable elements as described in sec 3.3. All of the displacement measurements in the FEA simulations are measured with respect to the reference point of the indenter rigid plate (just like the experimental measurements) unless otherwise specified.

Figures 5 and 6 show the measured load-displacement curves for all of the analyzed indentation experiments using the sharp and blunt indenters, respectively. The first contact points in the experimental data were adjusted slightly as described in sec 4.2 and the dotted lines indicate the major pop-in location for each experiment. The FEA-simulated elastic deformation curves (with deformable AFM-measured Berkovich indenters) are plotted for comparison. The simulated load-displacement curves and all of the experimental plots were in good agreement with each other for both the sharp and blunt indenters. The deviation between some of the measured and simulated curves in Fig. 6 at large indentation depth (above 12 nm) is caused by small amounts of plastic deformation. In some cases, this occurred slightly before the major pop-in events which are the main emphasis in this paper.

The FEA-simulated Berkovich load-displacement data for both the sharp and blunt indenters are replotted in Fig. 7. These data may be analyzed as if they came from a physical measurement, and Hertzian fits (using Eq. 3) to these data are shown as smooth curves. The resulting Hertzian-estimated radii are 163 nm for the sharp indenter and 1160 nm for the blunt indenter. These radii were then used for additional FEA simulations using deformable parabolic indenters (designated FEA_Hertz simulations) and the results are shown as black open circles for the blunt indenter and black open diamonds for the sharp indenter. All of the load displacement curves are in good agreement for the blunt indenter as shown in Fig. 7. However, the FEA_Hertz simulation for the sharp indenter exhibited a stiffer contact than that predicted from the Hertz equation; by trial and error, an FEA_HERTZ simulation using a radius of 125 nm (plotted using open squares in Fig. 7) was found to match the simulated Berkovich *P-h* relationship very well. A similar discrepancy between a rigid-sphere FEA simulation and the Hertz solution was described previously (Ma et al., 2009).

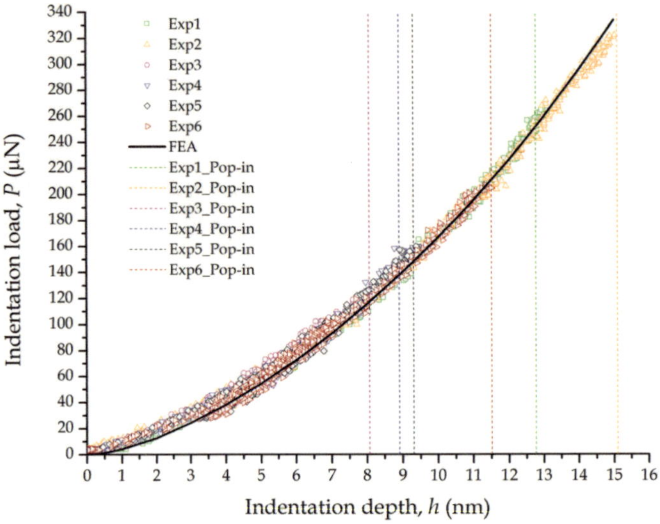

Figure 5. Measured load-displacement curves for all analyzed indentation experiments using the sharp indenter and the corresponding FEA-simulated elastic deformation curves using the AFM-measured indenter. The dotted lines indicate the major pop-in location for each experiment.

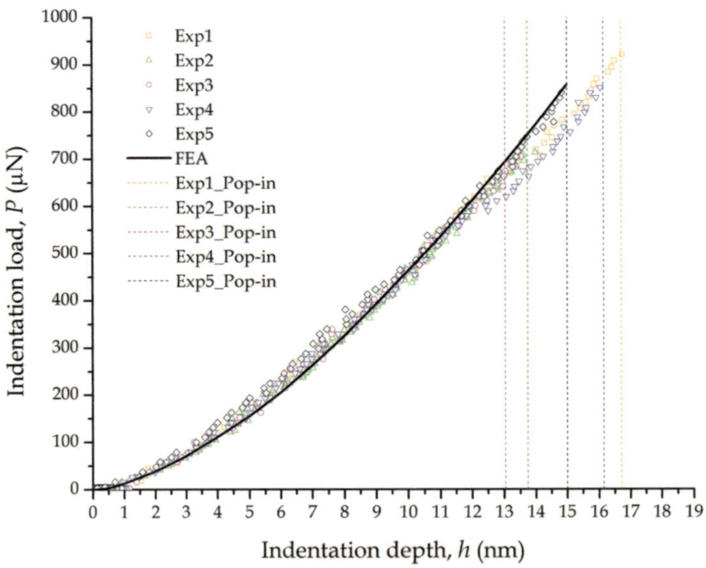

Figure 6. Measured load-displacement curves for all analyzed indentation experiments using the blunt indenter and the corresponding FEA-simulated elastic deformation curves using the AFM-measured indenter. The dotted lines indicate the major pop-in location for each experiment.

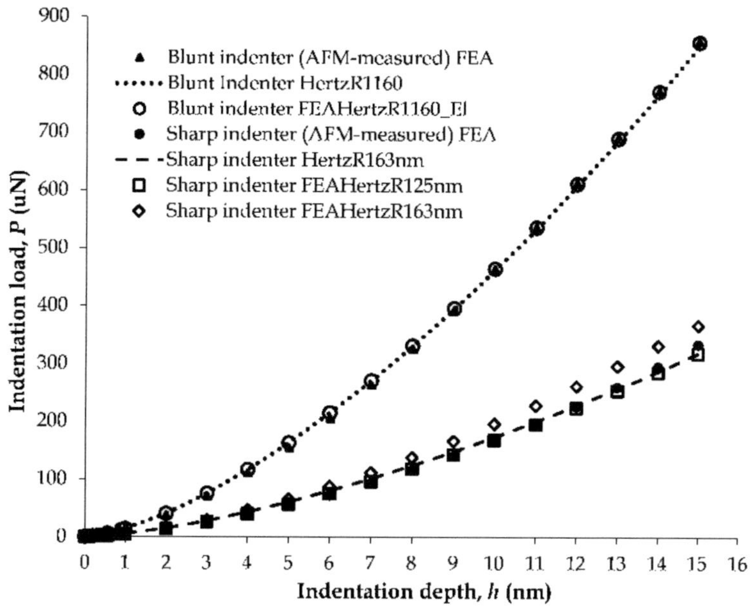

Figure 7. All of the FEA-simulated and Hertzian load-displacement curves for both the sharp and blunt indenters.

Figures 5 and 6 demonstrate that the simulated load-displacement relationships for the Berkovich-probe indentation agree with the experimental indentation results, and Fig. 7 shows that the analytic Hertz approximation and simulations using a deformable parabolic indenter can produce comparable load-displacement results. The next step is to compare the resulting simulated Tresca stresses, which is defined as the maximum difference between principal shear, throughout the specimen for all of these analysis methods.

Table 1 lists the radii and maximum shear stresses derived from the experimental data and the FEA simulations for both the sharp and blunt indenters. Figures 8a and 8b show the contact-surface shear-stress contours for the sharp indenter (Fig. 8a) at an indentation depth of 11 nm, and the blunt indenter (Fig. 8b) at an indentation depth of 15 nm, from the FEA simulations using the 3D deformable Berkovich indenter. These depths correspond to the average indentation depths where pop-ins occurred in the experiments. The contours show the shapes of the indenter tips at small scale. Figures 8c and 8d are the corresponding perspective views that show the distributions of shear stress around the locations of maximum shear stress ($\tau_{Tr}/2$). The stress contours are set so that shear stresses larger than those predicted by the Hertz approximation are shown in grey. Thus, the grey-region for the sharp indenter in Fig. 8c has $\tau_{Tr}/2 \geq 16.3$ GPa; for the blunt indenter in Fig 8d, this limit is $\tau_{Tr}/2 \geq 7.1$ GPa.

The shear-stress contours for both Berkovich probes in Fig. 8 exhibit pronounced irregularities that correspond to the non-spherical probe shape. For the sharp indenter (see Figs. 8a and 8c), the average indentation depth for the first pop-in is deep enough so that the three-sided pyramid region affects the shapes of the stress contours. Also, the innermost contour (gray color in Fig. 8c) encompasses a large volume where the shear stress exceeds the Hertzian estimate for the same indentation load and depth. Finally, although the point of maximum shear stress is located below the sample surface, in agreement with Hertz contact theory, it is shifted slightly from the central axis and towards the surface. For the blunt indenter (see Fig. 8b and 8d), the stress contours do not exhibit a clear three-sided shape such as that seen from the sharp indenter. However, the pattern of high shear stress within the sample is extremely irregular and is strongly influenced by the irregular shape of the indenter. The sample volumes where the shear stress exceeds the Hertzian estimate (gray color in Fig. 8d) are concentrated in several small local regions directly underneath the contact surface. Thus, the irregular shape of the measured indenter produces numerous volumes of highly localized shear stress that are higher than those produced by a spherical probe with an indistinguishable load-displacement relationship. In addition, the locations of these highly-stressed regions are not generally on the indentation axis and they occur much closer to the sample surface than predicted by the Hertz theory.

	Sharp Indenter	Blunt Indenter
Hertzian radius (nm)	163 ± 13	1160 ± 22
AFM measured radius (nm)	121 ± 13	1150 ± 14
FEA simulated radius (nm)	125	1160
Average depth for 1st pop-in (nm)	11	15
Load at average pop-in depth (μN)	233 ± 90	851 ± 131
τ_{MAX} (Hertz-estimated) (GPa)	16.3 ± 1.6 ≈ G/10	7.1 ± 0.4 ≈ G/20
τ_{MAX} (FEA) (GPa)	23.0 ± 2.0 ≈ G/7	14.0 ± 0.5 ≈ G/12

Table 1. Lists the radii and maximum shear stress derived from the experiment and FEA simulation for both the sharp and blunt indenters.

Figure 9a plots the maximum shear stress in the sample as a function of indentation load for the 3D Berkovich FEA simulations, the Hertz predictions, and the FEA Hertz simulation for the sharp and blunt indenters. Figure 9b plots the percent deviation of the Hertzian and FEA Hertzian predictions from the more rigorous 3D Berkovich FEA simulations. It can be seen that the maximum shear stress from all of the Hertzian and FEA Hertzian estimates are much smaller than those found from direct FEA simulation of the real indenter probe.

For the sharp indenter, all of the shear stress curves exhibit a smooth increase with increasing indentation load, consistent with the generally smooth character of the indenter tip shape. However, the FEA Hertz result deviates from the 3D Berkovich simulation by about 5 % to 8 % at the experimentally found pop-in load of P_{CRIT} = 233 μN ± 90 μN. At smaller loads, this deviation is even greater, increasing to around 20 % at an indentation load of about 6 μN. Meanwhile, the Hertzian estimate of the radius (R=163 nm) from the experimental data is much larger than those obtained from the direct AFM measurement (R = 121 nm ± 13 nm) and FEA Hertz fitting of the experimental load displacement curves (R =

125 nm). This overestimation of the sharp indenter tip radius introduces a substantial bias to the Hertz estimate of the maximum shear stress. The maximum shear stress is underestimated by more than 25 % at the pop-in loads and increases to more than 30 % at lower loads.

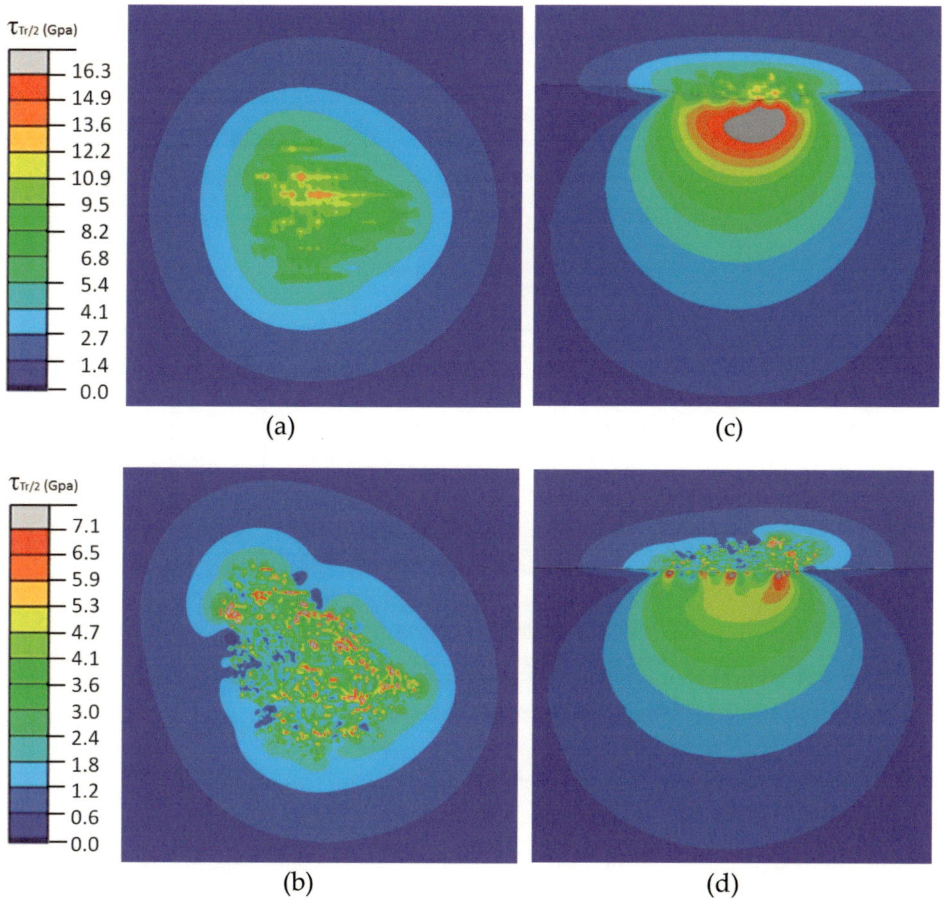

Figure 8. FEA Contact surface shear stress contours for both the sharp (a) and blunt (b) indenters and their perspective views ((c) sharp and (d) blunt indenters). The indentation depths were 11 nm for the sharp indenter and 15 nm for the blunt indenter.

For the blunt indenter, the tip radii obtained from the AFM measurement, the Hertzian estimate and the FEA Hertz simulation are all in good agreement (see Table 1). Thus, it is not surprising that the maximum shear stresses predicted by the Hertz approximation and the FEA Hertz simulation are nearly identical, as shown Fig. 9. However, the irregularity of the real indenter tip shape produces large stress excursions away from the more macroscopic stress contours. Thus, the Hertzian approximation and the FEA Hertz simulation both underestimate the maximum shear stress by more than 50 % at the pop-in loads (P_{CRIT} = 851 μN ± 131 μN) and this error increases to more than 70 % at smaller loads. This deviation is much larger than that found for the sharp indenter.

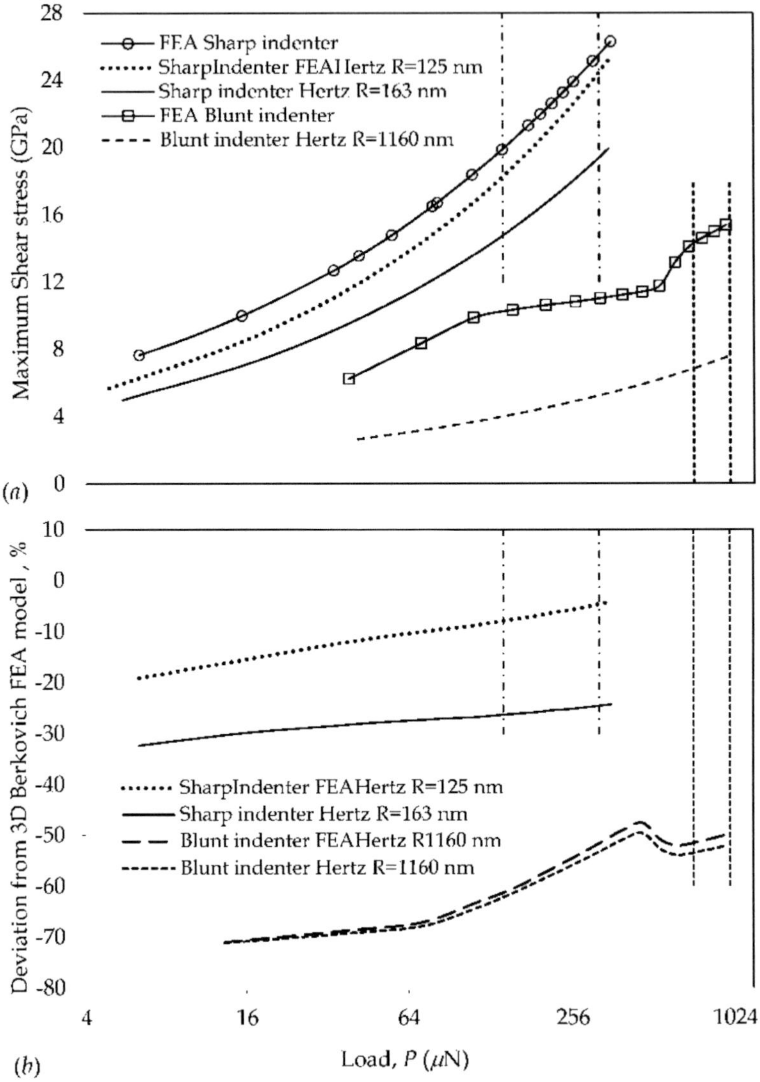

Figure 9. a) The maximum shear stress in the sample as a function of indentation load for the 3D Berkovich FEA simulations, the Hertz predictions, and the FEA Hertz simulations for the sharp and blunt indenters. B) The corresponding percent deviation of the Hertzian and FEA Hertzian predictions from the more rigorous 3D Berkovich FEA simulations.

The obvious irregularity of the blunt indenter is reflected in the distribution of maximum shear stresses within the sample (see Fig. 8b and 8d) and the behavior of τ_{MAX} with increasing indentation load (see Fig. 9). In Fig. 9a, the maximum shear stress under the blunt 3D Berkovich indenter (square with line) shows a distinct two step behavior. Thus, the stress increases fairly rapidly up to an indentation load of about 110 μN (about 4 nm indentation depth) where it abruptly slows. At an indentation load of about 538 μN (about 11 nm indentation depth) τ_{MAX} again increases rapidly with a tapering slope.

Figure 10 shows two perpendicular cross-section profiles (passing through the apex) of the 3D AFM-measured blunt indenter as shown in Fig. 2d. These cross sections show pronounced changes in curvature at a variety of distances from the apex. It is probable that such local shape changes are largely responsible for the irregular behavior of the blunt 3D Berkovich τMAX-load plot.

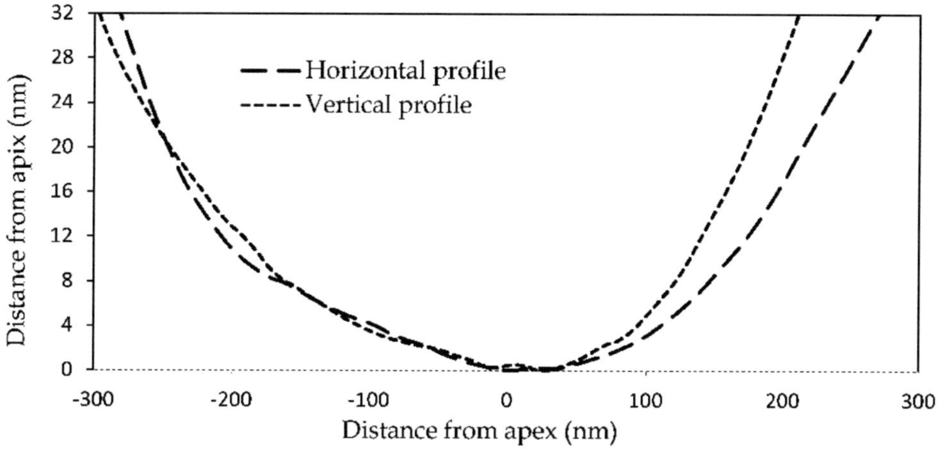

Figure 10. Two perpendicular cross-section profiles (passing through the apex) of the 3D AFM-measured blunt indenter as shown in Fig. 2d.

As shown in Fig. 2, new diamond Berkovich indenters have a faceted-pyramid shape with an approximately spherical end cap. Theoretically, the transition from the spherical cap to the pyramid geometry can be approximated using a simple geometrical model, where the spherical region is tangentially enclosed within an open-cap conical geometry, giving:

$$h^*/R = 1 - \sin\theta, \tag{4}$$

where h^* is the transition depth and θ is the semi-angle of the conical part (Cheng & Cheng, 1998; VanLandingham et al., 2005; Constantinides et al., 2007). For an axisymmetric approximation of a Berkovich probe, $\theta = 70.3°$ and $h^* = 0.0585\ R$. Thus, the sharp indenter with $R = 125$ nm has a predicted transition at $h^* = 7.3$ nm, which is smaller than any of the measured pop-in depths and explains the pronounced 3-fold symmetry of the surface-stress contours in Fig. 8a. The predicted transition for the blunt indenter with $R = 1160$ nm is considerably larger at $h^* = 67.86$ nm. In reality, after frequent indention, the diamond indenter tip will be worn on the apex (see Fig. 2), decreasing the transition depth. Nevertheless, a transition to the pyramidal region was not observed for the blunt indenter.

The approximate sphere-cone load-displacement relationship also follows a generalized power-law relationship,

$$P = Ch^m, \tag{5}$$

where C and m are constants. If fitted to Eq. 5, the P-h relationships from the 3D Berkovitch simulations have a best-fit parameter of $m = 1.59$ for the sharp indenter and $m = 1.54$ for the blunt indenter. Because $m = 1.5$ (Hertz) for a sphere, and $m = 2$ for a cone, an intermediate m is generally interpreted to mean that the spherical and conical parts of the indenter were both in contact with the material at the point of pop-in. This is certainly true for the sharp indenter as described above.

Next the contact radius, a, was estimated by the Hertzian relationship (Johnson, 1999),

$$a = \left[\frac{3PR}{4\bar{E}_R}\right]^{\frac{1}{3}} \tag{6}$$

Johnson (1999) observed that "... doubt must be cast on the Hertz results if the ratio a/R becomes too large. With metallic bodies, this restriction is ensured by the small strains at which the elastic limit is reached...". Yoffe (1984) shows that the surface displacement field, including the effect of lateral displacements in the contacted zone, for a Hertizan contact begins to deviate significantly from that of a sphere at normalized contact radius $a/R \approx 0.2$ when the Poisson's ratio of the indented material is $\upsilon \approx 0.3$. The normalized contact radius at the average pop-in position for the blunt indenter is about $a/R \approx 0.11$ for the estimated radius of 1160 mm. However, the sharp indenter shows roughly $a/R \approx 0.25$ for the Hertz estimated radius of 163 mm and $a/R \approx 0.34$ for the AFM measured radius $R =$ 121 nm. This significantly exceeds Yoffe's rue for the limits of Hertizian theory (Li et al., 2009).

Finally, it is well known that diamond indenters will exhibit substantial elastic deformation when indenting materials like tungsten that have an appreciable fraction of the diamond stiffness. This elastic deformation can be quantified by subtracting the displacement of the indenter tip from the displacement of the indenter base, here referred to as the elastic depth difference. Since almost all of the indenter deformation occurs near the tip where the cross-sectional area is small, the elastic depth difference is a well-defined quantity as long as the simulated base is far enough away from the tip as described in section 3.3. Figure 11 shows the elastic depth difference plotted as a function of indentation load for the 3D Berkovich FEA simulations and the FEA Hertz simulations for both the sharp and blunt indenters. At a given indentation load, the sharp indenter deforms the most because of its smaller cross section. This difference between the sharp and blunt indenters increases nonlinearly with the indentation force. The 3D Berkovich indenters deform more than the parabolic FEA Hertz indenters, but this difference is very small.

The amount of elastic deformation exhibited by the sharp indenter in Fig. 10 is an appreciable fraction of the total indentation depth. For example, the average pop-in depth for the sharp indenter was about 11 nm, with a corresponding load of about 196 μN. At this load, the elastic deformation depth from Fig. 11 is almost 3.7 nm, demonstrating that the shape of the indenter is being severely modified by the elastic stresses.

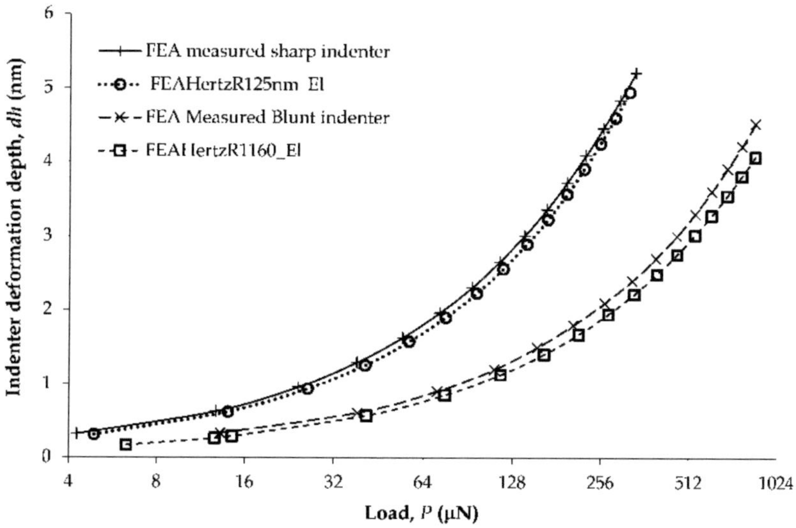

Figure 11. The elastic depth difference as a function of indentation load for the 3D Berkovich FEA simulations and the FEA Hertz simulations for both the sharp and blunt indenters.

5. Conclusion

Hertzian elastic contact theory is broadly used by the nanoindentation community to estimate the stresses required for interesting events such as pop-ins. However, the Hertzian approximation makes a number of assumptions that may not be valid for real systems; these include assumptions concerning the sphericity of the indenter tip, the flatness of the specimen, and a low ratio of indentation contact radius to indenter tip radius. In real systems, the indenter tip can be markedly irregular at nanometer length scales, the sample surface can have an irregular shape, and the indentation depth for the first pop-in can be deep enough so that the non-spherical sides of the indenter can contact the specimen. Additional problems that can occur include quantitative calibration of the nanonewton-level loads and determination of the actual indentation depth and contact area.

In this review chapter, we described some of the challenges and limitations for extracting the initial plastic yield stress from nanoindentation with the spherical indenter tip assumption, and we assessed possible errors and pitfalls of the Hertzian estimation of initial plastic yield at the nanoscale. Ultimately, solving these problems will require the development of new standard test methods and standard reference materials and this chapter describes some of our going work at NIST in that direction, including 1) measurements of the sample and indenter tip shapes using an AFM system that uses laser interferometry (traceable to NIST primary reference standards) for all distance measurements, 2) development of a custom nanoindentation instrument for nanoindentation where load and displacement are again traceable to primary NIST reference standards, 3) 3D, submicrometer resolution, synchrotron X-ray measurements of the local elastic strain tensor, crystallographic orientations, and dislocation density throughout of indentation volume both before and after indentation, and 4) combined FEA and molecular dynamics modeling of the experiments using the experimental measurement as direct inputs.

A preliminary combined experimental and FEA study was described where <100>-oriented single-crystal tungsten was used to examine the role of the indenter tip shape on nanoindentation determinations of the maximum shear stress at the dislocation nucleation point (plastic yield stress). The near-apex shape of two real Berkovich indenters, one slightly and another heavily used, were measured using AFM. These shapes were then used as a "virtual" indentation probe in a 3D FEA simulation of nanoindentation experiments on <100>-oriented single crystal tungsten. Independently, instrumented nanoindentation experiments were carried out with the real indenters on <100>-oriented single crystal tungsten. The agreement between the FEA and experimental load-displacement curves was excellent, and the FEA simulations provide a validated point-of-comparison for more conventional model predictions.

Analytic Hertzian fits and FEA simulated Hertzian simulations were also carried out, and the resulting fits to the load displacement data were excellent, even though there were substantial deviations from a spherical or parabolic geometry. These results demonstrate that even a highly irregular probe will produce an elastic load-displacement relationship that could be perceived as having originated from a spherical contact. That is, although the load-displacement data could reasonably be described by a fit using the Hertzian relationship, the contacting surfaces certainly did not fit the Hertzian criteria (specifically, that of contacting paraboloids) and the fitted radius for the small indenter was off by 35 %. Not surprisingly, the simulations showed that the irregular probes generated shear stresses in the sample that were significantly different, both in magnitude and location, from those generated with a true spherical probe. The effect of asphericity and other irregularities were dominant, focusing the indentation force into a smaller volume, and thereby increasing the maximum shear stress in the body over that produced by an ideal spherical contact. Hertzian estimates of the maximum shear stress in tungsten at the pop-in loads were smaller than those determined from FEA by over 25 % for the lightly used indenter and over 50 % for the heavily used indenter.

Author details

Li Ma, Lyle Levine, Ron Dixson and Douglas Smith
National Institute of Standards and Technology, USA

David Bahr
Washington State University, USA

Acknowledgement

The authors would like to thank Dr. Dylan Morris for measuring the Berkovich indenter tip shapes, Ms. Stefhanni Jennerjohn for conducting nanoindentation experiments and Dr.

Yoonkap Kim, whose work was supported by the National Science Foundation under grant number DMR-0907378, for the rate dependent experiments. This research work is funded and supported by the National Institute of Standards and Technology.

6. References

Abaqus/Standard (2011). *Theory and User's Manual,Version 11.1* , Dassault Systemes Simulia

Aldrich-Smith, G.; Jennett, N. M. & Hangen, U. (2005). Direct Measurement of Nanoindentation Area Sunction by Setrological AFM, *Z. Metallkd.* Vol. 96, No. 11, pp. 1267-1271, ISSN 0044-3093

Bahr, D. F.; Kramer, D. E. & Gerberich, W. W. (1998). Non-linear Deformation Mechanisms During Nanoindentation, *Acta Materialia,* Vol. 46, No. 10, pp. 3605-3617, ISSN 1359-6454

Bei, H., George, E.P., Hay, J.L., Pharr, G.M., 2005. Influence of Indenter Tip Geometry on Elastic Deformation During Nanoindentation, *Physics Review Letters,* Vol. 95, pp. 045501-045501-4

Bradby, J. E. & Williams, J. S. (2004). Pop-in Events Induced by Spherical Indentation in Compound Semiconductors, *Journal of Materials research,* Vol. 19, No. 1, pp. 380-386, ISSN 0884-291

Bradby, J. E.; Williams, J. S., Wong-Leung, J., Swain, M. V. & Munroe, P. R. (2002). Nanoindentation-induced Deformation of Ge, *Applied Physics Letters,* Vol. 80, No. 15, pp. 2651-2653

Chang, L. & Zhang, L. C. (2009). Deformation Mechanisms at Pop-out in Monocrystalline Silicon under Nanoindentation, *Acta Materialia,* Vol. 57, No. 7, pp. 2148-2153, ISSN 1359-6454

Cheng, Y. T. & Cheng, C. M. (1998). Further Analysis of Indentation Loading Curves: Effects of Tip Tounding on Mechanical Property Measurements. *Journal of Materials research,* Vol. 13 No. 4, pp. 1059-1064, ISSN 0884-2914

Chiu, Y. L. & Ngan, A. H. W. (2002). Time-dependent Characteristics of Incipient Plasticity in Nanoindentation of a Ni3Al Single Crystal, *Acta Materialia,* Vol. 50, No. 6, pp. 1599-1611, ISSN 1359-6454

Constantinides, G.; Silva, E. C. C. M., Blackman, G. S. & Vliet, K. J. V. (2007). Dealing with Imperfection: Quantifying Potential Length Scale Artefacts From Nominally Spherical Indenter Probes, *Nanotechnology,* Vol. 18, No. 30, pp. 305503 1-14, ISSN 0957-4484

Cross, G. L; Schirmeisen, W., A., Grutter, P. & Durig, U. T. (2006). Plasticity, Healing and Shakedown in Sharp-asperity Nanoindentation, *Natual Materials,* Vol. 5, No. 5, pp. 370-376, ISSN 1476-1122

Doerner, M. F. & Nix, W. D. (1986). A Method for Interpreting the Data from Depth-sensing Indentation Instruments, *Journal of Materials research,* Vol. 1, No. 04, pp. 601-609, ISSN 0884-2914

Fisher-Cripps, A. C. (2002). *Nanoindentation,* ISBN 1441998713, 9781441998712, New York: Springer-Verlag.

Gane, N. & Bowden, F. P. (1968). Microdeformation of Solids, *Journal of Applied Physics,* Vol. 39, No. 3, pp. 1432-1435, ISSN 0021-8979

Gerberich, W. W.; Nelson, J. C., Lilleodden, E. T., Anderson, P. & Wyrobek, J. T. (1996). Indentation Induced Dislocation Nucleation: The Initial Yield Point, *Acta Materialia,* Vol. 44, No. 9, pp. 3585-3598, ISSN 1359-6454

Gouldstone, A.; Koh, H.-J., Zeng, K.-Y., Giannakopoulos, A. E. & Suresh. S. (2000). Discrete and Continuous Deformation during Nanoindentation of Thin Films, *Acta Materialia,* Vol. 48, No. 9, pp. 2277-2295, ISSN 1359-6454

Haq, A. J.; Munroe, P. R., Hoffman, M., Martin, P. J. & Bendavid. A. (2007). Deformation Behaviour of DLC Coatings on (111) Silicon Substrates, *Thin Solid Films,* Vol. 516, No. 2–4, pp. 267-271, ISSN 0040-6090

Hasche, K.; Herrmann, K., Pohlenz, F. & Thiele, K. (1998). Determination of the Geometry of Microhardness Indenters with a Scanning Force Microscope, *Measurement Science and Technology,* Vol. 9, pp. 1082-1086, ISSN 0057-0233

Herrmann, K.; Hasche, K., Pohlenz, F. & Seeman, R. (2001). Characterisation of the Geometry of Indenters Used for the Micro- and Nanoindentation Method, *Measurement,* Vol. 29, pp. 201-207, ISSN 0263-2241

Jang, J.; Lance, M. J., Wen, S., Tsui, T. Y. & Pharr, G. M. (2005). Indentation-Induced Phase Transformation in Silicon: Influences of Load, Rate and Indenter Angle on the Transformation Behavior, *Acta Materialia,* Vol. 53, pp. 1759-1770, ISSN 1359-6454

Johnson, K. L. (1999). *Contact Mechanics,* ISBN-13, 978-0521347969, Cambridge university Press.

Juliano, T.; Domnich, V. & Gogotsi, Y. (2004). Examining Pressure-induced Phase Transformations in Silicon by Spherical Indentation and Raman Spectroscopy: A Statistical Study, *Journal of Materials research,* Vol. 19, No. 10, pp. 3099-3108, ISSN 0884-2914

Jungk, J. M.; Boyce, B. L., Buchheit, T. E., Friedmann, T. A., Yang, D. & Gerberich, W. W. (2006). Indentation Fracture Toughness and Acoustic Energy Release in Tetrahedral Amorphous Carbon Diamond-like Thin Films, *Acta Materialia,* Vol. 54, No. 15, pp. 4043-4052, ISSN 1359-6454

Kelchner, C. L.; Plimpton, S. J. & Hamilton, J. C. (1998). Dislocation Nucleation and Defect Structure during Surface Indentation, *Physical Review B,* Vol. 58, No. 17, pp. 11085-11088

Kiely, J. D. & Houston, J. E. (1998). Nanomechanical Properties of Au (111), (001), and (110) Surfaces, *Physical Review B,* Vol. 57, No. 19, pp. 12588-12594

Kiener, D.; Durst, K., Rester, M. & Minor, A.M. (2009). Revealing Deformation Mechanisms with Nanoindentation, *JOM,* Vol. 61, No. 3, pp. 14–23, ISSN 1047-4838

Lee, W. S. & Fong. F. J. (2008). Microstructural Study of Annealed Gold–silicon Thin Films under Nanoindentation, *Materials Science and Engineering: A,* Vol. 475, No. 1–2, pp. 319-326, ISSN 0291-5093

Levine, L. E.; Larson, B. C., Yang, W., Kassner, M. E., Tischler, J. Z., Deloas-Reyer, M. A., Fields R. J. & Liu, W. J. (2006). X-ray Microbeam Measurements of Individual Dislocation Cell Elastic Strains in Deformed Single Crystal Copper, *Nature Materials,* Vol. 5, pp. 619-622, ISSN 1476-1122

Levine, L. E.; Geantil, P., Larson, B. C., Tischler, J. Z., Kassner, M. E., Liu, W. J., Stoudt, M. R. & Tavazza, F. (20011). Disordered Long Range Internal Stresses in Deformed Copper and the Mechanisms Underlying Plastic Deformation. *Acta Materialia,* Vol. 59, pp. 5803-5811, ISSN 1359-6454

Li, J.; Van Vliet, K. J., Zhu, T., Yip, S. & Suresh, S. (2002). Atomistic Mechanisms Governing Elastic Limit and Incipient Plasticity in Crystals, *Nature,* Vol. 418, No. 6895, pp. 307-310, ISSN 0028-0836

Li, T. L; Gao, Y. F., Bei, H., George, E. P. (2011). Indentation Schmid Facor and Orientation Dependence of Nanoindentation Pop-in Behavior of NiAl Single Crystals. *Journal of the Mechanics and Physics of Solids,* Vol. 59, pp. 1147-1162, ISSN 0022-5096.

Lide, D. (2007). *CRC Handbook of Chemistry and Physics,* 87th Edition edition.

Lorenz, D.; Zeckzer, A., Hilpert, U., Grau, P., Johansen, H. & Leipner, H. S. (2003). Pop-in Effect as Homogeneous Nucleation of Dislocations during Nanoindentation. *Physica B: Condensed Matter,* Vol. 67, No. 17, pp. 172101-1-172101-4, ISSN 0163-1829

Ma, L. & Levine, L. E. (2007). Effect of the Spherical Indenter Tip Assumption on Nanoindentation, *Journal of Materials Research,* Vol. 22, No. 6, pp. 1656-1661, ISSN 0884-2914

Ma, L.; Morris, D. J., Jennerjohn, S. L., Bahr, D. F. & Levine. L. E. (2009). Finite Element Analysis and Experimental Investigation of the Hertzian Assumption on the Characterization of Initial Plastic Yield, *Journal of Materials Research,* Vol. 24, No. 3, pp. 1059-1068, ISSN 0884-2914

Ma, L.; Morris, D. J., Jennerjohn, S. L., Bahr, D. F. & Levine, L. E. (2012). The role of probe shape on the initiation of metal plasticity in nanoindentation. *Acta Materialia,* (in press, http://dx.doi.org/10.1016/j.actamat.2012.05.026), ISSN 1359-6454

Mason, J.K.; Lund, A.C. & Schuh, C.A. (2006). Determining the Activation Energy and Volume for the Onset of Plasticity During Nanoindentation, *Physics Review B,* Vol. 73, No. 5, pp. 054102 1-14, ISSN 1098-0121

Mao. W. G.; Shen, Y. G. & Lu, C. (2011). Deformation Behavior and Mechanical Properties of Polycrystalline and Single Crystal Alumina During Nanoindentation, *Scripta Materialia,* Vol. 65 pp. 127-130, ISDN 1359-6462

Nix, W.D., Greer, J.R., Feng, G. & Lilleodden, E.T. (2007). Deformation at the Nanometer and Micrometer Length Scales: Effects of Strain Gradients and Dislocation Starvation, *Thin Solid Films,* Vol. 515, pp. 3152–3157

McElhaney, K. W.; Vlassak, J. J. & Nix, W. D. (1998). Determination of Indenter Tip Geometry and Indentation Contact Area for Depth-sensing Indentation Experiments, *Journal of Materials Research,* Vol. 13, pp. 1300–1306, ISSN 0884-2914

McMinis, J.; Crombez, R., Montalvo, E. & Shen, W. (2007). Determination of the Cross-sectional Area of the Indenter in the Nano-indentation Tests, *Physica B,* Vol. 391, pp. 118-123, ISSN 0921-4526

Minor, A. M.; Lilleodden, E. T., Stach, E. A. & Morris, J. J. W. (2004). Direct Observations of Incipient Plasticity during Nanoindentation of Al, *Journal of Materials research,* Vol. 19, No. 1, pp. 176-182, ISSN 0884-2914

Minor, A. M.; Syed Asif, S. A., Shan, Z., Stach, E. A., Cyrankowski, E., Wyrobek, T. J. & Warren, O. L. (2006). A New View of the Onset of Plasticity During the

Nanoindentation of Aluminium, *Natual Materials,* Vol. 5, No. 9, pp. 697-702, ISSN 1476-1122

Misra, R. D. K.; Zhang, Z., Jia, Z., Somani, M. C. & Karjalainen, L. P. (2010). Probing Deformation Processes in Near-defect Free Volume in High Strength–high Ductility Nanograined/Ultrafine-grained (NG/UFG) Metastable Austenitic Stainless Steels, *Scripta Materialia,* Vol. 63, No. 11, pp. 1057-1060, ISSN 1359-6462

Morris, D. J. (2009). Cleaning of Diamond Nanoindentation Probes with Oxygen Plasma and Carbon Dioxide Snow, *Review of Scientific Instruments,* Vol. 80, No. 12, pp. 126102 1-3, ISSN 0034-6748

Morris, D. J.; Myers, S. B. & Cook, R. F. (2004). Sharp Probes of Varying Acuity: Instrumented Indentation and Fracture Behavior, *Journal of Materials research,* Vol. 10, No. 01, pp. 165-175, ISSN 0884-2914

Munoz-Paniagua, D. J.; McDermott, M. T., Norton, P. R. & Tadayyon, S. M. (2010). Direct Tip Shape Determination of a Berkovich Indenter: Effect on Nanomechanical Property, *IEEE Transactions on Nanotechnology,* Vol. 9, No. 4, pp. 487-493, ISSN 1536-125X

Ogata, S.; Li, J., Hirosaki, N., Shibutani, Y. & Yip, S. (2004). Ideal Shear Strain of Metals and Ceramics, *Physics Review B,* Vol. 70, pp. 104104, ISSN 1098-0121

Oliver, W. C. & Pharr, G. M. (1992). An Improved Technique for Determining Hardness and Elastic Modulus Using Load and Displacement Sensing Indentation Experiments, *Journal of Materials research,* Vol. 7, No. 6, pp. 1564-1583, ISSN 0884-2914

Oliver, W. C. & Pharr, G. M. (2004). Measurement of Hardness and Elastic Modulus by Instrumented Indentation: Advances in Understanding and Refinements to Methodology, *Journal of Materials research,* Vol. 19, No. 1, pp. 3-20, ISSN 0884-2914

Page, T. F.; Oliver, W. C. & McHargue, C. J. (1992). The deformation Behavior of Ceramic Crystals Subjected to Very Low Load (nano)Indentations, *Journal of Materials research,* Vol. 7, No. 2, pp.450-473, ISSN 0884-2914

Pratt, J. R.; Kramar, J. A., Newell, D. B. & Smith, D. T. (2005). Review of SI Traceable Force Metrology for Instrumented Indentation and Atomic Force Microscopy. *Measurement Science and Technology,* Vol. 16, No. 11, pp. 2129-2137, ISSN 0291-5093

Rajulapati, K. V.; Biener, M. M., Biener, J. & Hodge, A. M. (2010). Temperature Dependence of the Plastic Flow Behavior of Tantalum, *Philosophical Magazine Letters,* Vol. 90, No. 1, pp. 35-42, ISSN 0950-0839

Ruffell, S.; Bradby, J. E., Williams, J. S. & Munroe, P. (2007). Formation and Growth of Nanoindentation-induced High Pressure Phases in Crystalline and Amorphous Silicon, *Journal of Applied Physics,* Vol. 102, No. 6, pp. 063521-063528, ISSN 0021-8979

Schuh, C. A. & Lund, A. C. (2004). Application of Nucleation Theory to the Rate Dependence of Incipient Plasticity during Nanoindentation, *Journal of Materials research,* Vol. 19, No. 7, pp. 2152-2158, ISSN 0884-2914

Schuh, C. A.; Mason, J. K. & Lund, A. C. (2005). Quantitative Insight into Dislocation Nucleation from High-temperature Nanoindentation Experiments, *Nature Materials,* Vol. 4, No. 8, pp. 617-621, ISSN 1476-1122

Shim, S.; Bei, H., George, E. P. & Pharr, G. M. (2008). A Different Type of Indentation Size Effect, *Scripta Materialia,* Vol. 59, No. 10, pp. 1095-1098, ISSN 1359-6462

Smith, D. T.; Pratt, J. R. & Howard, L. P. (2009). A Fiber-optic Interferometer with Subpicometer Resolution for dc and Low-frequency Displacement Measurement, *Review of Scientific Instruments,* Vol. 80, pp. 035105, ISSN 0034-6748

Song, J.-F.; Low, S., Pitchure, D., Germak, A., DeSogus, S., Polzin, T., Yang, H.-Q., Ishida, H., & Barbato, G. (1997). Establishing a World-wide Unified Rockwell Hardness Scale with Metrological Traceability, *Metrologia,* Vol. 34, pp. 331-342, ISSN 0026-1394

Suresh, S.; Nieh, T.G. & Choi, B.W. (1999). Nano-indentation of Copper Thin Films on Silicon Substrates, *Scripta Materialia,* Vol. 41, pp. 951–957, ISSN 1359-6462

Szlufarska, I.; Kalia, R. K., Nakano, A. & Vashishta, P. (2007). A Molecular Dynamics Study of Nanoindentation of Amorphous Silicon Carbide, *Journal of Applied Physics,* Vol. 102, No. 2, pp. 023509 1-9, ISSN 0021-8979

Tschopp, M.A. & McDowell, D.L. (2005). Influence of Single Crystal Orientation on Homogeneous Dislocation Nucleation Under Uniaxial Loading, *Journal of the Mechanics and Physics of Solids,* Vol. 56, pp. 1806–1830, ISSN 0022-5096

Tschopp, M.A.; Spearot, D.E. & McDowell, D.L. (2007). Atomistic Simulations of Homogeneous Dislocation Nucleation in Single Crystal Copper. *Modelling and Simulation in Materials Science and Engineering,* Vol. 15, pp. 693–709, ISSN 0965-0393

Tyulyukovskiy, E. & N. Huber. (2007). Neural Networks for Tip Correction of Spherical Indentation Curves from Bulk Metals and Thin Metal Films, *Journal of the Mechanics and Physics of Solids,* Vol. 55, No. 2, pp. 391-418, ISSN 0022-5096

Vadalakonda, S.; Banerjee, R., Puthcode, A. & Mirshams, R. (2006) Comparison of Incipient Plasticity in bcc and fcc Metals Studied Using Nanoindentation, *Materials Science and Engineering A,* Vol. 426, No. 1-2, pp. 208-213, ISSN 0921-5093

Van Vliet, K. J.; Li, J., Zhu, T., Yip, S. & Suresh, S. (2003). Quantifying the Early Stages of Plasticity through Nanoscale Experiments and Simulations, *Physical Review B,* Vol. 67, No. 10, pp. 104105, ISSN 1098-0121

Vander Voort, G. F. (1984). *Metallorgraphy:Principles and Practice*. New York: McGraw-Hill.

VanLandingham, M. R.; Juliano, T. F. & Hagon, M. J. (2005). Measuring Tip Shape for Instrumented Indentation Using Atomic Force Microscropy, *Measurement Science and Technology,* Vol. 16, pp. 2173-2185, ISSN 0291-5093

Vlassak, J. J.; Ciavarella, M., Barber, J. R. & Wang, X. (2003). The Indentation Modulus of Elastically Anisotropic Materials for Indenters of Arbitrary Shape, *Journal of the Mechanics and Physics of Solids,* Vol. 51, No. 9, pp. 1701-1721, ISSN 0022-5096

Wagner, R. J.; Ma, L., Tavazza, F. & Levine, L. E. (2008). Dislocation Nucleation During Nanoindentation of Aluminum, *Journal of Applied Physics,* Vol. 104, No. 11, pp. 114311-1-114311-4, ISSN 0021-8979

Wang, L.; Bei, H., Li, T. L., Gao, T. F., George, E. P. & Nieh, T. G. (2011). Determining the Activation Energies and Slip Systems for Dislocation Nucleation in Body-Centered Cubic Mo and Face-Centered Cubic Ni Single Crystals, *Scripta Materialia,* Vol. 65, No. 3, pp. 179-182, ISSN 1359-6462

Wang, W.; Jiang, C. B. & Lu, K. (2003). Deformation Behavior of Ni3Al Single Crystals During Nanoindentation, *Acta Materialia,* Vol. 51, pp. 6169-6180, ISSN 1359-6454

Warren, O. L.; Downs, S. A. & Wyrobek, T. J. (2004). Challenges and Interesting Observations Associated with Feedback-controlled Nanoindentation. *Zeitschrift Fur Metallkunde,* Vol. 95, No. 5, pp. 287-296

Yoffe, E. H. (1984). Modified Hertz Theory for Spherical Indentation, *Philosophy magazine A* Vol. 50, No. 6, pp. 813-828, ISSN 0141-8610

Zbib, A. A. & Bahr, D. F. (2007). Dislocation Nucleation and Source Activation during Nanoindentation Yield Points, *Metallurgical and Materials Transactions A-Physical Metallurgy and Materials Science,* Vol. 38A, No. 13, pp. 2249-225, ISSN 1073-5623

Chapter 3

Uncertainties and Errors in Nanoindentation

Jaroslav Menčík

Additional information is available at the end of the chapter

http://dx.doi.org/10.5772/50002

1. Introduction

Nanoindentation provides information on mechanical properties by calculation from indenter load and displacement. The basic formulae for these computations are based on certain assumptions on the material response, which are valid only under some conditions. If a factor is neglected that is important at these conditions, the results can be wrong. The user of nanoindentation techniques thus should be aware of possible influences and errors, in order to prepare the tests and process the data so as to avoid or reduce them.

Various works deal with errors in nanoindentation measurements, for example (Menčík & Swain, 1995), Chapter 4 in (Fischer-Cripps, 2004), or some of the papers quoted in the following text. This chapter provides an overview of sources of uncertainties and errors in instrumented testing. The individual cases will be discussed and formulae given, showing the relationships between the errors in input quantities and in the calculated ones. This can help in the assessment whether a certain influence should be considered, and in the formulation of demands on the tests and the accuracy of input data.

This chapter is divided according to the following sources of uncertainties and errors:

- Properties of the tested material
- Models used for data evaluation
- Properties of the indenter
- Device properties
- Specimen properties
- Initial depth of penetration
- Temperature changes, drift
- Contact profile, pile-up
- Indentation size effect
- Surface forces and adhesion
- Scatter of measured values

2. Properties of the tested material

Errors can arise if the measurement is not done under appropriate conditions. The tests can be arranged in a more effective way if more is known about the tested material in advance: whether it is elastic-plastic or brittle, and if the deformations depend on time. It is also important to know something about its microstructure, basic components, their size and the influence of possible anisotropy. Until now, most of the papers on nanoindentation were devoted to isotropic materials, and so also is this chapter. The study of anisotropic properties is beyond its scope.

The first information on mechanical properties is obtained from a simple load-unload test, which provides a general idea and basic characteristics such as hardness and elastic modulus and also informs about irreversible processes (Fig. 1). All this can serve as the starting point for more detailed study. A pointed indenter is often used, but spherical indenter may be advantageous, as it enables continuous change of the mean contact pressure and characteristic strain. It can be used for the study in elastic regime, with reversible deformations, but also in the elastic-plastic regime, and enables the construction of stress-strain curves and determination of the yield strength and strain-hardening parameters of a ductile material, important also for the assessment of pile-up effects (see Section 9). In brittle materials, spherical indenter enables the determination of the limit stresses for crack formation or other irreversible phenomena, such as stress-induced phase transformations, densification, or generation of dislocations.

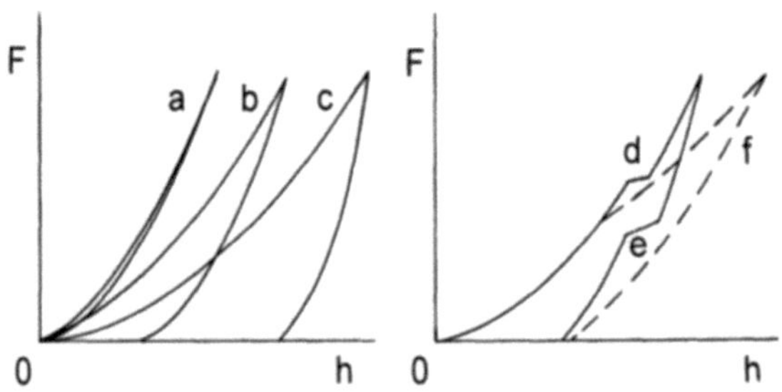

Figure 1. Typical load-depth curves of indentation into elastic-plastic materials – a schematic (after Menčík, 1996). a – highly elastic material, b – hard elastoplastic material, c – soft elastoplastic material, d – formation of cracks during loading, e – phase transformation during unloading, f – delamination of the indented coating from substrate

For components with coatings, also the coating strength and adhesion to the substrate are important. If delamination or another kind of coating damage occurs, the measured values of hardness or elastic modulus will be wrong. Nevertheless, nanoindentation can be used also for characterization of these phenomena.

Many materials exhibit time-dependent response, for example polymers or biological tissues, but also metals under certain conditions. Examples are shown in Figs. 2 and 3: the dotted curve in Fig. 2 illustrates distortion of unloading curve due to delayed deforming, Fig. 3 shows gradual indenter penetration into a polymer under constant load. Instrumented indentation enables the determination of parameters for characterization of viscoelastic response. A very important question is whether the deformations will be only delayed-reversible, or if also irreversible creep will be present. A role can be played by the magnitude of acting stresses. Low stress sometimes causes only reversible delayed elastic deforming, but if it exceeds some limit value, also irreversible viscous flow can occur, sometimes together with plastic (time-independent) deformations. One should be aware that pointed indenters always cause high stresses at the tip. Thus, if a low-stress response is to be studied, a spherical indenter must be used to ensure a sufficiently low stress level.

After the first information on the time-dependent response has been obtained, one can better specify the conditions of further exploration: the time course of the tests: velocity of the initial load increase, the dwell under constant load and the velocity of unloading, or the frequency if harmonic loading is used.

Properties of some materials depend on the environment (temperature, humidity). The measurements should be done in an environment corresponding to the assumed use of the material. This is especially important for biological materials, such as tissues or bones, where the properties of a dry specimen can significantly differ from those existing in human body. The tests, therefore, are often done on samples immersed in a suitable solution. Some materials also undergo slow changes during long time. For example, the properties of a child bone differ significantly from those of an old person. Gradual changes occur also in wood, concrete, plastics and some other materials.

It should be known in advance what structural details of the material will be studied, as some phenomena are evident only at certain loads and imprint size. Information on the material microstructure and the size of its building units is very important. Most materials are inhomogeneous on microscale. If indentation is done under so high load that many structural units are contained in the indentation-influenced volume, the indenter response reflects the average property of this volume. If properties of the individual phases are to be studied, appropriately lower load must be used. For more, see Section 12.

Under low loads (mN), with imprints smaller than several μm, hardness of some materials increases with decreasing load. This, so-called indentation size effect is treated in Section 10. Under even lower loads, tens of μN and less, another material-related problem can appear. Surface forces and adhesion between the indenter and sample are no more negligible, and can dramatically change the load-depth curves, especially for soft materials such as gels or biological tissues. This must be taken into account in the preparation of measurement and their evaluation, otherwise significant errors can appear; cf. Section 11.

The test conditions must also respect if the measured property depends on temperature. In any case, the temperatures of the specimen, device and environment should be constant during the test, otherwise additional errors appear due to thermal dilatations; see Section 8.

3. Models used for data evaluation

The quality and accuracy of results depend substantially on the quality and adequacy of the models used for data evaluation. A well known example is the increase of accuracy in the determination of hardness and elastic modulus, which was attained by calculating the contact stiffness via a power-law approximation of unloading curve (Oliver & Pharr, 1992) instead of the originally suggested linear approximation. The models are often only empirical and a matter of our choice. Examples are the function for approximation of unloading curve, of a creep curve for a viscoelastic material, or of the response during indentation into a coated sample. Sometimes, the same response can be described reasonably well by several models. When planning a test or evaluating the data, one should be aware that the models for determination of material properties are valid only under some conditions and within certain range of parameters (load magnitude, test duration, etc.). The purpose of the test should therefore always be known in advance, as it is important for the choice of a model with suitable accuracy.

In this chapter, the following sources of model-related errors will be discussed: 1. Calculation of contact depth, 2. Time-dependent response, 3. Indentation into coated specimens, 4. Others.

3.1. Models for contact depth

The most often determined mechanical properties are hardness H and elastic modulus E, calculated in depth-sensing indentation tests as:

$$H = F / A, E_r = \pi^{1/2} S / \left(2A^{1/2}\right). \tag{1}$$

F is the indenter load, E_r is the reduced (= sample + indenter) modulus, defined by Eq. (18), S is the contact stiffness, determined as the slope of the unloading curve at the beginning of unloading ($S = \mathrm{d}F/\mathrm{d}h$), and A is the contact area, calculated from the contact depth h_c (Fig. 2), which is related to the total indenter displacement h as

$$h_c = \beta h - \varepsilon F / S; \tag{2}$$

β and ε are constants. Usually, $\varepsilon = 0.75$ is used for a spherical or pointed indenter, and $\beta = 1$ for spherical and 1.05 for Berkovich indenter. These values are suitable for highly elastic materials, such as hard metals, glasses and ceramics. In substances with easy plastic flow, the material sometimes piles-up around the indenter, so that the above values of β and ε can lead to underestimated values of h_c. For such cases, Bec et al. (1996) and Fujisawa & Swain (2006) recommend $\beta = 1.2$ and $\varepsilon = 1$. (For pile-up, see Section 9.) S is mostly calculated from a power-law approximation of the unloading curve proposed by Oliver & Pharr (1992), $F = A(h - h_f)^m$, with empirical constants m and h_f.

Equations (1) and (2) are valid for elastic as well as elastic-plastic indentation and any indenter shape. For elastic contact with a spherical indenter (of radius R) also Hertz′ formulae (Johnson, 1985) may be used:

$$h = \sqrt[3]{\frac{9F^2}{16RE_r{}^2}}, \quad h_c = h/2. \tag{3}$$

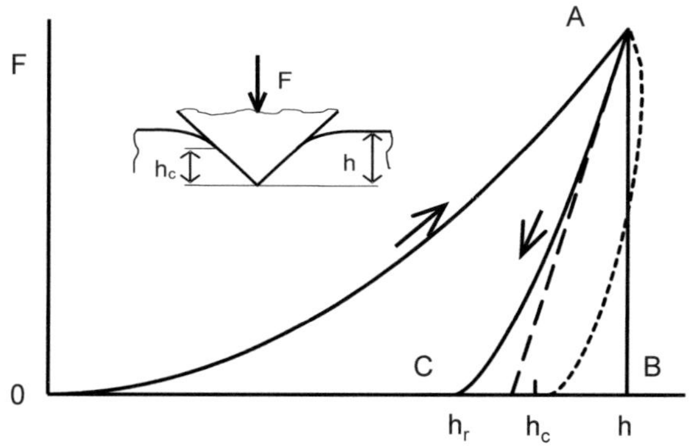

Figure 2. Load-depth curve of an indentation test. *F* – indenter load, *h* – displacement (penetration), h_c – contact depth. Dotted curve – distortion due to delayed viscoelastic response during unloading

3.2. Models for time dependent response

The load response of viscoelastic materials depends on the load and its duration. The indenter often continues penetrating into the specimen even under constant load (Fig. 3). As a consequence, the apparent elastic modulus and hardness decrease with time. This should be considered in the characterization of mechanical properties. Nevertheless, they are often determined also from common load-unload tests. Here, one more error can occur. Due to delayed response, the unloading part of the *F–h* curve is often distorted – more convex than for elastic materials (dotted curve in Fig. 2). As a consequence, the apparent contact stiffness *S*, determined from the slope of the upper part of unloading curve, is higher than the actual value. This could lead to errors in the determination of contact depth and all derived quantities, such as *H* and *E*.

The influence of viscoelastic after-effects can be reduced in various ways. Often, a dwell is inserted between the loading and unloading. According to Chudoba & Richter (2001), the effect of delayed deforming on the unloading curve may be neglected if the creep velocity has decreased so that the depth of penetration grows less than 1% per minute. The influence of creep at the end of dwell can also be mitigated by using the effective contact stiffness *S*, as proposed by Feng & Ngan (2002):

$$\frac{1}{S} = \frac{1}{S_{app}} + \frac{\dot{h}_d}{|\dot{F}_u|}; \tag{4}$$

S_{app} is the apparent stiffness, obtained from the unloading curve by the standard procedure; $\dot{h}_d$ is the indenter velocity at the end of dwell, and $\dot{F}_u$ is the load decrease rate at the beginning of unloading. A disadvantage remains that the indenter depth at the beginning of unloading after the dwell is larger than at the end of loading. This results in larger contact area and lower apparent hardness. Therefore, it is recommended to use relatively fast loading followed immediately by fast unloading (Cheng & Cheng, 2005). Nevertheless, it is generally insufficient to characterize materials, which flow under load, only by a single value of hardness or elastic modulus. Better are universal rheological models, created from springs and dashpots, which may also be used in the finite element analysis.

The parameters in these models can be obtained by fitting the time course of indenter penetration (Fig. 3) by a suitable creep function, generally in the form

$$h^{\mathrm{m}}(t) = K\psi(F, J, t) ; \tag{5}$$

m and K are constants related to the indenter shape. For pointed indenters, $m = 2$ and $K = \pi/(2 \tan\alpha)$; α is the semiangle of the equivalent cone (70.3° for Berkovich and Vickers). For a spherical indenter, $m = 3/2$ and $K = 3/(4\sqrt{R})$; R is the tip radius. $\psi(F, J, t)$ is a response

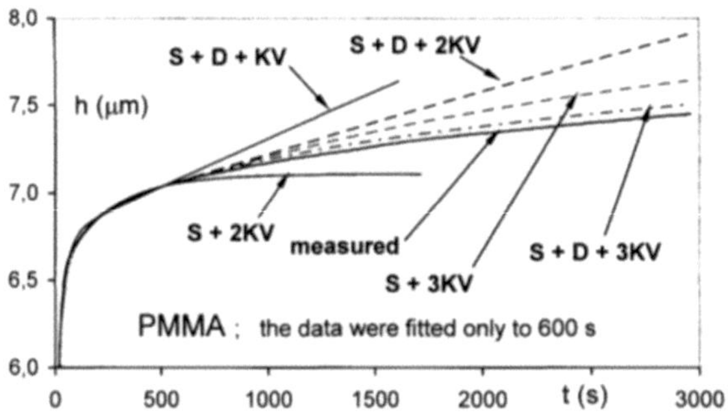

Figure 3. Time course of indenter penetration h into a polymer. Measured data and curves for various rheologic models (after Menčík, 2010). t – time, S – spring, D – dashpot, KV – Kelvin-Voigt body (the digits express the number of these bodies)

function depending on the load, material and time (Menčík et al., 2009, 2010). J is the creep compliance function, which expresses the material response to the unit step load. For monotonic load and instant deformation accompanied by reversible delayed deforming,

$$J(t) = C_0 + \mathrm{S}C_j[1 - \exp(-t / \tau_j)], \mathrm{j} = 1, \ldots \mathrm{n} , \tag{6}$$

with constants C_0, C_j and τ_j. The instantaneous compliance C_0 corresponds to a spring, and the terms with C_j correspond to Kelvin-Voigt bodies (a spring in parallel with a dashpot). For harmonic loading, trigonometrical series can be used (Menčík et al., 2004, 2011a).

The parameters in a model for viscoelastic response are usually determined from the time course of indenter penetration under constant load (Oyen, 2006, Menčík et al., 2009, 2010). However, the displacement was also influenced by the initial period of load-increase. This can be taken into account by correction factors. If the rate of load increase is constant ($\mathrm{d}F/\mathrm{d}t$ = const), the creep compliance function (6) can be modified as follows (Oyen, 2006):

$$J(t) = C_0 + \Sigma C_j[1 - \rho_j \exp(-t / \tau_j)] , \tag{7}$$

with ramp correction factors

$$r_j = (\tau_j / t_R)[\exp(t_R / \tau_j) - 1];\tag{8}$$

τ_j is so-called retardation time, t_R is the duration of load increase to the nominal value F. This correction is more reliable for shorter times t_R, where ρ_j are not very different from 1.

The practical procedure is as follows. The indenter is loaded to the nominal load F, which is then held constant. The time course of its penetration under constant load is fitted by a function of type (Menčík, 2011c, Menčík et al., 2009, 2010)

$$h^m(t) = FK[B_0 - \Sigma D_j \exp(-t / \tau_j)],\tag{9}$$

following from Eqs. (5) and (7). The constants B_0, D_j and τ_j are obtained using the least squares method. Then, the correction factors ρ_j are calculated via Eq. (8). The constants C_j in the creep function (7) are obtained as $C_j = D_j/\rho_j$. Finally, C_0 is determined as $C_0 = B_0 - \Sigma C_j$. If no permanent deformations occured, C_0 is related to the elastic modulus as $C_0 = 1/E_r$.

When choosing a model, one should know whether the time-dependent deformations are reversible or irreversible. Kelvin-Voigt elements are suitable for reversible deformations. If irreversible deformations appear, a plastic element in series with other elements must be added for time-independent deformations, or a dashpot for viscous flow; for details see Menčík et al. (2009, 2010) or Menčík (2011c). The information on the reversibility can be obtained from the unloading response – whether the deformations disappear or not. The indentation experiments should be arranged with respect to the assumed use of the material. If the stresses in operation will be low and the deformations reversible, also the test stress should be low, and a spherical indenter must be used.

In the choice of a model for time-dependent response, the following experience can be used (Menčík, 2010, 2011b, 2011c). Delayed deforming is modeled well by Kelvin-Voigt bodies. One K-V body is effective only within a limited time interval, roughly for $0.03\tau_j < t < 3\tau_j$, where τ_j is the retardation time – a material parameter of this body. For example, a K-V body with $\tau = 5$ s is effective from about 1.5 s to 15 s. If the response during a long time should be modeled, more K-V bodies must be arranged in the model, with retardation times τ_j scaled roughly by a factor of ten. It is thus useful to know approximately the duration of the process to be modeled, and the demanded accuracy. Models should not be complicated more than necessary. For long lasting processes, higher accuracy is usually needed for longer times, and lower for the initial period of load increase. The response can often be approximated similarly well by several models with different number of elements. Also the characteristic times τ_j in more complex models may vary in some range, with only small influence on the quality of the fit. Only sometimes it is possible to relate a particular K-V body (or the pertinent retardation time τ) with some process in the material. Experience with continuous stiffness measurement can be found in Pharr et al. (2009). Errors caused by the difference between harmonic and quasistatic stiffnesses in the evaluation of monotonic load-unload tests (in CSM mode) were discussed in (Menčík et al., 2005).

Generally, it is not difficult to find a good approximation for the response of a viscoelastic material within certain time interval. However, caution is necessary if the model should be used for predictions of behavior in longer times, especially if the indenter movement has not ended during the test. Example of differences in extrapolation for various models are shown in Fig. 3.

3.3. Models for indentation response of coated components

The response is influenced by the properties of the coating and substrate. The influence of substrate is negligible only if hardness of a relatively soft coating is to be determined and the indentation depth is smaller than 10% of the coating thickness. Otherwise it must be considered, especially if elastic modulus is measured, because the substrate influences the elastic response beginning from the minimum load. The determination of the coating property X_c (Young modulus, hardness, etc.) consists of measuring this property for several depths h, fitting the obtained $X(h)$ values by a suitable function (Fig. 4), and extrapolating to zero depth; $X_c = X(0)$ is found as one constant in the regression function. If CSM or DMA test mode (small harmonic loading superimposed on monotonously growing load) is used, all $X(h)$ data can be obtained in one test. General form of the fitting function is

$$X(\xi) = X_c \Phi(\xi) + X_s[1 - \Phi(\xi)] ; \tag{10}$$

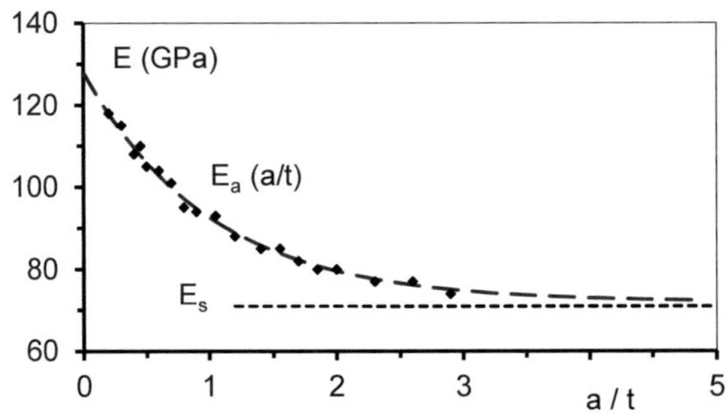

Figure 4. Apparent elastic modulus for various indenter penetration into a coating (a schematic). E_s – substrate modulus, a – contact radius, t – coating thickness

X_c and X_s is the pertinent property of the coating and the substrate, and $\Phi(\xi)$ is a non-dimensional weight function of parameter ξ, defined as the ratio of the penetration depth to the coating thickness, $\xi = h/t$, or the ratio of contact radius to the coating thickness, $\xi = a/t$. Function $\Phi(\xi)$ decreases from 1 for $\xi = 0$ (or $h = 0$) to zero for $h >> t$. The advantage of the dimensionless form of Eq. (10) is its universality: the data obtained on specimens of various thicknesses can be combined together, so that the fitting is based on more measured values.

Various functions for obtaining the coating elastic modulus are compared in Menčík et al. (1997). As most universal appeared the function proposed by Gao et al. (1992). Later, several other models were proposed. Gao´s function is based on the analytical solution of penetration of a punch into a semiinfinite elastic body with a layer on its surface. Some models, e.g. by Doerner & Nix (1986), Bhattacharya & Nix (1988), Rar, Song & Pharr (2001), Saha & Nix (2002) or Hay & Crawford (2011) correspond to the series arrangement of springs, representing the coating and substrate. All these models have theoretical substantiation and can lead to relatively reliable results. But even simple empirical models, based on exponential function $\Phi = \exp(-k\xi)$, can yield good results if the measurements are done for very small depths of penetration compared to the coating thickness. Extrapolation of the measured $X(\xi)$ values to zero depth ($\xi = 0$) should be as small as possible.

The indentation response varies between that for the coating and the substrate. The error in the determination of coating modulus will be smaller if it is close to that of the substrate. Sometimes it is possible to find a suitable substrate. It may also be useful to make the measurements for the coating deposited on two substrates: one stiffer than the coating, and the other more compliant (Menčík et al., 1997). The coating property will lie between the values obtained by extrapolating the $X(\xi)$ data for both samples to $\xi = 0$.

When fitting the measured $X(\xi)$ data by the $\Phi(\xi)$ function, one usually inserts the substrate property, X_s, as a known constant into Eq. (10). However, also this value is associated with some uncertainty and can contribute to the error in the determination of coating modulus. The pertinent substrate property thus should be known as accurately as possible. Some role is also played by the Poisson´s ratio of the coating and substrate.

The principle of determination of coating hardness is similar. This topic is treated in literature, e.g. Jönsson & Hogmark (1984), Burnett & Rickerby, 1987, Korsunsky et al., 1998). Four combinations of properties can appear: hard coating on a hard substrate, hard coating on a compliant (and ductile) substrate, soft coating on a compliant substrate, and a soft coating on a hard substrate. Approaches and functions Φ for various cases are also described in Menčík et al. (1997) and Menčík (1996). As it follows from Eq. (10), the accuracy of the determination of coating hardness depends also on the accuracy of determination the substrate hardness and the coating thickness t, appearing in the terms h/t or a/t. Sometimes, errors arise if the original (nominal) coating thickness is used in the evaluation of the $X(\xi)$ data, especially for soft coatings. High pressure under indenter can make the coating thinner. A correction for the thickness reduction can bring significant increase in accuracy of hardness as well as elastic modulus (Menčík et al., 1997).

Other sources of errors exist, as well. If a relatively soft coating was deposited on a hard and stiff substrate, the coating material can pile-up around the indenter. The contact area and stiffness are thus larger and do not correspond to the above models (cf. Section 9). Another problem occurs if the coating debonds from the substrate or if cracks arise in the coating or substrate during the indentation. In such cases the contact compliance becomes higher. These events are indicated by changes in the load-unload $F - h$ curves (Fig. 1d,f), whose shape is thus an important source of information. Also, indentation tests leading to coating delamination or nucleation of cracks can be used for the determination of material fracture toughness, interface strength or some sdhesion parameters. More about delamination by indentation can be found in Rother & Dietrich (1984), Marshall & Evans (1984), Matthewson (1986) or Li & Siegmund (2004). Determination of fracture toughness by indentation is treated in Niihara (1983), Niihara et al. (1982), Lawn et al. (1980) and Zhang et al. (2010).

3.4. Other cases

In the study of elastic-plastic behavior, the model for strain-hardening can be important; see, e.g., Field & Swain (1993), Herbert et al. (2001) and Cao & Lu (2004). If a material, consisting of several phases, is studied, the results depend on the indenter load and the number of phases considered in the model for statistical analysis; cf. Section 12. At very low loads, also surface (adhesive) forces can play a role, as explained in Section 11.

4. Indenter properties

Models for data evaluation assume certain shape and mechanical properties of the indenter. However, the actual parameters can deviate from the nominal values, used in the formulae.

4.1. Shape

The formulae for determination of mechanical properties usually work with contact area A. The indenters are mostly pointed or spherical. For an ideal pointed (conical) indenter,

$$A_{id} = kh_c^{\ 2}; \tag{11}$$

$k = \pi\,(\tan\alpha)^2$; α is the semiapical angle. For pyramidal indenters, an equivalent cone is worked with. The angle of the equivalent cone for Berkovich indenter is $\alpha \approx 24{,}5^{\circ}$.

The shape of a real indenter often deviates from the ideal one. The tip of a pointed indenter is never ideally sharp, but somewhat blunt, with the radius several tens of nm or more (Fig. 5). This deviation is important especially for very small depths of penetration. The simplest correction uses the addition of a small constant δ to the measured depth, $h_{corr} = h + \delta$. The constant δ can be obtained by fitting the series of $F(h)$ points by the expression $F = c(h + \delta)^2$. However, also the actual angle α can slightly differ from the nominal value, and even vary with depth. More reliable is indenter calibration. It provides so-called indenter area function, which expresses the projected contact area as a function of contact depth, $A = A(h_c)$. Especially for very small depths of penetration (less than several tens of nm), the difference between A_{true} and A_{id}, and the corresponding errors in the determination of A or E can amount several tens of percent.

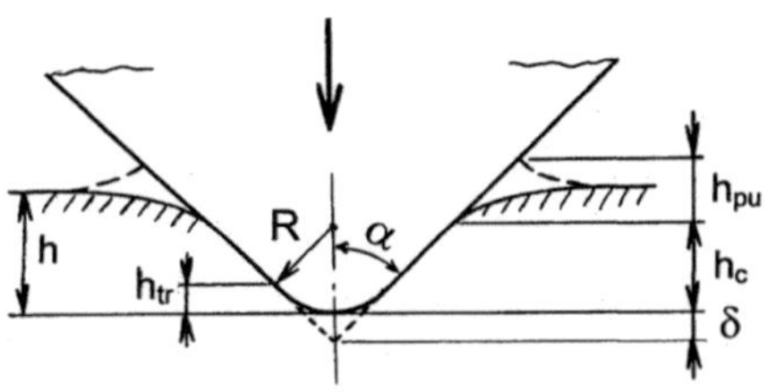

Figure 5. Shape of a real pointed indenter (a schematic). Contact profile without pileup (solid line) and with pileup (dashed line). h_{tr} – transition between the spherical tip and conical part

One method of indenter calibration determines the contact area via imaging the imprints by scanning- or atomic force microscope. More often, the indenter area function is calculated from the contact stiffness, obtained by indenting a sample of known elastic modulus:

$$A(h_c) = \pi[S(h_c)]^2 / \left(4E_r^{\ 2}\right). \tag{12}$$

The measurements must be done for a sufficient number of depths. The stiffness S can be obtained either in quasistatic manner from unloading curves, or in dynamic way, using small oscillations superimposed on the basic load (CSM or DMA mode). In this case, the area function $A(h_c)$ can be obtained in one test. The measured $A(h_c)$ values are then approximated by a suitable analytical expression, e.g. a polynomial (Oliver & Pharr, 1992):

$$A(h_c) = C_0 h_c^{\ 2} + C_1 h_c + C_2 h_c^{\ 1/2} + C_3 h_c^{\ 1/4} + C_4 h_c^{\ 1/8} + \ \dots \tag{13}$$

Indenter calibration is often done on fused silica, as its surface around the indenter deforms similarly to theoretical elastic models, without pile-up.

The contact area under a spherical indenter of radius R is

$$A_{\text{id}} = \pi a^2 = \pi\left(2Rh_c - h_c^{\ 2}\right) \approx 2\pi Rh_c. \tag{14}$$

If the deformations are elastic, Hertz′ formulae may be used as well. However, spherical indenters of small radius are never perfectly spherical, a role being sometimes played by anisotropy of properties (Menčík, 2011d). The indenters must thus be calibrated. Again, the indenter area function $A(h_c)$ can be determined via the unloading contact stiffness and formula (12). Sometimes, Hertzian formulae are used, which work with indenter radius R. In such case, the effective indenter radius, corresponding to certain contact depth, $R_{eff}(h_c)$, should be used. R_{eff} can be calculated from the actual contact area as

$$R_{\text{eff}}(h_c) = A(h_c) / (2\pi h_c)], \tag{15}$$

or from the elastic displacement of the indenter:

$$R_{\text{eff}}(h) = (9/16)F^2 E_r^{-2} h^{-3}. \tag{16}$$

This R_{eff} represents the average radius for the indenter penetration h. If the indenter shape is not exactly spherical, various R_{eff} values are obtained for various depths of penetration. In this way it is possible to construct the calibration curve $R_{eff}(h)$. Figure 6 shows calibration curve for a spherical indenter of nominal radius 200μm. The curve was obtained from the indenter penetration into fused silica, whose reduced modulus was assumed E_r = 70 GPa. The maximum load was F = 50 mN and the corresponding depth h = 130 nm. One can see that the effective radius for this depth (R_{eff} = 134 μm) differs significantly from the nominal value 200 μm. It is obvious that this indenter cannot be used for the evaluation of data for penetration less than 10 nm, and the results for depths up to 20 nm are not very reliable. Other examples of indenter calibration can be found in Oliver & Pharr (2004) or Nohava & Menčík (2012).

With pointed or spherical indenters, one should be aware that both shapes are, in principle, spheroconical. The tip is approximately spherical, but later it turns gradually into a cone or a pyramid. Thus, if one wants to use the formulae for spherical indenters, the contact depth should not be larger than the transition point between the sphere and cone (h_{tr} in Fig. 5). On the other hand, the formulae for a conical or pyramidal indenter are sufficiently accurate only if the round part at the tip is negligible compared to the total depth of penetration. In other cases, indenter area function or some correction must be used. The distance of the transition point from the tip is

$$h_{tr} = R(1 - \cos\alpha). \tag{17}$$

Sometimes, a cylindrical indenter („rigid punch") is used, for example for very compliant materials. Its advantage is constant contact area, independent of the penetration. However, the edge between the planar face and cylindrical part is not ideally sharp, and at extremely small depths of penetration the contact area slightly changes with depth. Moreover, high stress concentration at the edges can cause irreversible changes in the specimen, and thus some nonlinearity in the response.

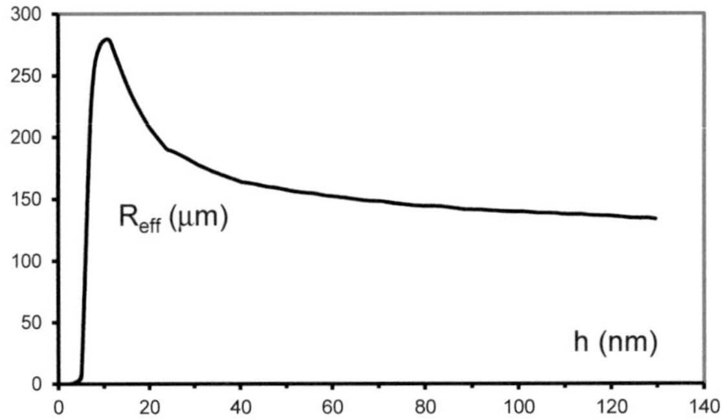

Figure 6. Effective radius R_{eff} of the spherical indenter as a function of penetration h. Nominal radius $R = 200$ μm. (After Menčík & Nohava, 2011)

The indenter roughness influences the measurement similarly as the specimen roughness (Section 6). Despite of high hardness, indenters gradually get worn, especially if very hard materials are repeatedly tested. Therefore, it is necessary to check their actual shape from time to time, and recalibrate it, if necessary.

4.2. Mechanical properties

Nanoindentation is very suitable for the determination of elastic modulus in small volumes. However, it is so-called reduced modulus E_r, which is obtained from Eq. (1). This modulus is influenced by the Young modulus of the specimen (E_s) and the indenter (E_i):

$$\frac{1}{E_r} = \frac{1}{E_s'} + \frac{1}{E_i'}; \tag{18}$$

E' means $E/(1 - \nu^2)$; ν is Poisson´s ratio. The specimen modulus is calculated from E_r as

$$E_s = (1-\nu_s^2) \Big/ \left(\frac{1}{E_r} - \frac{1-\nu_i^2}{E_i} \right). \tag{19}$$

Its value thus depends on the composite modulus E_r, but also on the modulus of the indenter and Poisson's ratio of the indenter and sample. These values, however, are often only assumed, for example E_i = 1000 GPa and ν_i = 0,07 for diamond, and the actual values can be different. (According to literature, Young modulus of diamond can vary between 800 and 1100 GPa, and indenters are made also from other materials, SiC, tungsten carbide, steel, etc.). Thus, if incorrect value of E_i is inserted into Eq. (19), the calculated sample modulus E_s is also wrong. The pertinent error will be non-negligible especially if relatively stiff materials are tested. The relative error in the sample modulus, $E_s'/E_{s,a}'$, expressed as the function of the relative error of indenter modulus $E_{i,a}'/E_i'$ and the ratio of the specimen's modulus and indenter modulus, $E_{s,a}'/E_{i,a}'$, is:

$$\frac{E_s'}{E'_{s,a}} = 1 \Big/ \left[1 + \frac{E'_{s,a}}{E'_{i,a}} \left(1 - \frac{E'_{i,a}}{E'_i} \right) \right]. \tag{20}$$

The subscript a denotes apparent value, i.e. the value obtained from the test ($E_{s,a}'$), or assumed ($E_{i,a}'$). Actually, we do not get the true specimen value from the test; we only try to obtain it from the apparent values. The situation is depicted in Fig. 7. One can see that the error in the determination of sample modulus, caused by the use of incorrect value E_i, is negligible for $E_s/E_i \leq 0.01$, but can amount several percent for $E_s/E_i > 0.1$, more for larger error in E_i and higher modulus of the tested material.

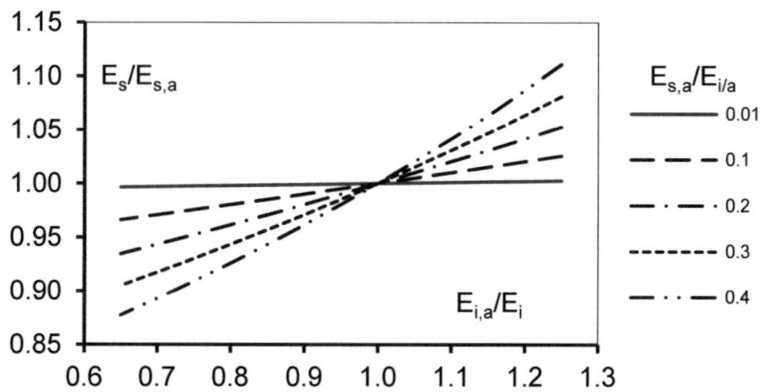

Figure 7. Relative error of sample modulus ($E_s/E_{s,a}$) as a function of relative error of indenter modulus $E_{i,a}/E_i$) and the ratio of specimen and indenter modulus E_s/E_i.
Subscripts: s – sample, i – indenter, a – apparent

The specimen modulus depends also on Poisson's ratio. As it follows from Eq. (19), the relative error is

$$\frac{E'_{s,a}}{E'_s} = \frac{1-\nu^2_{s,a}}{1-\nu^2_s}. \tag{21}$$

For example, if the Poisson's ratio $\nu_{s,a} = 0.2$ were used for the specimen instead of the correct value 0.3, the error in the determination of Young modulus would be about 5%.

5. Device properties

The results of measurement are influenced to some extent by the compliance of the device. The force acting on the indenter causes not only its penetration into the specimen, but also the deformation of the indenter shaft, the specimen holder and other parts of the device, all denoted together as the instrument frame. The total indenter displacement, measured by a sensor, h_m, is thus larger than the depth of penetration, h, by the frame deformation $h_f = C_f F$, where F is the load and C_f is the frame compliance. The actual depth of penetration is

$$h = h_m - C_f F. \quad (22)$$

C_f is one of the input quantities in the software for nanoindentation data processing. If an incorrect apparent value $C_{f,a}$ is used instead of the actual value C_f, it will cause an error in the depth of penetration h_a, contact stiffness S_a and all quantities calculated from them. The contact stiffness S is the reciprocal of the contact compliance, defined as $C = 1/S = dh/dF$. With h given by Eq. (22), the apparent stiffness is

$$S_a = \left[dh_m / dF - C_{f,a} \right]^{-1}, \quad (23)$$

If the value $C_{f,a}$ is (for example) smaller than the actual frame compliance C_f, the apparent depth h_a (= $h_m - C_{f,a}F$) will be larger than the actual depth h, the apparent contact compliance C_a will be higher and the apparent contact stiffness S smaller. The apparent contact depth,

$$h_{c,a} = h_a - \varepsilon C_a F = h_m - \varepsilon F dh_m / dF - C_{f,a} F(1 - \varepsilon), \quad (24)$$

will be larger than the actual depth. As a consequence, the apparent contact area will be larger and hardness lower than their actual values. Also the apparent elastic modulus, Eq. (1), will be lower than its actual value. For example, the results for a silicate glass (Berkovich indenter, $F = 50$ mN) were: $h = 694$ nm, $h_c = 527.9$ nm, $S = 225785$ N/m, $H = 7.32$ GPa and $E_r = 72.94$ GPa. If, instead of the correct frame compliance $C_f = 0.521$ nm/mN a wrong value $C_{f,a} = 0.3$ nm/mN would be used, the corresponding apparent values would be $h_a = 705$ nm, $h_{c,a} = 530.6$ nm, $S_a = 215054$ N/m, $H_a = 7.25$ GPa and $E_{r,a} = 69.11$ GPa. A higher value of frame compliance, for example $C_{f,a} = 1.0$ nm/mN, would give $h_a = 670$ nm, $h_{c,a} = 521.9$ nm, $S_a = 253165$ N/m, $H_a = 7.49$ GPa and $E_{r,a} = 82.72$ GPa. One can see that the wrong compliance has influenced the elastic modulus more than hardness. Generally, the differences between H_a and H and E_a and E are larger for higher loads and stiffer materials.

A wrong C_f value can also distort the trend in a series of $E(h)$ values, measured for various depths. This can be dangerous if the properties of a coating are determined by extrapolation of the data. If an unexpected increase or decrease in elastic modulus with increasing depth of penetration is observed, it is recommended to check the instrument compliance (Menčík & Swain, 1995).

The compliance C_f can be found by calibration. Several methods exist. A simple procedure is based on the general relationship between the load and depth. For a spherical indenter:

$$h_m = C_s F^{2/3} + C_f F; \tag{25}$$

C_s characterizes the compliance of contact with a spherical indenter. (For purely elastic deformations, $C_s = [(9/16)R/E_r^2]^{1/3}$.) Fitting the series of $h_m(F)$ data by Eq. (25) will yield the specimen-related compliance C_s and the frame compliance C_f. As Eq. (25) is nonlinear, a suitable fitting tool must be used (e.g. the solver in Excel is possible), with the optimization criterion „minimum of the sum of squared differences between the measured $h_m(F)$ and calculated $h_{calc}(F)$ values". As the instrument compliance is usually very low, it is recommended to make the calibration measurements up to relatively high loads. On the other hand, the deformations should be only elastic, so that the calibration block must be stiff and hard, and the indenter radius R must be at least several hundreds of μm. As Eq. (25) is based on Hertz formula for a spherical contact, the tip radius R should be constant within the used range of penetration depths. This is better achievable with larger radius.

Another method uses the fact that the derivative of Eq. (25) with respect to F,

$$dh_m / dF = (2/3) C_s F^{-1/3} + C_f, \tag{26}$$

represents a straight line in coordinates dh_m/dF vs. $F^{-1/3}$. The frame compliance C_f can be obtained as its intercept for $F^{-1/3} \to 0$. A disadvantage is the necessity of calculating the derivatives dh_m/dF. Equation (26) can also be expressed as a function of depth. More about calibration can be found in Fischer-Cripps (2004) or in manuals for nanoindenters.

The results of measurement can also be influenced by the internal noise in the measuring system, by vibrations, and by the way the indenter movement is controled. In tests done under harmonic load, also the dynamic properties of the device, its resonant frequency and damping ability can play a role. For details, the reader is referred to (e.g.) TestWorks (2002).

6. Specimen properties

The results of nanoindentation measurements are also influenced by the surface roughness and by the changes in the surface layer, caused by the specimen preparation. Reliable determination of the properties of a coating needs also a good knowledge of the substrate properties and coating thickness.

6.1. Surface roughness

No surface is ideally smooth. Even the best polished specimens have surface undulations of height from several nanometers to several tens of nanometers, and with the distances between individual hills or crests tens of nanometers and more. As a consequence, the indenter comes first into contact with the highest asperities (Fig. 8). During the load increase, the tips of the highest asperities become compressed and the indenter gradually comes into contact with lower asperities, etc. The specimen surface conforms more and more with the indenter and the response gradually approaches to that for the perfect contact. The overall response resembles that for a specimen with a thin layer of more compliant material on its surface.

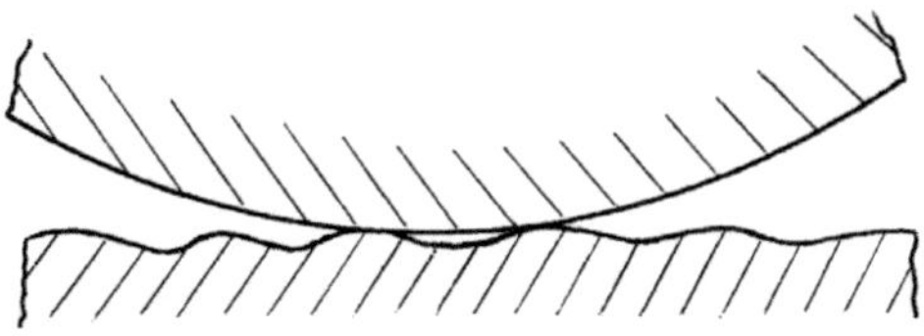

Figure 8. Indenter in contact with a surface of non-negligible roughness

A role is played not only by the shape of the asperities and by the distribution of their heights, but also by their density, by the radius and shape of the indenter tip and also by the elastic-plastic properties of the specimen. The situation resembles two nonlinear springs in series, one corresponding to the contact macrogeometry and the other to the microscopic asperities. At a first approximation, elastic contact and Hertz theory may be assumed. The influence of surface roughness can be assessed by the nondimensional roughness parameter, defined by Johnson (1976, 1985) as

$$\rho = \sigma_s R / a^2; \tag{27}$$

σ_s is the standard deviation of the height of surface asperities, R is the indenter radius, and a is the contact radius. As $a^2/R \approx 2\, h_c$, the parameter ρ is closely related with the ratio of the standard deviation of asperities height to the depth of penetration h_c. According to Johnson (1976), the influence of surface roughness on the elastic contact stiffness is small for $\rho < 0.05$. With ductile materials, also plastic flow must be considered.

6.2. Change of surface properties

The measured values depend also on the changes, undergone by the specimen surface during its preparation. In specimens of ductile materials, the properties of surface layer can be changed by polishing or another method of processing. Due to plastic flow, residual stresses can arise in this layer, and some materials also strain-harden. If brittle materials are prepared for nanoindentation, a very thin surface layer can be damaged by grinding and polishing. Hard abrasive grains cause irreversible deformations on the surface and below it and can nucleate submicroscopic cracks here, which reduce the contact stiffness.

Sometimes, the initial part of the load-depth curve looks like that in Fig. 9. The reason is a very thin hard oxidic layer, formed on the specimen surface during its preparation. It has properties different from the genuine specimen properties, and complicates also the identification of the first contact between the indenter and specimen.

7. Initial depth of penetration

One source of uncertainty in nanoindentation tests is the indenter penetration into the sample at the minimum load. The depth measurement can start only after the first contact has been made. This is at the minimum measurable load F_i, but the corresponding depth h_i (Fig. 9) is unknown yet. Nevertheless, it should be added to the measured depth h_m, so that the actual depth is

$$h = h_m + h_i. \tag{28}$$

The approximate magnitude of the initial penetration, h_i, can be obtained by fitting the initial part of the *F–h* curve and extrapolating it to zero load. Every indenter has a less or more rounded tip, and a reasonable approximation is

$$h = k\, F^n, \tag{29}$$

where *k* and *n* are constants. A suitable shape of the fitting curve is

$$h_f = k\, F^n - h_i \left(= h_{m,calc}\right). \tag{30}$$

Only the lower part (about 10 – 20 %) of the *h–F* curve should be used for the fitting.

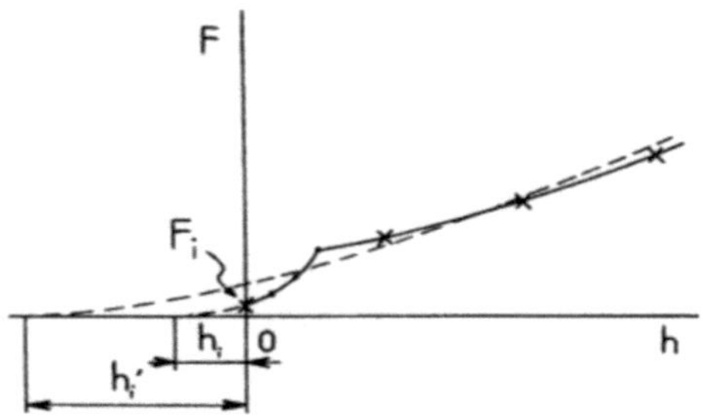

Figure 9. Determination of initial depth of penetration (a schematic). The indenter response at low loads was influenced by a thin hard oxidic layer on the surface.
F_i – initial load, h_i – initial depth of penetration, h_i' – initial depth obtained by extrapolation of measured depths (crosses) without considering the surface layer.

Three notes must be made here: 1) Sometimes, the shape of the load-depth curve at very low loads is influenced by a very thin oxidic layer on the specimen surface, formed here during its preparation (Fig. 9). The decision about what part of the *h–F* curve should be used for the determination of h_i in such case needs some experience. 2) The question of the initial depth of penetration is less important for advanced indentation devices with very high load and depth sensitivity, enabling the use of very low loads. 3) The above approach for the determination of the initial depth of penetration is not suitable for very compliant materials such as gels, where also adhesive forces play a role; see Section 11.

8. Temperature changes, drift

Errors can be caused by a gradual change of specimen or instrument dimensions due to temperature changes. If the specimen temperature at the test beginning is higher, it decreases during the measurement, and its dimensions become smaller. The sample surface moves in the same direction as the indenter tip, so that the apparent depth of indenter penetration, h_a, will be larger (Fig. 10), and so also the apparent contact depth, $h_{c,a}$. Thus, the apparent hardness, H_a, will be lower. The situation with elastic modulus is more complex. If the specimen temperature decreases only during the loading period, the apparent modulus (1) will be lower because of larger apparent contact depth and area. If, however, the temperature decreases during unloading, the indenter withdrawal from the contracting specimen lasts a shorter time and the seeming contact stiffness S_a will be higher and the apparent modulus E_a can be higher (cf. the higher slope of the unloading part of the $F-h$ diagram in Fig. 10); an opposite role is played by the larger depth. The effect of temperature decrease during the unloading is thus more complex than during loading. The opposite situation occurs if the specimen temperature during the test increases. Similar effects to the sample temperature changes can be caused by the device temperature changes.

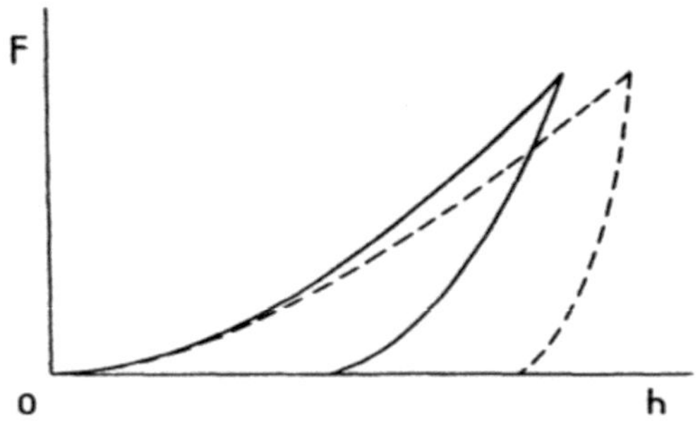

Figure 10. Load – penetration diagram without drift (solid lines) and with drift (dashed lines)

Several approaches are used to mitigate the thermal drift effects. The simplest way is to make the measurements only if all temperatures (specimen, nanoindenter) are constant during the test. Sufficient time before measurements is necessary for temperature equalization. The device is usually placed in a cabinet in order to minimize the random temperature variations. Another approach corrects the measured h_a values by subtracting the displacement caused by thermal drift. In this case, a dwell several tens of seconds is inserted into the unloading part of the test cycle (usually when the indenter load has dropped to 10% of the maximum value) to measure the velocity of the indenter movement due to the drift. This velocity is then used in a model for drift correction.

The influence of temperature is also reduced by making the critical parts of the device from materials with extremely low coefficient of thermal expansion. One device of this kind (UNHT made by CSM Instruments) uses also a reference shaft with a spherical tip, parallel with the indenter and close to it (several mm). This reference shaft, with its own actuator and sensor, is in contact with the specimen surface, so that the indenter tip displacement is measured not from the nanoindenter body, but from the specimen surface. With all these sophisticated tools, the drift in very advanced devices is lower than 1 nm per minute.

The influence of drift is also smaller if the indenter is loaded by a small harmonic signal. If the loading cycle does not last more than about 0.01 – 0.1 s, the creep or drift effects are negligible. On the other hand, if viscoelastic materials are tested, their response depends on the frequency of loading, and the harmonic contact stiffness S_h (and thus also elastic modulus) usually increases with the excitation frequency. However, if small harmonic oscillations are superimposed on the slowly growing load (CSM mode), the total indenter displacement is the same (and influenced by the drift in similar way) as in the common quasistatic load-unload cycle (Menčík et al., 2005). With harmonic loading also a danger exists that the dissipated energy can change the temperature and viscosity in the loaded volume, and thus also its mechanical properties.

9. Contact profile, pile-up

In depth-sensing tests, contact depth h_c and area A are not measured, but calculated from the total indenter displacement h via Eqs (2) and (11, 12, 13). In Equation (2) one usually assumes $\varepsilon = 0.75$, which follows from the analysis of elastic contact with a pointed or spherical indenter. According to this model, the surface around the indenter sinks down. If plastic deformations appear, ductile material is sometimes squeezed out upwards around the indenter, Fig. 5. If this pile-up occurs, the actual contact area is larger than the apparent values, and the material seems to be stiffer. If no correction for pileup is done, the calculated (apparent) hardness and elastic modulus are higher than in reality.

Important for pileup is the ratio of the elastic modulus and yield strength of the material, its strain-hardening behavior, and also the indenter shape. Pile-up around a pointed indenter is typical of materials, which have high ratio of elastic modulus to yield stress and do not strain-harden. Examples are relatively soft metals, which were cold-worked. High E/Y ratio (low hardness) means relatively large plastic zone, and thus a large volume of material around the indenter, which could be displaced by plastic flow upwards. In materials with strain-hardening, the yield stress increases during plastic deforming. The intensively deformed surface layer around the indenter becomes harder and constrains the upward flow of the material. For hard and highly elastic materials, such as glasses and ceramics, the E/H ratio is lower, the size of the plastic zone is small and the volume of the material displaced by the indenter can be accommodated by the elastic deformations of the material around. In this case, pile-up is not observed.

The error in the calculated contact area and hardness due to incorrect use of elastic formulae for materials with significant pile-up can be as high as 60% (Oliver & Pharr, 1992). For elastic modulus it is lower, but it can also amount up to several tens of percent, as it follows from the role of contact area A in Eqs. (1). The situation with pointed indenters was studied by comparison of the measured $h_c(h)$ and $A(h)$ values with those obtained by direct imaging of the imprints (Mc Elhaney et al., 1998) and by the finite element modeling (Bolshakov & Pharr, 1998, Hay et al., 1998). Several conclusions can be made from these studies. Pileup can be expected for $E/H = 10^2$ and more. The stress field around a pointed indenter is self-similar and the degree of pileup does not depend on the depth of penetration. A simple quantity to assess the material propensity to pile-up is the ratio of the final depth of the imprint after unloading to the total depth under load, h_f/h. Pileup is large if h_f/h is close to 1.0 and the work-hardening is small. If $h_f/h < 0.7$, none or only very small pileup is observed regardless whether the material work-hardens or not (Oliver & Pharr, 2004, Bolshakov & Pharr, 1998).

The situation with spherical indenters is more complex. The mean contact pressure and the stresses and strains increase with the depth of penetration. The representative strain is usually expressed as the ratio of the contact radius *a* and indenter radius *R*. At relatively low loads, the deformations are only elastic, and the material around the indenter sinks. Plastic flow starts at the mean contact pressure $p_m = 1.1\ \sigma_Y$, but the Hertz formulae for the displacement $h(F)$ are approximately valid till $p_m = 1.6\ \sigma_Y$, when the volume of the plastically deformed zone beneath the indenter tip becomes so large that this material is pushed upwards around the indenter (Mesarovic & Fleck, 1999). The tendency to pile-up increases with increasing ratio E/Y and the relative depth h/R of indenter penetration. From a finite element study (Mesarovic & Fleck, 1999) it follows that for a non-strain-hardening material with $E/Y = 200$, the pile-up is negligible for $h/R < 0{,}001$, and should be considered for $h/R >$ 0,01. For materials exhibiting strain-hardening, the tendency to pile-up is smaller.

Pile-up can also occur if a relatively soft coating on a hard substrate is indented: the coating material is displaced upwards from the contact zone around the indenter more than in a homogeneous specimen. The oposite case are materials, which densify under high compressive stresses. Here, sinking-in of the surface is more pronounced than in common elastic materials.

More about pile-up can be found in (Fischer-Cripps, 2004) and (Oliver & Pharr, 1992); some methods for its corrections were proposed in (Cheng & Cheng, 1998). In all cases with a suspicion of errors due to pile-up or sink-in, direct observation of the imprint (for example via AFM) is recommended for obtaining more reliable values of the contact area.

10. Indentation size effect

It is well known that hardness of metals, measured by a pointed indenter at small depths of penetration, often increases with decreasing depth of penetration (Fig. 11). This so-called indentation size effect (ISE) becomes observable for imprint depths smaller than several μm. ISE can appear also with a spherical indenter, though here the apparent hardness does not depend on depth, but increases with decreasing radius. Even some ceramics exhibit indentation size effect to a small extent.

There are several reasons for depth dependence of hardness in metals. Sometimes, a thin hard oxidic layer has been formed on the specimen surface during its preparation (Fig. 9). Surface layer can also become harder due to strain hardening or residual stresses caused by plastic flow during polishing. Another reason is the bluntness of the tip of every real indenter. For example, the tip of a Berkovich pyramid is roughly spherical with the radius 50 – 100 nm. As a consequence, the actual contact area for very small distances from the tip is much larger than at the same distance for an ideal indenter (Fig. 5). For a given load, the penetration of a real indenter is thus smaller, and the hardness appears to be higher. With increasing depth, the influence of the tip bluntness becomes smaller and the measured hardness values approach gradually to those for the ideal shape. However, this is no actual indentation size effect, but an error, which can be avoided by a proper indenter calibration.

Nevertheless, even if these errors are removed, there still may be some dependence of hardness on indentation depth. According to the classical continuum mechanics, hardness of a homogeneous material, measured by a self-similar indenter, should be constant. However, plastic deformation of real crystalline materials is facilitated by the presence of dislocations, more with their higher density. If the indentation depth is large, the influenced volume contains many dislocations, and the plastic flow is insensitive to this depth. If the size of the indented volume becomes comparable with the distance between dislocations, less dislocations are present, and the average contact stress for the initiation of plastic flow must be higher. The role of dislocations also explains why the size effect is more significant in annealed samples than in cold-worked ones of the same basic material, as the latter contain more dislocations created during the mechanical treatment.

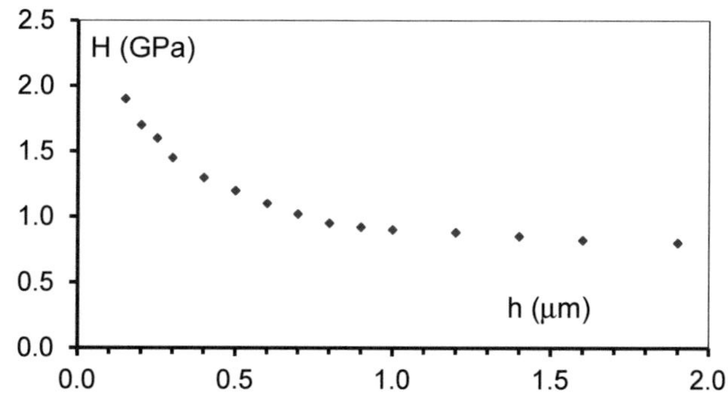

Figure 11. Depth dependence of hardness (ISE) of a single crystal Cu (after Nix & Gao, 1998)

A semiempirical formula has been proposed for the depth dependence of hardness (Nix & Gao, 1998):

$$H = H_0\left[1+\left(h^* / h\right)\right]^{1/2}; \tag{31}$$

H is the hardness for a given depth *h* of indentation, H_0 is the macro-hardness (corresponding to large indents), and h^* is a characteristic length that depends on the shear modulus, on H_0 and on indenter shape. The constant h^* (and H_0) for the tested material can be obtained by fitting a series of $H(h)$ values. It appears that a correct description of material response in micro or nano scale needs – in addition to „macro" parameters such as elastic modulus or yield strength – also a parameter, describing characteristic dimension of the microstructure. This material length scale may be defined, e.g., as (Nix & Gao, 1998):

$$l^* \approx cL^2 / \mathbf{b} = c / (\mathbf{b}\rho), \tag{32}$$

where c is a non-dimensional constant, L (m) is the mean spacing between dislocations, **b** (m) is the Burgers vector, and ρ (m^{-2}) is the average density of dislocations, calculated as the number of dislocation lines crossing a unit area (or as the total length of dislocation lines in a certain volume, divided by this volume). Material length scale l^* can be related to the characteristic length h^*. Also other formulations for l^* exist.

Equation (31) indicates that hardness is less dependent on depth for smaller values of h^* (and l^*), i.e. for higher density of dislocations. Nix & Gao (1998) have shown that hardness does not depend much on the depth of penetration if the material is intrinsically hard. An example is fused quartz, whose hardness is essentially independent of depth of indentation.

Indentation size effect also occurs if hardness is measured by a spherical indenter. In this case, however, it depends on the indenter radius rather than on the indentation depth. The contact-influenced volume, corresponding to a certain depth of penetration, is smaller for smaller indenter radius R, so that it will contain less pre-existing dislocations. Beginning from certain small R, hardness will increase with decreasing tip radius. Formula (31) may also be used for a spherical indenter, if the depth h is replaced by the curvature radius R_p of the residual surface impression and h^* is replaced by $R^* = c'/\mathbf{b}\rho$ (Swadener et al., 2002).

The relationship between the imprint size and the measured hardness is influenced by several factors and can be more complex. If the size of the influenced volume is smaller than the distance between the dislocations, the plastic flow starts only when the local stress reaches a value high enough to generate new dislocations. A role is also played by the fact that the strain field under the indenter is strongly inhomogeneous. This is considered in the strain gradient plasticity theory (Nix & Gao, 1998). Nevertheless, Gerberich et al. (2002) presume that gradient plasticity models do not apply at very shallow depths. Moreover, for very small depths of indentation by a spherical tip, the surface-related phenomena become important, such as friction or consumption of energy for the creation of larger surface as the indenter penetrates into the originally flat specimen.

In brittle materials, contact compliance is influenced by the presence of cracks in the contact region. Cracks can be present here from the manufacture or can be created during the test. Glass objects contain many tiny surface cracks, nucleated in mutual sliding with other bodies. Ceramics contains weak places from manufacture (broken crystalline grains, cracks from machining or polishing, etc.). Cracks are also formed during indentation (Zhu et al., 2008), for example at the imprint edges. For larger contact area or volume a higher probability exists that a larger pre-existing defect will be present in the stressed region, making the contact compliance higher. If the contact area is smaller, less cracks are present (or even none), and higher contact pressure is necessary for their nucleation or growth.

More about indentation size effect can be found in the above cited references and also in (Sargent, 1989, Huang et al., 2006, Spary et al., 2006, Rashid & Al-Rub, 2007, Sangwal, 2009). The most important conclusion is that for small imprints, the information on the indent size must always be given together with the hardness value and other details of the test.

11. Surface forces and adhesion

The models for instrumented indentation usually consider only the action of indenter load. However, this is correct only if this load is higher than several tens of μN. If it is lower, also adhesive forces should be considered, otherwise the results can be erroneous. Atoms of solids attract mutually, but the range of action of these attractive forces is very short, about 1 nm. Atoms in a very thin surface layer act also onto any other matter close to the surface, and this can result in adhesive joining. If the distance between two surfaces drops below some very small value, the bodies sometimes cling spontaneously and remain in close contact even without action of external force, and separate suddenly only after they are pulled apart by a certain tensile force (Fig. 12). In contact of relatively stiff solids, adhesion is usually not observed. Due to surface roughness, both bodies come into contact only at a limited number of matching asperities. The area of close contacts is minute, and the resultant attractive force is insufficient to compress the "hills". If, however, both surfaces are very smooth (roughness of the order of nm) or very compliant, the surfaces adjust one another better and the adhesion is stronger.

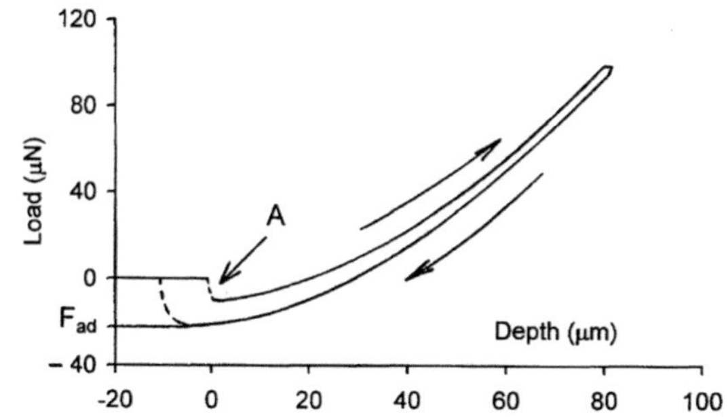

Figure 12. Indentation into a soft material influenced by adhesion and viscoelasticity (a schematic). F_{ad} – adhesion force at removal, A – adhesion at approach. After Menčík (2012)

Adhesion can also influence nanoindentation measurements, especially for compliant materials and very low loads, from μN to mN. Here, surface forces change the character of load-displacement curves (Gupta et al, 2007) and must be considered in the preparation and evaluation of the tests.

Several theories of adhesion exist; for an overview see Israelachvili (1992). Derjaguin, Muller and Toporov (1975) have developed theory of contact of stiff bodies (DMT), which is used, for example, in AFM indentation. For common compliant materials, the "Johnson–Kendall–Roberts" (JKR) model, proposed by Johnson et al. (1971), is used most often. It assumes spherical shape of contacting bodies and is based on Hertz′ theory of elastic contact, but considers also adhesive forces. Due to them, the contact deformations differ from those by Hertz. The actual contact radius a is larger and during unloading it remains finite even to zero load ($F = 0$), for which:

$$a_0 = [9\pi R^2 W / (2E_r)]^{1/3}; \tag{33}$$

R and E_r are the equivalent radius and elastic modulus of contacting bodies, and W is the specific surface energy of the contact of both bodies. The solids keep together up to a small tensile force, and suddenly separate at

$$F_{ad} = (3/2)\pi RW, \tag{34}$$

with the corresponding contact radius $a_{ad} = a_0/4^{1/3}$. For example, adhesion force F_{ad} of a glass sphere of $R = 100$ μm on rubber ($W = 62$ mJ/m^2) is 29 μN.

The resultant adhesive force depends also on the roughness of the surfaces in contact, on the viscosity of one or both bodies and also on the presence of a liquid at the contact. The influence of surface roughness was studied by Johnson (1976) and Tabor (1977) on asperities of spherical shape and randomly distributed heights. Each asperity adheres until its distance from the counter-surface exceeds some critical value, δ_c. The ratio of the standard deviation σ of asperity heights and the distance δ_c at separation, called elastic adhesion index $\alpha = \sigma/\delta_c$, can be related with the ratio of the force $F(\sigma)$, required to compress an asperity by σ, and the force of adhesion for that asperity, F_{ad}. Figure 13 shows how the apparent adhesion force decreases with increasing roughness. For $\alpha = 0$, this force is maximum and corresponds to the JKR model. For $\alpha = 1$ it drops to 50%, for $\alpha = 2$ it drops to 10%, and becomes negligible for $\alpha > 3$.

Adhesion is stronger if one or both bodies are very compliant so that the surfaces can easily adjust one another. Examples are gels, soft elastomers, or biological tissues. Their response, however, is visco-elastic, which also influences adhesion. Generally, the force at which two viscoelastic bodies get adhered during mutual approaching, is lower than the force needed for their separation on unloading (Fig. 12). A role is played by the decreasing apparent modulus $E(t)$ and growing contact area with time; the force for separation thus corresponds to a larger area. Moreover, this force also depends on the rate of separation – it is higher for faster separation. Greenwood & Johnson (1981) came to the conclusion that during the mutual catching or separation, the surface energy W in the JKR model should be modified by the ratio of instantaneous and asymptotic modulus, E_0/E_∞. A fracture mechanics approach to adhesion was presented by Maugis & Barquins (1978). Recently, an alternative model was proposed by Kesari et al. (2010).

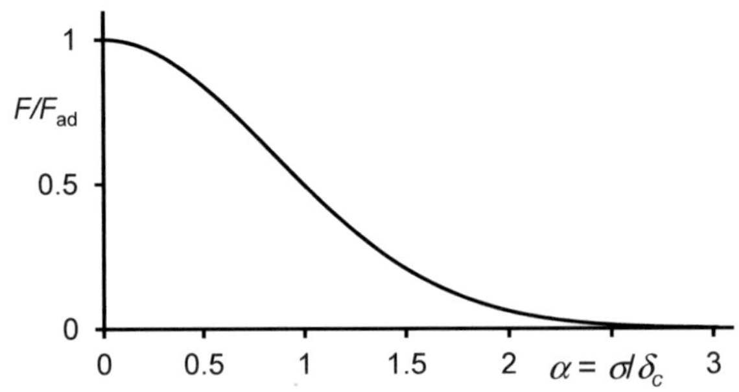

Figure 13. Relative decrease of adhesive force (F/F_{ad}) with growing relative surface roughness $\alpha = \sigma/\delta_c$; σ – standard deviation od asperity heights, δ_c – critical distance at the loss of adhesion. (A schematic, after Johnson, 1976)

If a thin layer of liquid is present between contacting bodies, it contributes to their mutual attraction by its surface tension. This is also the case in nanoindentation (Fig. 14):

1. Tensile surface stress γ acts along the circumference of the liquid ring around the indenter-sample contact and attracts the indenter to the specimen by the force $F_\gamma = 2\pi a\gamma$, where a is the contact radius and γ is the surface tension between the liquid and the air or another surrounding environment.
2. If the liquid wets the indenter and specimen, its shape is torroidal and the surface tension causes underpressure in it, $p = \gamma/r$, which attracts the indenter to the specimen by the force

$$F_{\mathrm{up}} = -\pi a^2 p = \pi a^2 \gamma / r; \tag{35}$$

r is the meniscus radius (Israelachvili, 1992). This capillary force is much bigger than F_γ and can also be expressed by the contact angle φ between the liquid and indenter or specimen. With some simplifications, the attractive force is (Israelachvili, 1992)

$$F_{\mathrm{up}} \approx 4\pi\, \gamma R \cos\varphi. \tag{36}$$

For example, water ($\gamma = 73$ mN/m) between the sample and indenter with $R = 100$ μm (and $\varphi = 20°$) can increase the adhesion force by 86 μN.

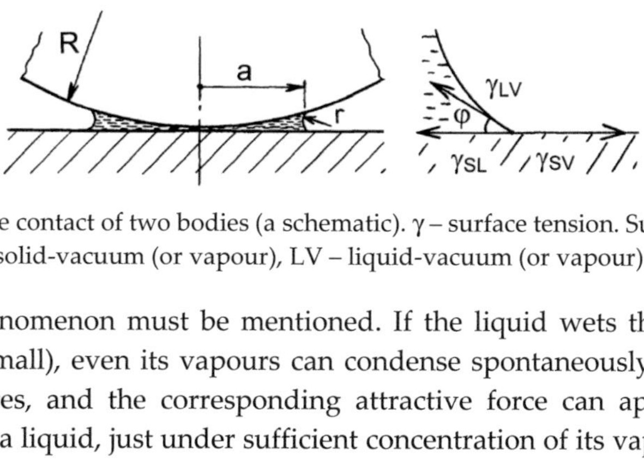

Figure 14. Liquid at the contact of two bodies (a schematic). γ – surface tension. Subscripts: SL – solid-liquid, SV – solid-vacuum (or vapour), LV – liquid-vacuum (or vapour)

One interesting phenomenon must be mentioned. If the liquid wets the surface well (the contact angle φ is small), even its vapours can condense spontaneously in a narrow space between two surfaces, and the corresponding attractive force can appear even without obvious presence of a liquid, just under sufficient concentration of its vapours. This also can happen between a spherical indenter and a specimen. For more, see Israelachvili (1992).

The presence of adhesive forces influences the load-depth curves and must be considered in the evaluation of low-load indentation tests. Elastic modulus is usually determined from a load-unload test by the Oliver & Pharr (1992) procedure. With a spherical indenter and elastic deformations, E can also be calculated from the expression (Gupta et al., 2007):

$$E_{\mathrm{r,\,H}} = \left[S^3 / (6RF) \right]^{1/2}, \tag{37}$$

which follows from the Hertz formula, relating the indenter load F and displacement h. R is the indenter radius and S is the contact stiffness at the nominal load. The advantage of Eq. (37)

is that one does not need to know the contact area A nor the instant of the first contact, necessary for the determination of indenter penetration h. If the surface forces are not negligible, the relationship between load and displacement is not proportional to $h^{3/2}$, but nearly linear for very low loads (Gupta et al., 2007). If the JKR model for adhesion is assumed, it is possible to determine the reduced elastic modulus using a spherical indenter and formula (37), but modified as follows (Carrillo et al, 2005, 2006, Gupta et al., 2007, Menčík, 2012):

$$E_{r,\,JKR} = E_{r,\,H} \times \Phi; \tag{38}$$

$$\Phi = [(2/3)(1+\psi)^{-1/2} + 1]^{3/2} / \{1 + 2\psi^{-1}[1+(1+\psi)^{1/2}]\}^{1/2}; \tag{39}$$

Φ is the correction factor (Fig. 15), and ψ is the ratio of indenter load and adhesion force, F/F_{ad}. The determination of E needs (in addition to F and R) the knowledge of adhesion force F_{ad} and contact stiffness S. They both can be ascertained in one load-unload test.

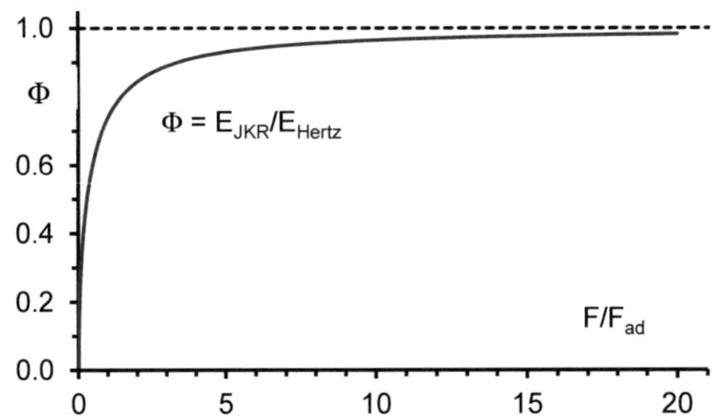

Figure 15. Ratio Φ of E-modulus calculated with the consideration of adhesive forces (E_{JKR}) and without (E_{Hertz}), plotted as a function of the ratio of the load F and adhesive force F_{ad}

12. Scatter of measured values

In principle, there are three kinds of errors: random, systematic, and gross. The gross errors occur due to extreme conditions in a particular test, such as random mechanical shock or sudden fluctuation of temperature. It usually does not appear if the test is repeated. Systematic errors are caused by inadequate conditions of measurement or data processing.

They can be revealed by the use of another method. Random errors are observable in repeated measurements, and can be reduced by processing the data by statistical methods.

Random scatter of individual values is caused by the noise in the device, by random variations of test conditions, and, especially, by the variability of material properties and inhomogeneities. At low loads, the results are influenced more by the random surface roughness of the indented place. Sometimes, porosity plays a role. Common also is the variability of coating properties along the surface; the range of values measured over a large area can be much wider than for the area of size less than 1×1 mm². Statistical data processing is necessary, and the number of tests should be adjusted to the scatter of the data.

If a certain material parameter X is to be determined, for example elastic modulus, one usually wants to ascertain it as accurately as possible. The uncertainty due to random scatter can be reduced by repeating the measurements. The confidence interval, which contains the true mean value μ_x with probability α, is

$$\mu_x = m_x \pm t_{\alpha,n-1} s_x / \sqrt{n}, \tag{40}$$

where m_x is the average value of x (calculated as $\Sigma x_i / n$), s_x is the standard deviation, $t_{\alpha,n-1}$ is the α-critical value of the Student's distribution for n–1 degrees of freedom, and n is the number of measurements. For $n > 30$, the width of the confidence interval is roughly indirectly proportional to $\sqrt{n}$. In this way, the parameter X could, in principle, be determined with any accuracy, just by increasing n. In reality, the number of tests is a compromise between the demanded accuracy and the costs of measurements.

The scatter of measured values often reflects the inherent variability of the pertinent property. For example, strength of a brittle material, such as ceramics, is given by the size of internal defects, which is random and depends on the raw materials and manufacturing conditions. Large scatter is also typical of biological samples. In such cases, there is no single value of the parameter X, and the quantity should be described by its probability distribution. The knowledge of this distribution allows the determination of confidence interval, containing certain percentage of all possible values or probable extreme value, for example the minimum strength that will not be atained only with some allowable probability, e.g. 1%. Analytical distributions, common in materials sciences (normal, log-normal, Weibull, etc.) are described by their parameters; the procedures for their determination can be found in textbooks on statistics. The number of measurements must be so high as to allow reliable determination of the distribution parameters. In some cases, the histogram of measured values cannot be approximated by a simple analytical function, and the use of order statistics is more useful.

Nanoindentation is also used for the investigation of heterogeneous materials consisting of several phases. Examples are concrete, ceramics, polycrystalline metals, plastics with hard particles dispersed in soft matrix, etc. Here, the measured values depend on the imprint size and thus also on the load. Large imprints, caused by high load, encompass many particles and the matrix between them, and give the average properties of the compound. If, on the contrary, the imprint is small compared to the size of the individual particles, and is near the centre of a particle (or matrix) and sufficiently far from the interface with another phase, the property of the phase alone is obtained. However, reliable indenter positioning is possible only if the individual particles or phases are visible in the nanoindenter's microscope. Sometimes, the phases cannot be distinguished optically. In such case, a grid of several hundred indents is created, e.g. 20x20, distributed uniformly over an area. The histogram of the measured property (Fig. 16) often exhibits several, less or more pronounced, peaks. This indicates that the measured values are a mixture of two or more probability distributions, each for one phase (Constantinides et al., 2006). In such case, the decomposition of the resultant distribution can yield the distribution parameters of the individual phases. For a composite of N-phases, with Gaussian distribution each, the probability density of the measured quantity X is given as the sum of densities of the individual quantities,

$$f(X) = \sum_{j=1}^{N} \frac{p_j}{\sqrt{2\pi}\sigma_j} \exp\left[-\frac{(X-\mu_j)^2}{2\sigma_j^2} \right]; \sum_{i=1}^{N} p_j = 1, \tag{41}$$

where μ_j and σ_j are the mean and standard deviation of the j-th phase (j = 1, 2, ..N), and p_j is the relative proportion of this phase. The procedure for obtaining all parameters is described in (Constantinides et al., 2006, Sorelli et al., 2008) or the chapter by J. Němeček in this book; examples can be found in Němeček & Lukeš (2010), Nohava & Haušild (2010) and Nohava et al. (2012).

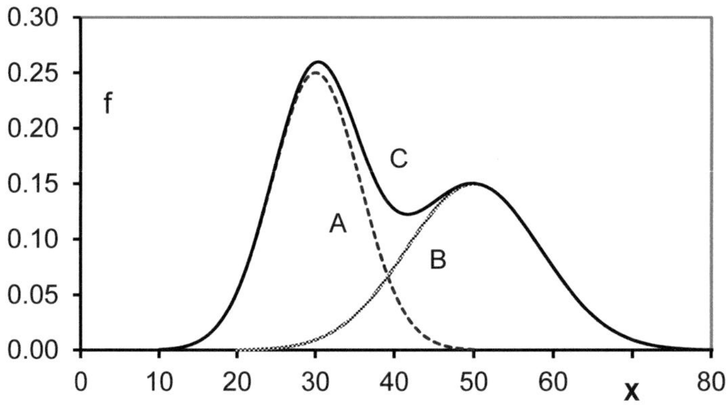

Figure 16. Distribution of property X in a two-phase material (a schematic). f – relative frequency (n_j/n; n is the total number of indents), A, B – distribution of property A or B alone, C – distribution of all X-values, obtained by grid indentation.

Until now, this method has been used mostly for systems consisting of two phases (N = 2). Unfortunately, some indents occur at or near the interfaces of both phases, and this influences the response (similarly like the response to indentation into a coating is influenced by the substrate). The interaction depends on the indent size and on the difference of properties of the phases; see, e.g., Buršík (2011) or Němeček & Lukeš (2010). Sometimes, therefore, trimodal Gaussian distribution is assumed, with the third fictitious „phase" pertaining to the influenced results. In this case, N = 3 and the genuine properties of the phases 1 and 2 can be determined better. Examples of application are described in (Nohava et al., 2012). In principle, this method can be used for materials consisting of more phases. However, the number of indents in such analysis must be very high, especially if the probability distributions of individual quantities overlap and the differences between the properties of individual phases are low. It can also be recommended to make several grids over a larger area.

13. Summary and conclusion

Accuracy of determination of mechanical properties by nanoindentation can be impaired for various reasons, such as inadequate model for data processing or omission of some specific properties or factors, such as pileup, surface forces or temperature changes. In this chapter, individual sources of errors were reviewed and explained and formulae or advice given for their avoiding or reducing.

Author details

Jaroslav Menčík
University of Pardubice, Czech Republic

Acknowledgement

The author highly appreciates the effort of InTech Publishing House to facilitate access to knowledge to everyone. He is also grateful for its help in the preparation of this chapter.

14. References

Bec, S., Tonck, A., Georges, J. M., Georges, E. & Loubet, J. L. (1996). Improvements in the indentation method with a surface force apparatus. *Phil. Mag. A*, 74, 1061 – 1072

Bhattacharya, A.K., & Nix, W. D. (1988). Analysis of elastic and plastic deformation associated with indentation testing of thin films on substrates. *Int. J. Solids Struct.*, Vol. 24, No. 12, 1287 – 1298

Bolshakov, A. & Pharr, G.M. (1998). Influences of pileup on the measurement of mechanical properties by load and depth sensing indentation techniques. *J. Mater. Res.*, Vol. 13, No. 4, 1049 – 1058

Burnett, P. J. & Rickerby, D. S. (1987). The mechanical properties of wear resistant coatings I, II: Modelling of hardness behaviour. *Thin Solid Films*, Vol. 148, 41 – 65

Buršík, J. (2011). Modelling of hardness distribution curves obtained on two-phase materials by grid indentation technique. *Chem. Listy*, Vol. 105, pp. s660 – s663

Cao, Y. P. & and Lu, J. (2004). A new method to extract the plastic properties of metal materials from an instrumented spherical indentation loading curve. *Acta Materialia*, Vol. 52, 4023 – 5032

Carrillo, F., Gupta, S., Balooch, M., Marshall, S. J., Marshall, G. W., Pruitt, L. & Puttlitz, Ch. M. (2005). Nanoindentation of polydimethylsiloxane elastomers: Effect of crosslinking, work of adhesion, and fluid environment on elastic modulus. *J. Mater. Res.*,Vol. 20, 2820 – 2830

Carrillo, F., Gupta, S., Balooch, M., Marshall, S. J., Marshall, G. W., Pruitt, L. & Puttlitz, Ch. M. (2006). Erratum: „Nanoindentation of polydimethylsiloxane elastomers: Effect of crosslinking, work of adhesion, and fluid environment on elastic modulus" [*J. Mater. Res.*, Vol. 20 (2005) 2820 - 2830] *J. Mater. Res.*, 21, 535 – 537

Cheng, Y. T. & Cheng, C. M. (2005). Relationships between initial unloading slope, contact depth, and mechanical properties for conical indentation in linear viscoelastic solids. *J. Mater. Res.*, Vol. 20, No. 4, 1046 – 1053

Cheng, Y.T. & Cheng, C.M. (1998). Effects of `sinking in' and `piling up' on estimating the contact area under load in indentation. *Phil. Mag. Lett.*, Vol. 78, No. 2, 115 – 120

Chudoba, T. & Richter, F. (2001). Investigation of creep behaviour under load during indentation experiments and its influence on hardness and modulus results. *Surf. Coat. Technol.*, Vol. 148, 191 – 198

Constantinides, G., Ravi Chandran, K.S., Ulm, F. J. & Van Vliet, K. J. (2006). Grid indentation analysis of composite microstructure and mechanics: Principles and validation. *Materials Science and Engineering*, A 430, 189 – 202

Derjaguin, B. V., Muller, V. M. & Toporov, Yu. P. (1975). Effect of contact deformations on adhesion of particles. *J. Colloid Interface Sci.*, Vol. *53*, pp. 314 – 326

Doerner, M.F. & Nix, W.D. (1986). A method for interpreting the data from depth-sensing indentation instruments. *J. Mater. Res.*, Vol. 1, No. 4, 601 – 609

Feng, G. & Ngan, A.H.V. (2002). Effects of creep and thermal drift on modulus measurement using depth-sensing indentation. *J. Mater. Res.*, Vol. 17, No. 3, 660 – 668

Field, J. & Swain, M.V. (1993). A simple predictive model for spherical indentation. *J.Mater.Res.*, Vol. 8, No. 2, 297 – 306

Fischer-Cripps, A. C. (2004). *Nanoindentation* (2nd edition). Springer-Verlag, ISBN 0-387-22045-3, New York

Fujisawa, N. & Swain, M.V. (2006). Effect of unloading strain rate on the elastic modulus of a viscoelastic solid determined by nanoindentation. *J. Mater. Res.*, 21, No. 3, 708 – 714

Gao, H., Chiu, C.H. & Lee, J. (1992). Elastic contact versus indentation modelling of multilayered materials. *Int. J Solids Structures*, Vol. 29, 2471 – 2492

Gerberich, W. W., Tymiak, N. I., Grunian, J. C., Horstemeyer, M. F. & Baskes, M. I. (2002). Interpretations of indentation size effects. *J. Appl. Mech.*, Vol. 69, July 2002, 433 – 442

Greenwood, J. A. & Johnson, K. L. (1981). The mechanics of adhesion of viscoelastic solids. *Philosophical Magazine A*, 43, 697 – 711

Gupta, S., Carrillo, F., Li Ch., Pruitt, L. & Puttlitz, Ch. (2007). Adhesive forces significantly affect elastic modulus determination of soft polymeric materials in nanoindentation. *Materials Letters*, Vol. 61, Issue 2, 448 – 451

Hay, J. & Crawford, B. (2011). Measuring substrate-independent modulus of thin films. *J. Mater. Res.*, 26, No. 6, 727 – 738

Hay, J., Oliver, W.C., Bolshakov, A. & Pharr, G.M. (1998) Using the ratio of loading slope and elastic stiffness to predict pile-up and constraint factor during indentation. *Mat. Res. Proc. Symp.*, Vol. 522, 101 – 106

Herbert, E. G., Pharr, G. M., Oliver, W. C., Lucas, B. N. & Hay, J. L. (2001). On the measurement of stress-strain curves by spherical indentation. *Thin Solid Films*, Vol. 398 – 399, pp. 331 – 335

Huang, Y., Zhang, F., Hwang, K. C., Nix, W. D., Pharr, G. M. & Feng, G. (2006). A model of size effects in nanoindentation. *J. Mech. Phys. Solids*, Vol. 54, 1668 – 1686

Israelachvili, J. N. (1992). *Intermolecular and surface forces* (2nd ed.). Academic Press, London

Johnson, K. L.: Adhesion at the contact of solids (1976). pp. 133– 143 in: *Theoretical and Applied Mechanics* (W. T. Koiter, editor) North-Holland Publishing Comp., Amsterdam – New York – Oxford.

Johnson, K.L. (1985). *Contact mechanics*, Cambridge, Cambridge University Press.

Johnson, K. L., Kendall, K. & Roberts, A. D. (1971). Surface energy and the contact of elastic solids. *Proc. R. Soc. Lond. A*. Vol. 324, 301 – 313

Jönsson, B. & Hogmark, S. (1984). Hardness measurement of thin films. *Thin Solid Films*, Vol. 114, 257 – 269

Kesari, H., Doll, J. C., Pruitt, L. B., Cai, W. & Lew, A. J. (2010). Role of surface roughness in hysteresis during adhesive elastic contact. *Phil. Mag. Lett.*, Vol. 90, Issue 12, 891–902

Korsunsky, A. M., McGurk, M. R., Bull, S. J. & Page, T. F. (1998). On the hardness of coated systems. *Surf. Coat. Technol.*, Vol. 99, No. 1-2, 171 – 183

Lawn, B.R., Evans, A.G. & Marshall, D.B. (1980)- Elastic/plastic indentation damage in ceramics: The median/radial crack system. *J. Amer. Ceram. Soc.*, Vol. 63, No. 9-10, 574 – 581

Li, W. & Siegmund, T. (2004). An analysis of the indentation test to determine the interface toughness in a weakly bonded thin film coating – substrate system. *Acta Materialia*, Vol. 52, 2989 - 2999

Marshall, D. B. & Evans, A. G. (1984). Measurement of adherence of residually stressed thin films by indentation. *J. Appl. Phys.*, Vol. 56, No. 10, 2632 - 2644

Matthewson, M. J. (1986). Adhesion measurement of thin films by indentation. *Appl. Phys. Lett.*, Vol. 49, No. 21, 1426 - 1428

Maugis, D. & Barquins, M. (1978). Fracture mechanics and the adherence of viscoelastic bodies. *J. Phys. D: Appl. Phys.*, Vol. 11, 1989 – 2023

Menčík, J. (1996). *Mechanics of Components with Treated or Coated Surfaces*. Kluwer Academic Publishers, ISBN 0-7923-3700-X, Dordrecht

Menčík, J. (2010). Determination of parameters of viscoelastic materials by instrumented indentation. Part 3: Rheological model and other characteristics. *Chem Listy*, 104, pp. s275 – s278

Menčík, J. (2011a). Simple Models for Characterization of Mechanical Properties by Nanoindentation. Chapter 4 in: *Advances in Nanotechnology. Volume 5*, Bartul, Z. & Trenor, J. (editors), pp. 127 – 162. Nova Science Publishers, Inc., ISBN 978-1-61761-322-7, New York

Menčík, J. (2011b). Determination of parameters of viscoelastic materials by instrumented indentation. *Chem. Listy*, Vol. 105, pp. s115 – s119

Menčík, J. (2011c). Determination of parameters of viscoelastic materials by instrumented indentation. Part 2: Viscoelastic-plastic response. *Chem. Listy*, Vol. 105, s.143 – s145

Menčík, J. (2011d). Opportunities and problems in nanoindentation with spherical indenters. *Chem. Listy*, Vol. 105, pp. s680 – s683

Menčík, J. (2012). Low-load nanoindentation: Influence of surface forces and adhesion. Chem. Listy, Vol. 106 (special issue, Local Mechanical Properties 2011)

Menčík, J. & Swain, M. V. (1995). Errors associated with depth-sensing microindentation tests. *J. Mater. Res.*, Vol. 10, No. 6, 1491 – 1501

Menčík, J., Munz, D., Quandt, E., Weppelmann, E.R., & Swain, M.V. (1997). Determination of elastic modulus of thin layers using nanoindentation. *J. Mater. Res.*, Vol. 12, No. 9, 2475 – 2484

Menčík, J., Rauchs, G., Belouettar, S., Bardon, J. & Riche, A. (2004). Modeling of response of viscoelastic materials to harmonic loading. *Proceedings of Int. Conf. Engineering Mechanics 2004. Svratka, May 10-13,* Institute of Thermomechanics ASCR, ISBN 80-85918-88-9, Prague, p. 187 – 188, full paper on CD-ROM

Menčík, J., Rauchs, G., Bardon, J. & Riche, A. (2005). Determination of elastic modulus and hardness of viscoelastic-plastic materials by instrumented indentation under harmonic load. *J. Mater. Res.*, Vol. 20, No. 10, 2660 – 2669

Menčík, J., He, L.H. & Swain, M. V. (2009). Determination of viscoelastic-plastic material parameters of biomaterials by instrumented indentation. *J. Mech. Behav. Biomed.*, Vol. 2, 318 – 325

Menčík, J., He, L. H. & Němeček, J. (2010). Characterization of viscoelastic-plastic properties of solid polymers by instrumented indentation. *Polymer Testing*, 30, 101 – 109

Menčík, J. & Nohava, J. (2011). Nanoindentation into PMMA and fused silica by spherical and pointed indenters – a comparison. *Chem. Listy*, Vol. 105, Issue 17, s.834 – s835

Mesarovic, S. Dj. & Fleck, N.A. (1999). Spherical indentation into elastic-plastic solids. *Proc. R. Soc. Lond. A*, Vol. 455, 2707 – 2728

Němeček, J. & Lukeš, J. (2010). On the evaluation of elastic properties from nanoindentation of heterogeneous systems. *Chem Listy*, Vol. 104, pp. s279 – s282

Niihara, K., Morena, R. & Hasselman, DPH (1982). Evaluation of KIC of brittle solids by the indentation methods with low crack to indent ratios. *J Mater Sci Lett*, Vol. 1, No. 1, 13 – 16

Niihara, K. (1983). A fracture mechanics analysis of indentation-induced Palmqvist crack in ceramics. *J Mater Sci Lett*, Vol. 2 , No. 5, 221 – 223

Nix, W. D. & Gao, H. (1998). Indentation size effects in crystalline materials: a law for strain gradient plasticity. *J. Mech. Phys. Solids*, Vol. 46, No. 3, 411 – 425

Nohava, J. & Haušild, P. (2010). New grid indentation method for multiphase materials. Chem. Listy, 104, pp. S360 – s363.

Nohava, J., Houdková, Š. & Haušild, P. (2012). On the use of different instrumented indentation procedures for HVOF sprayed coating. Chem. Listy, Vol. 106 (special issue, Local Mechanical Properties 2011)

Nohava, J. & Menčík, J. (2012). A contribution to understanding of low-load spherical indentation – Comparison of tests on polymers and fused silica. *J. Mater. Res.*, Vol. 27, No. 1, 239 – 244

Oliver, W.C. & Pharr, G.M. (1992). An improved technique for determining hardness and elastic modulus using load and displacement sensing indentation experiments. *J. Mater. Res.*, Vol. 7, No. 6, 1564 – 1583

Oliver, W.C. & Pharr, G.M. (2004). Measurement of hardness and elastic modulus by instrumented indentation: Advances in understanding and refinements to methodology. *J. Mater. Res.*, Vol. 19, No. 1, 3 – 20

Oyen, M. L. (2006). Analytical Techniques for Indentation of Viscoelastic Materials. *Philosophical Magazine*, 86, No. 33-35, 5625 – 5641

Pharr, G. M., Strader, J. H. & Oliver, W. C. (2009). Critical issues in making small-depth mechanical property measurements by nanoindentation with continuous stiffness measurement. *J. Mater. Res.*, Vol. 24, No. 3, 653 – 666

Rar, A., Song. H. & Pharr, G. M. (2001). Assessment of New Relation for the elastic compliance of a film-substrate system. In: *Thin Films: Stresses and Mechanical Properties IX*, (C. S. Ozkan, L. B. Freund, R. C. Cammarata & H. Gao, editors). *Mat. Res. Soc. Symp. Proc.*, Vol. 695, MRS, Warrendale, Pennsylvania; paper L10.10

Rashid, K. & Al-Rub, A. (2007). Prediction of micro and nanoindentation size effect from conical or pyramidal indentation. *Mechanics of Materials*, Vol. 39, 787 – 802

Rother, B. & Dietrich, D. A. (1984). Evaluation of coating-substrate interface strength by differential load feed analysis of load-indentation measurements. *Thin Solid Films*, vol. 250, Issue 1 – 2, 181 - 186

Saha, R. & Nix, W. D. (2002). Effects of the substrate on the determination of thin film mechanical properties by nanoindentation. *Acta Materialia*, Vol. 50, 23 – 38

Sangwal, K. (2009). Review: Indentation size effect, indentation cracks and microhardness measurement of brittle crystalline solids – some basic concepts and trends. *Cryst. Res. Technol.*, Vol. 44, No. 10, 1019 – 1037

Sargent, P. M. (1989). Indentation size effect and strain hardening. *J. Mater. Sci. Letters*, Vol. 8, 1139 – 1140

Sorelli, L., Constantinides, G., Ulm, F. J. & Toutlemonde, F. (2008). The nano-mechanical signature of ultra high performance concrete by statistical nanoindentation techniques. *Cement and Concrete Research*, Vol. 38, 1447 – 1456

Spary, I. J., Bushby, A. J. & Jennett, N. M. (2006). On the indentation size effect in spherical indentation. *Phil. Mag.*, Vol. 86, Issue 33 – 35, pp. 5581 – 5593

Swadener, J. G., George, E. P. & Pharr, G. M. (2002). The correlation of the indentation size effect measured with indenters of various shapes. *J. Mech. Phys. Solids*, Vol. 50, 681 – 694

Tabor, D. (1977). Surface forces and surface interactions. *J. Colloid Interface Sci.*, Vol. 58, 2 – 13

TestWorks 4 Nanoindentation Manual (2002). Version No. 16. CSM Option – Theory: Introduction. 199 – 207

Zhang, T., Feng, Y., Yang, R., & Jiang P. (2010). A method to determine fracture toughness using cube-corner indentation. *Scripta Materialia*, Vol. 62, 199 – 201

Zhu, T. T., Bushby, A. J. & Dunstan, D. J. (2008). Size effect in the initiation of plasticity for ceramics in nanoindentation. *J. Mech. Phys. Solids*, Vol. 56, 1170 – 1185

Section 2

Nanomechanical Performance of Composite Materials

Chapter 4

Nanoindentation Based Analysis of Heterogeneous Structural Materials

Jiří Němeček

Additional information is available at the end of the chapter

http://dx.doi.org/10.5772/50968

1. Introduction

Nanoindentation is undoubtedly a powerful experimental technique, developed for more than a decade together with new device technologies, characterization of lower scale physical laws, theories and small scale numerical modelling. Nowadays, nanoindentation (Fischer-Cripps, 2002) is commonly used for investigation of local mechanical properties of mostly homogeneous materials modelled as isotropic (Oliver & Pharr, 1992) or anisotropic solids (Swadener & Pharr, 2001; Vlassak et al., 2003). On the other hand, many materials and especially structural ones exhibit phase heterogeneity and mechanical differences of the phases on different length scales (nanometers to meters). In order to model heterogeneous material systems, multiscale approach that allows for separation of scales based on some characteristic dimension of a material microscopic feature for each level is often utilized. Material micro-level of selected structural materials whose properties are accessible by nanoindentation (below one micrometer) will be analyzed in this chapter. Phase separation based on their different mechanical behaviour and intrinsic phase properties of the selected materials will be performed. Statistical grid indentation technique will be employed (Constantinides et al., 2006; Ulm et al., 2007; Němeček et al., 2011c). Finally, micromechanical framework (Zaoui, 2002) will be applied in the analysis of the effective composite properties for higher material levels.

2. Heterogeneity of structural materials

Structural materials exhibit several types of heterogeneity at microscale. The first type of their heterogeneity comes from mixing of components that do not react chemically in the composite like sand, fibres, and other additives. Such heterogeneity is usually known in advance and is given by the mixing proportions. The second type of heterogeneity comes from chemical reactions that are evolving after the mixing of basic components. As a result of these reactions, new phases are produced. In the case of structural materials, it is hard to rigorously define their volumes and microstructural distribution. Cement paste or alkali-activated materials can be given as typical examples. After many decades of research, exact microstructure development of these materials and the link between their basic components and their mechanical performance is still an open issue. Their microstructure is rather complex and it is, therefore, impossible to separate the phases from the composite and to prepare homogeneous-like samples suitable for mechanical testing.

Formation of the new phases in the structural composites includes fully or partly reacted matrix, unreacted grains of the raw material, interfacial zones with different chemical and also mechanical properties (e.g. Taylor, 2003; Bentz, 1999) and porosity. Structural materials based on cement (like cement paste, concrete, plasters) or waste materials (like fly-ash, furnace slag, etc.) usually include both types of the heterogeneity.

Complementary techniques to nanoindentation, such as optical imaging, electron microscopy (SEM) or atomic force microscopy (AFM) are often used to separate and to characterize the material phases. These techniques allow qualitative as well as quantitative investigation of individual material phases at small volumes near or at the sample surface. Measurement of intrinsic properties of individual material phases can be performed almost exclusively by nanoindentation that can directly access mechanical properties at small volumes starting from several tens of nanometers (depending on sample and probe). Complicated microstructures (e.g. trabecular bones, porous ceramics) are sometimes treated by a combination of nanoindentation with small scale mechanical testing (e.g. Jiroušek et al., 2011b). The microstructure of these materials can be reconstructed by X-ray microtomography in combination with the specimen loading (Jiroušek, 2011a).

The final step in the micromechanical analysis includes up-scaling of the properties to the higher material level. Multiple tools of classical analytical micromechanics or numerical approaches can be employed in this task (e.g. Zaoui, 2002; Moulinec & Suquet, 1994, 1998; Michel et al., 1999).

3. Testing strategies

In contrast to usual indentation on homogeneous materials (e.g. glass, films, coatings), structural materials (e.g. cement paste, alkali-activated materials, gypsum) are much more complex in their microstructure and mechanical performance. The situation is further complicated by their time-dependent load response (Němeček, 2009), aging and property fluctuations due to temperature or humidity (Beaudoin et al., 2010; Randall, 2009). The evaluation methodology however, is currently restricted mostly to homogeneous isotropic systems (Oliver & Pharr, 1992). The indentation response in the form of force-penetration (*P-h*) curves is characterized by two elastic constants, indentation modulus:

$$E_r = \frac{1}{2\beta} \frac{\sqrt{\pi}}{\sqrt{A_c}} \frac{dP}{dh}\bigg|_{P=P_{\max}} \tag{1}$$

and indentation hardness:

$$H = \frac{P}{A_c} \tag{2}$$

where $\left.\frac{dP}{dh}\right|_{P=P_{\max}}$ is the contact stiffness evaluated from the initial slope of the unloading branch of the force-penetration curve, P is the indentation force, A_c is the projected contact area and β is the correction factor for indenters with non-symmetrical shape (β=1.034 for Berkovich tip). Direct application of these equations to heterogeneous materials poses several difficulties, as the underlying analysis relies on the self-similarity of the indentation test which holds only for homogeneous materials (Borodich et al., 2003; Constantinides, 2007). The interaction of phases in heterogeneous materials is unavoidable but depending on the length scale it can be more or less important. Properties extracted from indentation data of a heterogeneous solid can be treated as averaged quantities dependent on the depth h. Therefore, the choice of an indentation depth directly determines the length scale of the tested material volume. For example, the effective volume affected by an indent can be estimated as 3×h for the Berkovich indenter (Constantinides et al., 2006).

Composite structural materials are multiphase materials in which distinct phases are intermixed spatially and chemically. Taking the microstructural heterogeneity into account one can formulate basically three testing strategies to obtain mechanical properties of a composite or its phase properties:

(i) Averaged (effective) composite properties can be found if the indentation depth is much larger than the characteristic phase dimension ($h \gg D$). In this case, a phase compound is indented and thus, physically averaged properties are obtained. This strategy does not give access neither to distinct phases' properties nor to their volume fractions.

(ii) Another possibility is to perform pointed indentation to a specific material phase with individual indent's dimension much smaller than the characteristic dimension of the tested phase ($h \ll D$). In this case, intrinsic properties of the distinct phase (which may also include a phase porosity smaller than the tested size h) are obtained. This strategy can be used, provided the material phase can be distinguished prior to indentation by some other means (e.g. optical microscope, SEM) which is not always the case. It gives access to the distinct phase properties but not to volume fraction of the phase compared to other phases.

(iii) The last one, but for structural materials probably the most powerful technique, is based on the statistical (massive grid) indentation in which small indents are produced over a large area to capture the sample heterogeneity, but the dimension of a single indent is kept still smaller than the characteristic dimension of an individual phase ($h \ll D$). In this case, the results provide information on all phases' properties as well as their volume ratios, but without any knowledge which indent belongs to which phase. The properties can be evaluated in terms of property histograms for which subsequent deconvolution techniques can be employed and individual phase properties assessed (Constantinides et al., 2006; Ulm et al., 2007; Němeček et al., 2011c).

The extraction of material properties of a heterogeneous system from nanoindentation in cases of (ii) and (iii) relies on the fact that the volume affected by an indenter is small enough not to mechanically interact with other phases. As a rule of a thumb, the indentation depth is usually chosen as 1/10 of the characteristic size of the measured inclusion or phase D (Durst, 2004). In the literature, the solution of a mutual influence in the matrix-inclusion system is rather rare. The situation of phases with different stiffness was studied for thin films placed on a substrate e.g. by Gao et al., 1992. It was shown by Gao et al. that the substrate effects are negligible if the stiffness mismatch ratio is:

$$\frac{E_{substrate}}{E_{film}} \in (0.2;5) \tag{3}$$

as long as the indentation depth is smaller than 10% of the film thickness. The layered substrate-film system is not completely equivalent to the disordered structural multiphase materials but it can be successfully used as the first estimate.

Applying (ii) strategy to cement paste, for example, where calcium-silica-hydrates of different densities (low and high) are intermixed with $Ca(OH)_2$ zones in hydrated cement matrix (Taylor, 2003; Thomas et al., 1998) is not an option due to impossible differentiation of the reaction products in optical microscope or SEM. Therefore, it is advantageous to perform massive grids (hundreds of indents) on large sample area containing all material phases. Then, the indentation offers statistical set of data which can be analyzed by the deconvolution technique (Ulm et al., 2007; Němeček, 2011c).

4. Identification of intrinsic phase properties by the deconvolution procedure

It is assumed in the algorithm that a large statistical set of independent events (i.e. measurements of elastic modulus or hardness at individual material points) is obtained from grid nanoindentation, i.e. by applying the (iii) testing strategy from the previous section. The analysis begins with the generation of experimental probability density function (PDF) or cumulative distribution function (CDF) for the data set. Using of PDF is more physically intuitive since significant peaks associated with mechanically distinct phases can be often distinguished in the graph. On the other hand, the construction of PDF requires the choice of a bin size. Application of CDF (Ulm et al., 2007) is more straightforward (does not require the choice of a bin size) and is more appropriate for cases where no clear peaks occur in the property histogram. The deconvolution algorithms based on PDF of CDF are analogous and thus the procedure will be explained just for the case of PDF.

Experimental PDF is firstly constructed from all measurements whose number is N^{exp}, using equally spaced N_{bins} bins of the size b. Each bin is assigned a frequency of occurrence $f_i^{\exp}$ that can be normalized with respect to the overall number of measurements as $\frac{f_i^{\exp}}{N^{\exp}}$. From that, one can compute the experimental probability density function (PDF) as a set of discrete values:

$$p_i^{\exp} = \frac{f_i^{\exp}}{N^{\exp}} \frac{1}{b} \quad i = 1,...,N_{bins} \tag{4}$$

The task of deconvolution into M phases represents finding of $r=1,...,M$ individual distributions related to single material phase. Assuming normal (Gauss) distributions, the single phase PDF can be written as:

$$p_r(x) = \frac{1}{\sqrt{2\pi s_r^2}} \exp \frac{-(x-\mu_r)^2}{2s_r^2} \tag{5}$$

in which μ_r and s_r are the mean value and standard deviation of the *r-th* phase computed from n_r values as:

$$\mu_r = \frac{1}{n_r} \sum_{k=1}^{n_r} x_k \quad s_r^2 = \frac{1}{n_r - 1} \sum_{k=1}^{n_r} (x_k - \mu_r)^2 \tag{6}$$

and x is the approximated quantity (i.e. elastic modulus or hardness). The overall PDF constructed from M phases is then:

$$C(x) = \sum_{r=1}^{M} f_r p_r(x) \tag{7}$$

where f_r is the volume fraction of a single phase defined as:

$$f_r = \frac{n_r}{N^{\exp}} \tag{8}$$

Individual distributions can be found by minimizing the following error function:

$$\min \sum_{i=1}^{N^{bins}} [(P_i^{\exp} - C(x_i))P_i^{\exp}]^2 \tag{9}$$

in which quadratic deviations between experimental and theoretical PDFs are computed in a set of discrete points. The function is weighted by the experimental probability in order to put emphasis on the measurements with a higher occurrence. The minimization in Eq. 9 can be based on the random Monte Carlo generation of M probability density functions satisfying the condition:

$$\sum_{r=1}^{M} f_r = 1 \tag{10}$$

As mentioned above, the bin size needs to be chosen prior the computation. Also, it is beneficial to fix the number of mechanically distinct phases M in advance to minimize the computational burden and to stabilize the ill-posed problem (Němeček et al., 2011c). Such knowledge can be supplied by some independent analyses (chemical composition, SEM or image analyses).

5. Assessment of effective material properties

5.1. Analytical homogenization methods

Continuum micromechanics will serve as fundamental tool in our assessment of effective material properties. A material is considered as macroscopically homogeneous with microscopically inhomogeneous phases that fill a representative volume element (RVE) with characteristic dimension l. The scale separation condition requires to be $d<<l<<D$, where d stands for a size of the largest microlevel inhomogeneity in the RVE (e.g. particles or phases), l is the RVE size and D stands for structural dimension of a macroscopically homogeneous material which can be continuously built from the RVE units. The characteristic structural dimension is usually at least 4-5 times larger than the RVE size (Drugan & Willis, 1996).

The RVE with substantially smaller dimensions than the macroscale body allows imposing homogeneous boundary conditions over the RVE (Hill, 1963, 1965; Hashin, 1983). Then, continuum micromechanics provides a framework, in which elastic properties of heterogeneous microscale phases are homogenized to give overall effective properties of the upper scale (Zaoui, 2002). A significant group of analytical homogenization methods relies on the Eshelby's solution (Eshelby, 1957) that uses an assumption of the uniform stress field in an ellipsoidal inclusion embedded in an infinite body. Effective elastic properties are then obtained through averaging over the local contributions. The methods are bounded by rough estimates based on the mixture laws of Voigt (parallel configuration of phases with perfect bonding) and Reuss (serial configuration of phases). The bounds are usually quite distant so that more precise estimates need to be used. Very often, the Mori-Tanaka method (Mori & Tanaka, 1973) is used for the homogenization of composites with continuous matrix (reference medium) reinforced with spherical inclusions. In this method, the effective bulk and shear moduli of the composite are computed as follows:

$$k_{eff} = \frac{\sum_r f_r k_r (1+\alpha_0(\frac{k_r}{k_0}-1))^{-1}}{\sum_r f_r (1+\alpha_0(\frac{k_r}{k_0}-1))^{-1}} \qquad \mu_{eff} = \frac{\sum_r f_r \mu_r (1+\beta_0(\frac{\mu_r}{\mu_0}-1))^{-1}}{\sum_r f_r (1+\beta_0(\frac{\mu_r}{\mu_0}-1))^{-1}}$$

$$\alpha_0 = \frac{3k_0}{3k_0+4\mu_0}, \beta_0 = \frac{6k_0+12\mu_0}{15k_0+20\mu_0} \qquad (11)$$

where f_r is the volume fraction of the rth phase, k_r is its bulk modulus, μ_r is its shear modulus, and the coefficients α_0 and β_0 describe bulk and shear properties of the 0th phase, i.e. the reference medium (Mori & Tanaka, 1973). The bulk and shear moduli can be directly linked with Young's modulus E and Poisson's ratio ν used in engineering computations as:

$$E = \frac{9k\mu}{3k+\mu} \quad \nu = \frac{3k-2\mu}{6k+2\mu} \tag{12}$$

Materials with no preference of matrix phase (i.e. polycrystalline metals) are usually modeled with the self-consistent scheme (Zaoui, 2002). It is an implicit scheme, similar to Mori-Tanaka method, in which the reference medium points back to the homogenized medium itself.

5.2. Numerical homogenization based on FFT

The homogenization problem, i.e. finding the link between microscopically inhomogeneous strains and stresses and overall behavior of a RVE can be solved e.g. by finite element calculations or by applying advanced numerical schemes that solve the problem using fast Fourier transformation (FFT), for example. The later was found to be numerically efficient in connection with grid indentation that serves as a source of local stiffness parameters in equidistant discretization points. The behavior of any heterogeneous material consisting of periodically repeating RVE occupying domain Ω can be described with differential equations with periodic boundary conditions and prescribed macroscopic load ($\boldsymbol{\varepsilon}^0$) as

$$\boldsymbol{\sigma}(\mathbf{x}) = \mathrm{L}(\mathbf{x}) : \boldsymbol{\varepsilon}(\mathbf{x}) \quad div\boldsymbol{\sigma}(\mathbf{x}) = \mathbf{0} \quad \mathbf{x} \in \Omega \tag{13}$$

$$\langle \boldsymbol{\varepsilon} \rangle := \frac{1}{|\Omega|} \int_{\Omega} \boldsymbol{\varepsilon}(\mathbf{x}) d\mathbf{x} = \boldsymbol{\varepsilon}^0 \tag{14}$$

where $\boldsymbol{\sigma}(\mathbf{x})$ denotes second order stress tensor, $\boldsymbol{\varepsilon}(\mathbf{x})$ second order strain tensor and $\mathrm{L}(\mathbf{x})$ the fourth order tensor of elastic stiffness at individual locations $\mathbf{x}$. The effective (homogenized) material tensor L_{eff} is such a tensor satisfying

$$\langle \boldsymbol{\sigma} \rangle = \mathrm{L}_{\mathrm{eff}} \langle \boldsymbol{\varepsilon} \rangle \tag{15}$$

Local strain tensor can be decomposed to homogeneous (macroscopic) and fluctuation parts which leads to the formulation of an integral (Lippmann–Schwinger type) equation:

$$\boldsymbol{\varepsilon}(\mathbf{x}) = \boldsymbol{\varepsilon}^0 - \int_{\Omega} \Gamma^0(\mathbf{x}-\mathbf{y}) : (\mathrm{L}(\mathbf{y}) - \mathrm{L}^0) : \boldsymbol{\varepsilon}(\mathbf{y}) d\mathbf{y} \tag{16}$$

where Γ^0 stands for a periodic Green operator associated with the reference elasticity tensor L^0 which is a parameter of the method (Moulinec & Suquet, 1998). The problem is further discretized using trigonometric collocation method (Saranen & Vainikko, 2002) which leads to the assemblage of a nonsymmetrical linear system of equations. The system can be resolved e.g. by the conjugate gradient method as proposed by Zeman et al. (Zeman et al., 2010). Elastic constants received from grid nanoindentation have been used as input parameters for this FFT homogenization with the assumption of plane strain conditions.

6. Experimental program

6.1. Sample preparation and microstructure

6.1.1. Cement paste

Typical heterogeneous structural materials were selected to illustrate the methodology described in previous sections. Firstly, effective elastic properties were studied for cement paste which is a basic component of a wide range of cementitious composites. Cement paste (i.e. hydrated cement clinker in hardened state) was prepared from Portland cement CEM-I 42,5 R (Mokrá, CZ) with water to cement weight ratio equal to 0.5 (Němeček, 2009). Samples were stored in water for two years. Once the cement powder is mixed with water, an exothermic reaction leading to the development of hydration products begins. The reaction kinetics is very rapid in early minutes and hours but slows down significantly after days. After a year, high degree of hydration (over 90%) could be anticipated in the samples. The microstructure of cement paste after hydration includes several major chemical phases, namely calcium-silica hydrates (C-S-H), calcium hydroxide $Ca(OH)_2$ called Portlandite, residual clinker and porosity. The cement paste microstructure is shown in Fig. 1. Very light areas in Fig. 1 can be attributed to the residual clinker, light grey areas are rich of Portlandite, dark grey zone belongs to C-S-H gels and black colour represents very low density regions or capillary porosity. Note, that C-S-H gel and Portlandite zones are spacially intermixed in small volumes (<<10 μm) and the resolution of SEM-BSE images does not allow for a direct separation of these phases from the image.

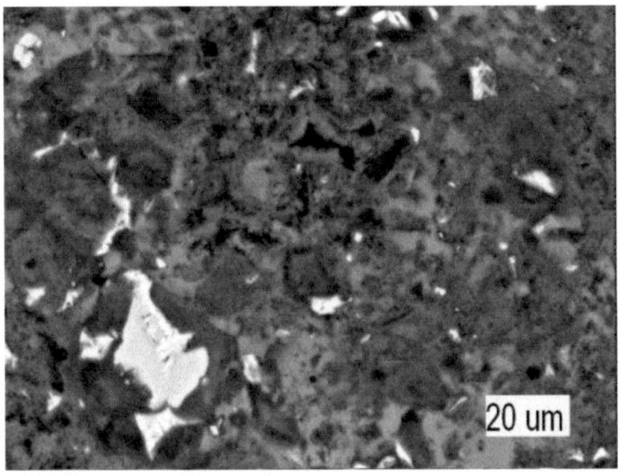

Figure 1. SEM image of cement paste microstructure.

The majority of the material volume mostly consists of poorly crystalline or amorphous phases (C-S-H) and partly of crystalline phases (Portlandite). Portlandite crystals are known for their anisotropy. Since their volume is not large in the sample and they can be mixed with C-S-H, all the phases will be supposed to be mechanically isotropic for simplification in the analysis.

Cement paste includes also wide distribution of pores. The He/Hg-porosimetry have been performed on the samples (Fig. 2). Majority of pores lies in nanometer range (<100 nm) and, on the other hand, large capillary pores are present in the scale above the indentation level (i.e. >>1 μm, not seen by the He/Hg-porosimetry). Therefore, the indentation depth was chosen so that the nanoporosity was included in the tested volume but the large capillary porosity was not. The depth range 100-400 nm was suitable for the analysis.

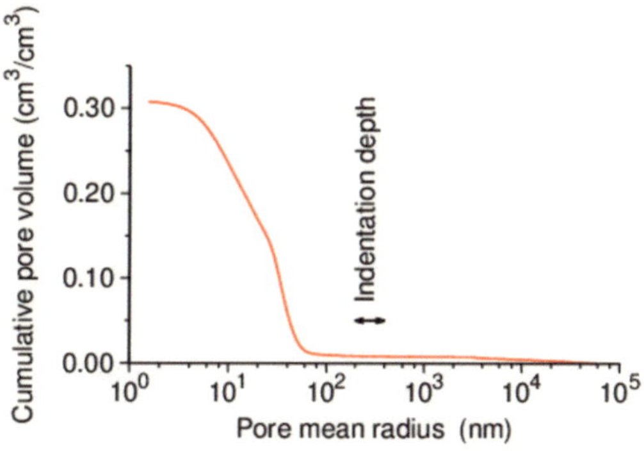

Figure 2. Cumulative pore volume on cement paste samples.

Samples for nanoindentation testing were cut from larger volume by a diamond saw and polished on series of SiC papers and diamond spray to achieve flat surface on a ~5 mm thick disk with diameter ~30 mm. The surface roughness was checked with AFM to be $R_q \approx 10$ nm on 50×50 μm area.

6.1.2. Gypsum

Secondly, dental gypsum (Interdent® Interrock New) was chosen as a model system for gypsum based materials (Tesárek & Němeček, 2010). From the chemistry point of view, every gypsum binder is composed of three main components – calcium sulphate anhydrite ($CaSO_4$), calcium sulphate hemihydrate ($CaSO_4 \cdot ½H_2O$) in two modifications: α- or β-hemihydrate, and calcium sulphate dihydrate ($CaSO_4 \cdot 2H_2O$). The gypsum binder consists also some impurities and additives in case of natural sources. The Interdent® gypsum is a low-porosity purified α-hemihydrate used for dental purposes.

From the micromechanics point of view, gypsum samples can be viewed as porous polycrystalline materials that are characterized with a macroscopically compact solid. The microstructure of the polished cross-section of the dental gypsum sample as seen in electron microscope is shown in Fig. 3. Dark areas in Fig. 3 can be attributed to the porosity, very light areas belong to low hydrated $CaSO_4$ grains or carbonates and the majority of the sample volume composes of hydrated crystalline mass.

Samples were prepared with water to binder weight ratio 0.2 and stored in ambient conditions for 20 days (full strength occurs after a day). After mixing, the material begins to crystallize with maximum heat development in the order of minutes. The hardened gypsum mass is a porous material with a relatively large internal surface consisting of interlocking crystals in the form of plates and needles (Singh & Middendorf, 2007). In case of β-hemihydrate hydration, the resulting sample porosity is typically very large (more than 50% for higher water to binder ratii) and crystals are interlocked very weakly. Therefore, ordinary gypsum systems used for building purposes which are based on β-hemihydrate are characterized with relatively low strengths (<10 MPa in compression). In contrast, hydration of our samples based on α-hemihydrate produced a dense matrix. Overall sample porosity reached just 19% as assessed by sample weighing. Since pore system plays a key mechanical role in gypsum materials, the pore distribution was monitored with Hg-porosimetry (MIP) as depicted in Fig. 4. The majority of pores lay in the range 0-1 μm (~12%) and virtually no pores appeared between 1 and 100 μm (<0.5%). The MIP technique does not allow to access large capillary/entrapped air porosity. It means that ~7% of pores was not accessible by MIP and can be attributed to large air voids.

Again, the indentation depth was chosen so that major nanoporosity was included in the tested volume whereas large air porosity was not. The indentation depths ~500 nm (i.e. $(3\times0.5)^3=1.5^3$ μm³ tested volume) were chosen as suitable for the analysis.

For nanoindentation testing, samples were cut and polished in the same way like cement paste to receive ~5×30 mm disk with a flat surface. The surface roughness checked with AFM was $R_q \approx 40$ nm on 50×50 μm area.

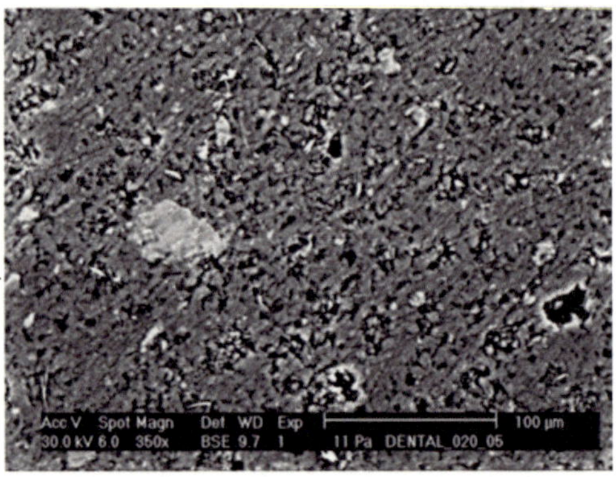

Figure 3. SEM image of gypsum.

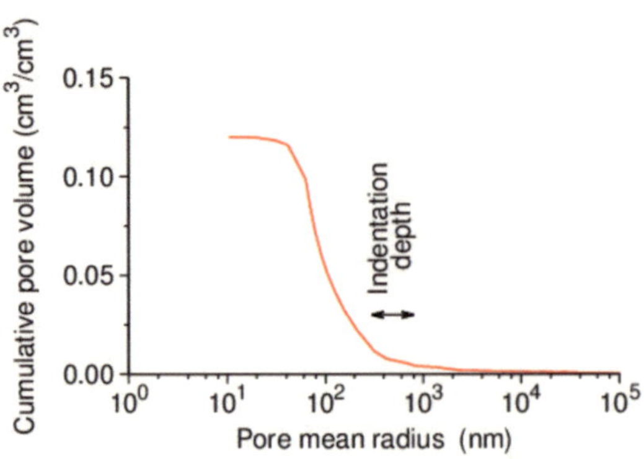

Figure 4. Cumulative pore volume on gypsum samples.

6.2. Grid nanoindentation

6.2.1. Cement paste

Nanoindentation measurements were performed in a load control regime using the CSM Nanohardness tester in the Micromechanics laboratory of the Czech Technical University in Prague. The trapezoidal loading diagram was prescribed for all tests (Fig. 5). Maximum force 2 mN was applied with constant load rate 12 mN/min. The loading lasted for 10 s. The holding period, in which the load was kept constant for 30 s, followed, allowing the material to creep (Němeček, 2009). The following unloading branch of 12 mN/min for 10 s was supposed to be purely elastic. The applied maximum load of 2 mN led to maximum penetration depths ranging from 100 nm to 400 nm (average 220 nm) depending on the hardness of the indented material phase.

The effective depth captured by the tip of the indenter can be roughly estimated as three times the penetration depth for the Berkovich indenter (Constantinides et al., 2006). It yields the affected volume of around ~0.7^3 μm^3 for this particular case. Nanoindentation response obtained for different material constituents is depicted in Fig. 5 and it clearly shows different deformations and stiffness of distinct phases if the same load is applied. Elastic properties were evaluated from nanoindentation tests according to the Oliver-Pharr methodology (Oliver & Pharr, 1992).

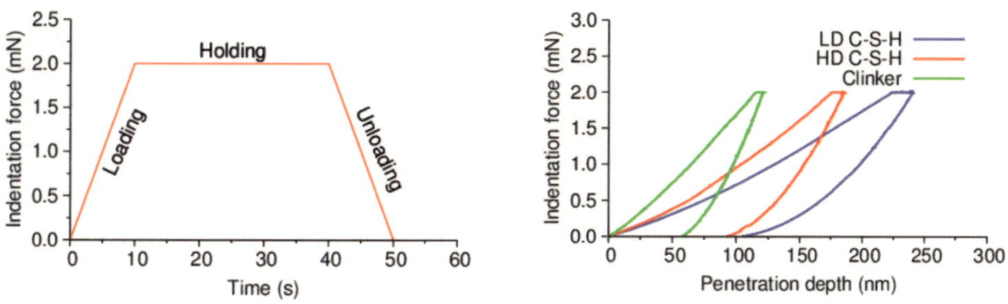

Figure 5. Prescribed loading diagram and examples of force-penetration curves for different cement paste constituents.

Due to the heterogeneity of the sample and uneven distribution of the phases in its volume, the results from individual indentation grids depend very much on the location and indents' spacing. As observed in SEM, the heterogeneity occurs on the scale of tens or even hundreds of μm (Němeček, 2009). This fact leads to the necessity to cover relatively large areas of the sample by large grids. Several sample locations containing the material phases in a sufficient content (~100×100 to 200×200 μm with 10-20 μm spacing) have been performed on the sample.

Results from ~400 indents have been obtained on the sample and Young's moduli evaluated from individual locations (assuming Poisson's ratio 0.2 for all phases). Results were merged and property histograms constructed. The experimental histogram covers all these material phases including their intrinsic nanoporosity. According to the cement chemistry, the phases can be denoted as:

a. Low stiffness phase (all phases with Young's modulus less than ~10 GPa)
b. Low density C-S-H (i.e. C-S-H gel with loose packing density)
c. High density C-S-H (i.e. C-S-H gel with high packing density)
d. Portlandite $Ca(OH)_2$
e. Clinker residue

The deconvolution into the mechanically distinct phases that correspond to the aforementioned chemical phases (A-D) has been carried out with the assumption that values of Young's modulus higher then ~50 GPa can be attributed to residual clinker grains (E) and were not considered in the deconvolution because of very low content. Instead, separate 'ex-situ' measurements of clinker stiffness were performed (Němeček, 2009). The reason for the separate clinker measurements was twofold. The first reason lies in the low clinker content and the second reason is a very high stiffness contrast to other phases which is more then 5 (Eq. 3). Due to phase interactions, the clinker stiffness is underestimated in 'in-situ' measurements.

Experimental histogram and the deconvolution of phases on cement paste are depicted in Fig. 6. Elastic constants for individual phases with their volume fractions are summarized in Tab. 1.

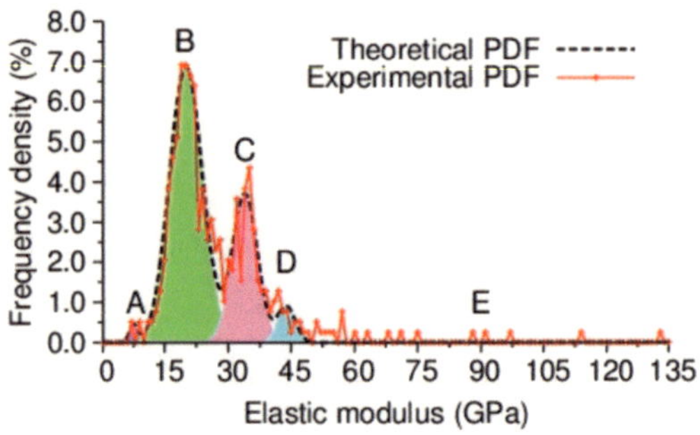

Figure 6. Deconvolution of the experimental histogram of Young's moduli to distinct material phases of cement paste.

Phase	E (GPa)	Volume fraction
Low stiffness (A)	7.45±0.98	0.0105
Low density C-S-H (B)	20.09±3.85	0.6317
High density C-S-H (C)	33.93±2.98	0.2634
$Ca(OH)_2$ (D)	43.88±2.15	0.0461
Clinker (E)	121.0±14.0*	0.0483

*Note: The clinker value was adjusted according to (Němeček, 2009).

Table 1. Phase properties from deconvolution on cement paste.

6.2.2. Gypsum

Similar experimental setup as for cement paste samples was used. Two locations were tested on gypsum samples. Each place was covered by 15×12=180 indents with 15 μm spacing.

Also, similar loading was used (i.e. load controlled test) but to maximum force 5 mN. Typical loading diagrams received on gypsum samples are depicted in Fig. 7. A bit wider range of final depths on indented phases (200-800 nm) was obtained due to larger differences in the polycrystalline stiffness. However, the majority of indents were performed to final depths ranging between 400 to 500 nm. Thus, the material volume affected by indentation can be estimated as $(3\times0.5)^3=1.5^3$ μm³. The RVE size defined by the tested area is ~200 μm in this case.

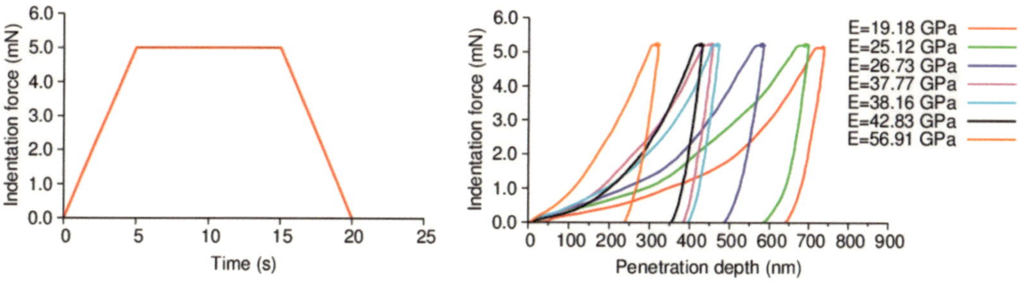

Figure 7. Prescribed loading diagram and examples of force-penetration curves at different locations on gypsum sample.

Again, results of Young's moduli (assuming Poisson's ratio 0.32), evaluated from all positions, were merged and property histogram constructed. The phase separation is not straightforward in this case, since the anisotropic gypsum crystals are squeezed in a dense polycrystalline matrix with random orientations. The response in nanoindentation is then measured on differently oriented crystals and also on a combination of differently oriented crystals located under the indenter in the affected volume ~1.5^3 μm³. The tested location can be viewed as a set of mechanically different phases that are physically averaged by an indenter. Apparent isotropic elasticity constants associated with the tested indentation volume are derived in this case.

Further, it is possible to hypothesize on the number of mechanically dissimilar groups of responses from the experimental frequency plot (Fig. 8). For example, three significant peaks related to the three symmetry axes of the gypsum crystal (monoclinic system) is one of the options. Deconvolution methodology can be used to separate the three phase distributions. One can also compute apparent elastic moduli of an isotropic solid from all responses in an ensemble (i.e. compute average value from all results). Both approaches have been tested and compared.

The numerical error (Eq. 9) of the two fits was very similar and they can both be treated as numerically equivalent. Resulting phase distributions are depicted in Fig. 8. The homogenized response computed from the three-phase fit by analytical homogenizations was also mechanically equivalent to the single phase fit ($E_{(3\text{-phases})}\approx33$ GPa, not shown here).

Therefore, the apparent Young's modulus of a single isotropic phase will be further used as effective stiffness of already physically homogenized gypsum matrix located in the tested RVE (~200 μm). This effective Young's modulus was found to be 33.90±10 GPa (Fig. 8, right).

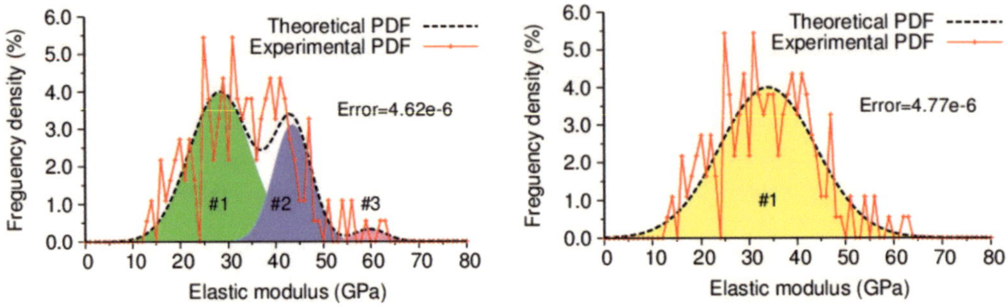

Figure 8. Experimental histogram of Young's moduli with the three-phase (left) and single–phase (right) Gaussian fits on gypsum.

7. Effective elastic properties

7.1. Cement paste

Two-scale micromechanical model was considered for cement paste. The majority of the specimen volume is occupied by C-S-H phases (in two significant densities) reinforced with nanocrystalline Portlandite particles (Thomas et al., 1998; Tennis & Jennings, 2000). Therefore, the first level of the model with the RVE size ~1 μm was considered. Effective elastic constants of the first level were obtained by analytical homogenization. Low density C-S-H phase (i.e. the phase with the highest frequency) was considered as the reference phase in the Mori-Tanaka method. Analytical estimates (Mori-Tanaka and self-consistent) are shown in Tab. 3. They both give very similar results bounded with very close Voigt and Reuss limits. The Mori-Tanaka result is considered in further computations.

The second level of the micromechanical model consisted of the homogenized C-S-H phase, Portlandite, low stiffness phase and clinker. Such arrangement defined the RVE size ~100 μm which corresponds to the size of indentation grids. The same analytical estimates were computed (Tab. 4).

Input	E (GPa)	Poisson's ratio	Volume fraction
Low density C-S-H (B)	20.09	0.2	0.6317
High density C-S-H (C)	33.93	0.2	0.2634
Output (B+C)			
Mori-Tanaka	**23.36**		
self-consistent	23.41	0.2	
Voigt bound	24.16		
Reuss bound	22.83		

Table 2. Analytical homogenization of the C-S-H level in cement paste.

Input	E (GPa)	Poisson's ratio	Volume fraction
C-S-H level (B+C)	23.36	0.2	0.8951
Low stiffness (A)	7.45	0.2	0.0105
$Ca(OH)_2$ (D)	43.88	0.3	0.0461
Clinker (E)	121.00	0.3	0.0483
Output			
Mori-Tanaka	**25.39**	0.207	
self-consistent	25.47	0.208	
Voigt bound	29.05	0.234	
Reuss bound	24.29	0.204	

Table 3. Analytical homogenization of the cement paste level.

Results of the analytical homogenization were checked with independent FFT-based method which takes into account all indents in a grid. Elastic constants in individual grid points are considered as input values related to one discretization point in the method. The homogenized result in the form of a stiffness matrix (in Mandel's notation) of cement paste was:

$$\mathbf{L}_{\mathbf{eff}}^{\mathbf{FFT}} = \begin{bmatrix} 26.177 & 6.778 & 0.068 \\ 6.778 & 26.224 & 0.014 \\ 0.068 & 0.014 & 19.818 \end{bmatrix} (\text{GPa}) \tag{17}$$

The off-axis terms in the matrix are almost zero which shows on isotropic character of the material. The result is directly comparable with the analytical result by assuming plane strain stiffness matrix for an isotropic material. Results from the Mori-Tanaka homogenization on cement paste level (E=25.37 GPa, ν=0.207) give:

$$\mathbf{L}_{\mathbf{eff}}^{\mathbf{A}} = \frac{E}{(1+\nu)(1-2\nu)} \begin{bmatrix} 1-\nu & \nu & 0 \\ \nu & 1-\nu & 0 \\ 0 & 0 & 1-2\nu \end{bmatrix} = \begin{bmatrix} 28.44 & 7.43 & 0 \\ 7.43 & 28.44 & 0 \\ 0 & 0 & 21.02 \end{bmatrix} (\text{GPa}) \tag{18}$$

The difference between the estimates can be computed by the matrix error norm:

$$\delta = \sqrt{\frac{\left(\mathrm{L}_{\mathrm{eff}}^{\mathrm{FFT}} - \mathrm{L}_{\mathrm{eff}}^{\mathrm{A}}\right)::\left(\mathrm{L}_{\mathrm{eff}}^{\mathrm{FFT}} - \mathrm{L}_{\mathrm{eff}}^{\mathrm{A}}\right)}{\left(\mathrm{L}_{\mathrm{eff}}^{\mathrm{FFT}}::\mathrm{L}_{\mathrm{eff}}^{\mathrm{FFT}}\right)}} \approx 0.08 \tag{19}$$

The 8% error shows good agreement of the methods and also near to isotropic nature of the cement paste.

7.2. Gypsum

As discussed earlier, the gypsum can be viewed as a composition of polycrystalline matrix and pore space. Therefore, two-scale micromechanical model can be defined for the gypsum sample. The lower level (i.e. the crystalline matrix) was studied by nanoindentation and effective isotropic properties were directly assessed as an average value (E=33.90 GPa, ν=0.32) for RVE~200 μm. This level also includes intrinsic porosity which is naturally included in the nanoindentation results received from the affected volume ~1.5^3 μm^3. This part of the porosity was determined by MIP to be 12%.

The assumption of effective properties on the matrix level was checked with the FFT-based homogenization by using elastic constants measured with nanoindentation in grid points as input. The homogenized stiffness matrix (in Mandel's notation) for the lower level is:

$$\mathbf{L}_{\mathbf{eff}}^{\mathbf{FFT}} = \begin{bmatrix} 45.302 & 21.185 & 0.101 \\ 21.185 & 45.497 & -0.008 \\ 0.101 & -0.008 & 24.396 \end{bmatrix} (\text{GPa}) \tag{20}$$

Comparison with the result received from nanoindentation (E=33.90 GPa, ν=0.32) using Eq. 18 gives:

$$\mathbf{L}_{\mathbf{eff}}^{\mathbf{NI}} = \begin{bmatrix} 48.51 & 22.84 & 0 \\ 22.84 & 48.51 & 0 \\ 0 & 0 & 25.69 \end{bmatrix} (\text{GPa}) \tag{21}$$

The difference between the results computed by the matrix error norm (Eq. 19) yields δ=0.07, i.e. 7%. Again, very good agreement between the results from different methods was achieved.

The upper model level (the sample level) includes the first level and a part of the sample porosity that is above the indentation level which was 7% (see section 6.1.2). Therefore, micromechanical homogenization was performed for the second level assuming additional 7% porosity. Several order of magnitude smaller stiffness (E=0.0001 GPa) compared to the matrix was prescribed to spherical air inclusions. Naturally, Voigt and Reuss bound are very distant in this case since the first represents parallel phase configuration and the second serial configuration which in case of one almost zero stiffness phase (the air) leads to almost zero composite stiffness. The real situation in the composite (matrix reinforced with spherical air inclusions) is better described with the Mori-Tanaka method. For the sake of comparison, the self-consistent scheme was also used. It yielded a similar result for the specific case. The results are summarized in Tab. 5.

8. Discussion

It has been shown that the grid indentation and the deconvolution give access to intrinsic properties of individual material phases that are mechanically dissimilar. Both analytical and numerical schemes used for homogenization on cement paste and gypsum resulted in very similar values of elastic constants. The estimated overall elastic moduli of cement paste

Method	E (GPa)	Poisson's ratio
Mori-Tanaka	**29.46**	0.31
self-consistent	29.14	0.31
Voigt bound	31.52	0.32
Reuss bound	0.0014	0.49

Table 4. Results of analytical homogenization on gypsum (the sample level).

(E=25.37 GPa) are in good agreement with those experimentally measured on larger material volumes. For example, similar values of Young's modulus can be found for the hydrated compound of cement paste in the literature, e.g. E~26.4±1.8 GPa (Němeček, 2009), E~22.8±0.5 GPa (Constantinides & Ulm, 2007; Constantinides & Ulm, 2004) or E~26.5 GPa (Hughes & Trtik, 2004).

Since the gypsum crystallizes in the monoclinic system, the elastic stiffness tensor contains 13 nonzero components. The measurement of these constants has been performed with acoustic measurements by Haussühl, 1960. Computation of an angular average from the elastic moduli tensor leads to an isotropic stiffness E~45.7 GPa and ν=0.33 (Meille and Garboczi, 2001). Sanahuja et al. elaborated a modified self-consistent scheme for elongated gypsum crystals and computed their equivalent isotropic stiffness E~45 GPa and ν=0.33 found for zero crystal porosity. If an inter-crystalline porosity ϕ=0.2 was taken into account the isotropic stiffness decreased to E~28 GPa and ν~0.3 (Sanahuja, 2010).

We performed measurements on large scale gypsum samples (40x40x160 mm prisms) on which Young's moduli were investigated by uniaxial compression and by the resonant method. Both measurements gave similar values of the mean Young's modulus: E~28.6±0.2 GPa in compression tests and E~26.6±0.4 GPa, ν=0.29±0.008 in the resonant method, respectively. Thus, good agreement with the homogenized result on our samples (E~29.46 GPa, ν=0.31) was achieved.

9. Conclusions

Effective elastic properties of selected heterogeneous structural materials were studied by nanoindentation and methods of micromechanics. Cement paste and gypsum (α-hemihydrate) were in focus but the methodology described in this chapter can be easily adopted for other heterogeneous materials, e.g. for high-performance concretes and mortars (Sorelli et al., 2008; Němecek et. al., 2011b), alkali-activated materials (Němeček et al., 2011c) or for metal alloys (Němeček et al., 2011a).

Firstly, nanoindentation was utilized for the assessment of elastic parameters of small material volumes. The size of individual indents was kept small enough (h≈200-500 nm) to represent individual phase behaviour. The mechanical response was received from the volume ~0.7^3-1.5^3 μm^3. As an unavoidable fact, intrinsic porosity of the constituents was included in nanoindentation results.

Complicated microstructures of the studied heterogeneous structural materials and impossible ex-situ preparation of individual material constituents lead to the use of statistical grid indentation to cover the response from all phases. Elastic constants of the basic constituents have been derived and further used in multi-scale models to receive effective composite properties within the RVE.

Deconvolution technique has proven to be an efficient tool for the separation of distinct phases in cement paste. On the other hand, local anisotropy of the polycrystalline gypsum matrix was not assessed and effective (average) matrix properties have been directly derived from the grid nanoindentation as mean values.

Effective RVE properties were successfully determined with analytical schemes (Mori-Tanaka, self-consistent) and verified with numerical FFT-based method. The performance of both approaches was in good agreement for the tested materials. Comparison with macroscopic experimental data also showed good correlation with the predicted values and validated the presented methodology.

Author details

Jiří Němeček

Czech Technical University in Prague, Faculty of Civil Engineering, Department of Mechanics, Czech Republic

Acknowledgement

Support of the Czech Science Foundation (P105/12/0824) is gratefully acknowledged.

10. References

Beaudoin, J.J.; Raki, L.; Alizadeh, R. & Mitchell, L. (2010). Dimensional change and elastic behavior of layered silicates and Portland, cement paste, *Cement & Concrete Composites* 32, pp. 25–33.

Bentz, D.P. & Garboczi, E.J. (1999). Computer Modelling of Interfacial Transition Zone: Microstructure and Properties, *Engineering and Transport Properties of the Interfacial Transition Zone in Cementitious Composites*, Rilem Report No. 20, RILEM Publ. s.a.r.l., Cachan, France, Part 5, Chapter 20, pp. 349-385.

Borodich, F.M.; Keer, L.M. & Korach, C.J. (2003). Analytical study of fundamental nanoindentation test relations for indenters of non-ideal shapes, *Nanotechnology* 14 (7), pp. 803-808.

Constantinides, G. & Ulm, F.-J. (2004). The effect of two types of C–S–H on the elasticity of cement-based materials: results from nanoindentation and micromechanical modeling, *Cement and Concrete Research* 34 (1), pp. 67–80.

Constantinides, G. & Ulm, F.-J. (2007). The nanogranular nature of C–S–H, *Journal of the Mechanics and Physics of Solids* 55, pp. 64–90.

Constantinides, G.; Chandran, F.R.; Ulm, F.-J. & Vliet, K.V. (2006). Grid indentation analysis of composite microstructure and mechanics: Principles and validation, Materials *Science and Engineering: A*, 430 (1-2), pp. 189-202.

Drugan, W.R. & Willis, J.R. (1996). A micromechanics-based nonlocal constitutive equation and estimates of representative volume element size for elastic composites, *Journal of the Mechanics and Physics of Solids* 44 (4), pp. 497–524.

Durst, K.; Göken, M. & Vehoff, H. (2004). Finite element study for nanoindentation measurements on two-phase materials, *J Mater. Res.* 19 , pp. 85–93.

Eshelby, J.D. (1957). The determination of the elastic field of an ellipsoidal inclusion and related problem, *Proc. Roy. Soc. London A* 241, pp. 376–396.

Fischer-Cripps, A.C. (2002). *Nanoindentation*, Springer Verlag, ISBN 0-387-95394-9.

Gao, H.J.; Chiu, C.H. & Lee, J. (1992). Elastic contact versus indentation modeling of multi-layered materials, *International Journal of Solids and Structures* 29 (20), pp. 2471–2492.

Hashin, Z. (1983). Analysis of composite materials - a survey. *ASME J. Appl. Mech.* 50, pp. 481–505.

Haussuhl, S. (1960). Elastische und thermoelastische Eingenschaften von $CaSO_4 2H_2O$ (Gips), *Zeitschrift fur Kristallographie* 122, pp. 311-314.

Hill, R. (1963). Elastic properties of reinforced solids - Some theoretical principles. *Journal of the Mechanics and Physics of Solids* 11, pp. 357–372.

Hill, R. (1965). Continuum micro-mechanics of elastoplastic polycrystals. *Journal of the Mechanics and Physics of Solids* 13, pp. 89–101.

Hughes, J.J. & Trtik, P. (2004). Micro-mechanical properties of cement paste measured by depth-sensing nanoindentation: a preliminary correlation of physical properties with phase type, *Mater Charact* 53, pp. 223-31.

Jiroušek, O.; Jandejsek, I. & Vavřík, D. (2011a). Evaluation of strain field in microstructures using micro-CT and digital volume correlation, *Journal of Instrumentation* 6 (1), art. no. C01039.

Jiroušek, O.; Němeček, J.; Kytýř, D.; Kunecký, J.; Zlámal, P. & Doktor, T. (2011b). Nanoindentation of trabecular bone-comparison with uniaxial testing of single trabecula, *Chemicke Listy* 105 (17), pp. s668-s671.

Meille, S. & Garboczi, E. J. (2001). Linear elastic properties of 2D and 3D models of porous materials made from elongated objects. *Modell. Simul. Mater. Sci. Eng.* 9(5), pp. 371–390.

Michel, J.C.; Moulinec, H. & Suquet, P. (1999). Effective properties of composite materials with periodic microstructure: a computational approach, *Comput. Methods Appl. Mech. Engrg.* 172, pp. 109-143.

Mori, T. & Tanaka, K. (1973). Average stress in matrix and average elastic energy of materials with misfitting inclusions, *Acta Metallurgica* 21 (5), pp. 571-574.

Moulinec, H. & Suquet, P. (1994). A fast numerical method for computing the linear and nonlinear properties of composites, *Comptes-Rendus de l'Académie des Sciences série II* 318, pp. 1417-1423.

Moulinec, H. & Suquet, P. (1998). A numerical method for computing the overall response of nonlinear composites with complex microstructure, *Computer Methods in Applied Mechanics and Engineering* 157 (1–2), pp. 69–94.

Němeček, J. (2009). Creep effects in nanoindentation of hydrated phases of cement pastes, *Materials Characterization* 60 (9), pp. 1028–1034.

Němeček, J.; Králík, V.; Vondřejc, J. & Němečková, J. (2011a). Identification of micromechanical properties on metal foams using nanoindentation, *Proceedings of the Thirteenth International Conference on Civil, Structural and Environmental Engineering Computiing* [CD-ROM]. Edinburgh: Civil-Comp Press, 2011, pp. 1-12. ISBN 978-1-905088-46-1.

Němeček, J.; Lehmann, C. & Fontana, P. (2011b). Nanoindentation on Ultra High Performance Concrete System, *Chemicke Listy* 105 (17), pp. 656-659.

Němeček, J.; Šmilauer, V. & Kopecký, L. (2011c). Nanoindentation characteristics of alkali-activated aluminosilicate materials, *Cement and Concrete Composites* 33 (2), pp. 163-170.

Oliver, W. & Pharr, G.M. (1992). An improved technique for determining hardness and elastic modulus using load and displacement sensing indentation experiments, *J. Mater. Res.* 7 (6), pp. 1564–1583, 1992.

Randall, N.X. (2009). Mechanical Properties of Cementitious Materials, *CSM Instruments Application Bulletin* No.29, available from: http://www.csm-instruments.com/

Sanahuja, J.; Dormieux, L.; Meille, S.; Hellmich, C. & Fritsch, A. (2010). Micromechanical Explanation of Elasticity and Strength of Gypsum: From Elongated Anisotropic Crystals to Isotropic Porous Polycrystals, *Journal of Engineering Mechanics* 136 (2), pp. 239-253.

Saranen, J., Vainikko, G. (2002), *Periodic Integral and Pseudodifferential Equations with Numerical Approximation*, Springer, Berlin.

Singh, N.B. & Middendorf, B. (2007). Calcium sulphate hemihydrate hydration leading to gypsum crystallization, *Progress in Crystal Growth and Characterization of Materials* 53 (1), pp. 57–77.

Sorelli, L.; Constantinides, G.; Ulm, F.-J. & Toutlemonde, F. (2008). The nano-mechanical signature of Ultra High Performance Concrete by statistical nanoindentation techniques, *Cement and Concrete Research* 38, pp.1447–1456.

Swadener J.G & Pharr, G.M. (2001). Indentation of elastically anisotropic half-spaces by cones and parabolae of revolution, *Phil. Mag. A* 81 (2), pp. 447-466.

Šmilauer, V.; Hlaváček, P.; Škvára, F.; Šulc, R.; Kopecký, L. & Němeček, J. (2011). Micromechanical multiscale model for alkali activation of fly ash and metakaolin, *Journal of Materials Science* 46 (20), pp. 6545-6555.

Taylor, H.F.W. (2003). *Cement Chemistry, Second Edition*, Thomas Telford, ISBN 0-7277-2592-0.

Tennis, P.D. & Jennings, H.M. (2000). A Model for Two Types of C-S-H in the Microstructure of Portland Cement Pastes, *Cement and Concrete Research* 30 (6), pp. 855-863.

Tesárek, P. & Němeček, J. (2011). Microstructural and Micromechanical Study of Gypsum, *Chemicke Listy* 105 (17), pp. 852-853.

Thomas, J.J.; Jennings, H.M. & Allen, A.J. (1998). The Surface Area of Cement Paste as Measured by Neutron Scattering – Evidence for Two C-S-H Morphologies, *Cement and Concrete Research* 28 (6), pp. 897-905.

Ulm, F.-J.; Vandamme, M.; Bobko, C. & Ortega, J.A. (2007). Statistical Indentation Techniques for Hydrated Nanocomposites: Concrete, Bone, and Shale, *J. Am. Ceram. Soc.* 90 (9), pp. 2677–2692.

Vlassak, J.J.; Ciavarella, M.; Barber, J.R. & Wang, X. (2003). The indentation modulus of elastically anisotropic materials for indenters of arbitrary shape, *Journal of the Mechanics and Physics of Solids* 51, pp. 1701 – 1721.

Zaoui, A. (2002). Continuum Micromechanics: Survey, *Journal of Engineering Mechanics* 128 (8), pp. 808–816.

Zeman, J.; Vondřejc, J.; Novák, J. & Marek, I. (2010). Accelerating a FFT-based solver for numerical homogenization of periodic media by conjugate gradients, *Journal of Computational Physics* 229(21), pp. 8065-8071.

Chapter 5

Application of Nanoindentation Technique in Martensitic Structures

L. Zhang, T. Ohmura and K. Tsuzaki

Additional information is available at the end of the chapter

http://dx.doi.org/10.5772/48327

1. Introduction

The instrumented nanoindentation technique with a capability of in situ scanning-probe microscopy (SPM) or atomic force microscopy (AFM) can measure the mechanical properties of a precise site in a microstructure with accuracy within a nanometer [1]. As a precise and quantitative measuring method, nanoindentation technique has been widely used in various materials systems. In this chapter, the application of nanoindentation technique in the martensitic structures (lath martensite [2, 3] and lenticular martensite [4] as illustrated in Fig. 1) will be introduced. Also, to understand the relationship between martensitic transformation with grain size, nanostructured NiTi shape memory alloy [5] with a graded surface nanostructure was produced by surface mechanical attrition treatment (SMAT) and tested by nanoindentation.

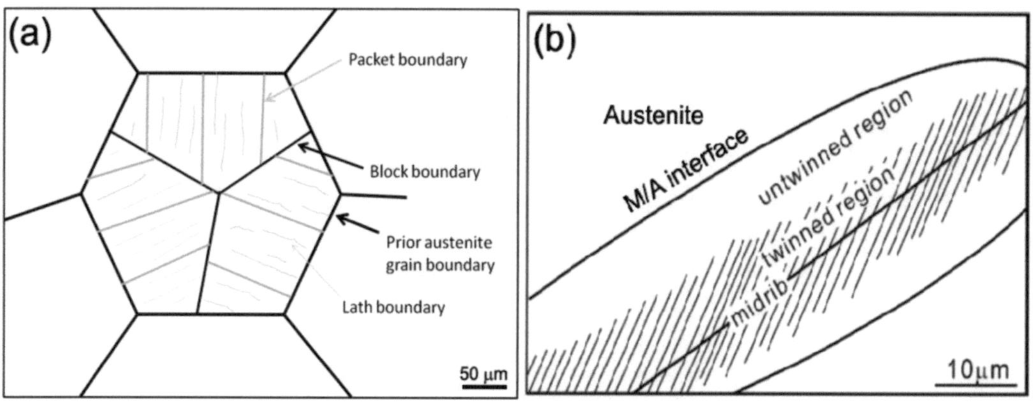

Figure 1. Schematic drawing of three packets of lath martensite within a prior austenite grain (a) and a lenticular martensite plate (b).

As one of the most important structural materials, the morphology and substructure of martensite in the Fe-Ni/Fe-C alloys have been investigated widely. The formation of various morphologies such as lath, butterfly, lenticular and thin plate has been shown to depend on the alloy composition or the transformation temperature (Ms Temperature). Each of them is unique in crystallographic features as well as in substructure [6]. Fig. 1 is schematic drawing of three packets of lath martensite within a prior austenite grain (a) and a lenticular martensite plate (b).

1.1. Lath martensite

Lath martensite, which forms in the highest temperature range, consists of parallel platelets or laths containing a high density of random dislocations [7, 8]. The lath martensite is mainly composed of four structural units: lath, block, packet and prior austenite grain (Fig. 1(a)). The prior austenite grain consists of some packets, the packet is divided into the blocks, and the block is subdivided into the lath structure. Only the lath boundaries are low-angle ones [9-12]. Hence the block structure can be recognized as an effective grain. Many studies have revealed that the macroscopic strength of the martensite in Fe-C-based alloys depends mainly on the carbon content [13] and tempering temperatures [14]. These dependences are understood as a matrix strengthening such as a carbon solid solution, dislocation interaction and fine precipitation strengthening. However, matrix strength (i.e. block matrix connecting solid solution, dislocation interaction and fine precipitation strengthening etc. of the matrix) without the contributions of high-angle boundaries has never been measured directly. Nanoindentation test with indent sizes much smaller than the block width is the key to obtain an exact measurement of the matrix strength without contributions of the high-angle boundaries, i.e. block boundaries. Simple alloy systems such as pure Fe-C binary alloy are suitable for conducting systematic investigations.

1.2. Lenticular martensite

On the other hand, lenticular martensite, which is plate-like with curved interfaces, forms at the intermediate temperature between lath martensite and thin plate martensite. The substructure of lenticular martensite is much more complicated than that of the other types. The martensite in the Fe-Ni alloy containing about 28.5% ~ 33 wt% Ni is lenticular with three regions, namely the mid-rib region, the twinned region and the untwined region [15]. The midrib region, where the martensitic transformation is thought to begin, has a substructure of fine transformation twins. The twin spacing is 6.9 nm in the Fe-31 wt% Ni alloy which was cooled just below the Ms temperature (223 K) [10, 16]. The twinned region is partially twinned, and the untwined region does not contain any twins but has many dislocations similar to that of the lath martensite [17, 18]. Since various lattice defects such as ultra-fine twins and dislocations exist simultaneously in one sample, the lenticular martensite in the Fe-Ni alloy is appropriate for understanding the strengthening mechanism in the structural materials.

As shown in Fig. 1, the microstructure of the martensite is complex and very fine on the submicron scale, so it is difficult to separate the strength contribution of each factor by the conventional macroscopic testing method. To understand the whole strengthening mechanism of martensite, it is important to estimate each factor independently and reveal the relationship between the microstructure and the deformation behavior.

1.3. NiTi shape memory alloy

Nanoindentation technique can not only be used to characterize mechanical properties in transformed martensite, but also to detect the martensitic transformation during deformation. In NiTi alloys the reversible thermoelastic martensitic transformation can be induced by either temperature or stress and leads to their well-known shape memory and superelastic properties [19]. This thermoelastic martensitic transformation indicates a close, intrinsic interaction between microscopic phase transformations and macroscopic mechanical properties, resulting in complex mechanical behavior and uncertainty in determining the mechanical properties of NiTi. The stress-induced martensitic transformation (SIM) may contribute to the apparent elastic strain in the stress–strain test, leading to a low apparent Young's modulus [20, 21]. Despite suggestions by previous studies, direct evidence for the effect of the SIM on the Young's modulus is still lacking. Recent transmission electron microscopy (TEM) observations have indicated that the martensitic transformation is suppressed in nanocrystalline NiTi because of grain size effects [22]. In a very recent study [23] it was also suggested that a higher stress is required for the SIM for a smaller grain size. Therefore, it is important to investigate the effect of grain size on the Young's modulus in nanostructured NiTi. To understand the relationship between martensitic transformation with grain size, nanostructured NiTi with a graded surface nanostructure was produced by surface mechanical attrition treatment (SMAT).

2. Experimental procedures

2.1. Fe-C lath martensite and Fe-Ni lenticular martensite

Same experimental procedures were applied in Fe-C lath martensite and Fe-Ni lenticular martensite samples. The specimen surfaces for nanoindentation were mechanically polished with care, and subsequently electropolished. Nanoindentation measurements were carried out using a Hysitron, Inc. Triboscope with a Berkovich indenter. The tip truncation of the indenter was calibrated using fused silica. The Oliver and Pharr method [16] was used for the tip calibration and the calculation of nanohardness. The probed sites and the shape of the indent marks on the specimen surface were confirmed before and after the indentation measurements with the AFM mounted with the Triboscope. The conducted peak load was 500 μN. The loading/unloading rate was 50 μN/s and was held at the peak load for 10 s. Conventional Vickers hardness (H_v) tests as a macroscopic strength evaluation were conducted with a load of 4.9 N.

2.2. TiNi shape memory alloy

For nanoindentation the SMAT surface was protected with an electroplated Ni coating. Cross-sectional samples were carefully polished with decreasing grit sizes. The final polishing agent was a mixture of OP-S (0.04 μm colloid Si, Struers Inc.), H_2O_2 and NH_3-H_2O solutions (50:1:1) to reduce possible surface damage. After polishing, the state of the sample surface was further confirmed by scanning electron microscopy and electron backscattering diffraction.

Nanoindentation measurements were carried out using a Hysitron, Inc. TriboIndenter with a Berkovich indenter. The tip truncation of the indenter was calibrated using fused silica. The Oliver and Pharr method [16] was used for the tip calibration and the calculation of nanohardness. The position and configuration of the indents were checked in situ by SPM. A peak load of 1000 μN was used with a loading/unloading rate of 50 μN/s and the indenter was held at peak load for 5 s before unloading.

The calculated modulus by an analysis of the unloading curve is reduced modulus E_r and its relationship with the Young's modulus is expressed as:

$$E_r^{-1} = \frac{1-\nu_s^2}{E_s} + \frac{1-\nu_i^2}{E_i}, \tag{1}$$

where E_s and ν_s are the Young's modulus and the Poisson's ratio for the specimen, and E_i and ν_i are the same parameters for the indenter (1140 GPa and 0.07, respectively). Using 0.35 for the Poisson's ratio of NiTi [5], the value of E for NiTi can be determined.

3. Nanoindentation in Fe-C lath martensite

We firstly start with high purity Fe-C binary alloys which contain as-quenched lath martensite. Five nominal carbon contents of 0.1, 0.2, 0.4, 0.6 and 0.8 wt% (shorted as Fe-0.1C, 0.2C, 0.4C, 0.6C and 0.8C) were employed. A Fe-23wt% Ni alloy with lath martensite (shorted for Fe-23Ni) was also used for comparison.

3.1. Microstructures of Fe-C lath martensite

Fig. 2 represents the optical micrographs of the Fe-C lath martensite with different carbon content [2]. The block width deceases with increasing carbon content. The microstructure of the Fe-0.1C martensite composes clearly packets and blocks but for higher carbon content specimens, the block structure is very fine and complicated so it is hard to identify the boundaries of the block and packet. The microstructure of the Fe-23Ni lath martensite consists of blocks with a relatively large width typically in a few 10 μm.

3.2. Conventional nanoindentation in Fe-C lath martensite

Fig. 3 shows the typical load-displacement curves of five Fe-C martensites [2]. The smaller penetration depths show higher nanohardness values on the load-depth curve under the same peak load conditions. Maximum penetration depths of nanoindentation are less than 60 nm at the 500 μN peak load for all of the specimens. There is no indent size effect when the contact depth is in the range of 20 nm to 100 nm [24]. The previous study showed that the typical size of lath width is about 200 nm for carbon contents higher than 0.05 %. On the hemispherical approximation [25], the plastic zone sizes corresponding to the depths are estimated to be submicron in diameter, comparable to or smaller than the typical block size because the martensite block consists of several laths with identical crystal orientation. Therefore, the measured nanohardness H_n can be considered to be dominated by the matrix strength. A grain boundary might exist just beneath the indenter or within the plastic zone in some cases. However, the number of the boundaries associated with the plastic zone may be only a few, and the effect of the grain boundary is considered to be relatively small. On the other hand, the data of H_v includes a significant effect of grain boundaries with the indent size around 50 μm (4.9 N load), which is much larger than the grain size.

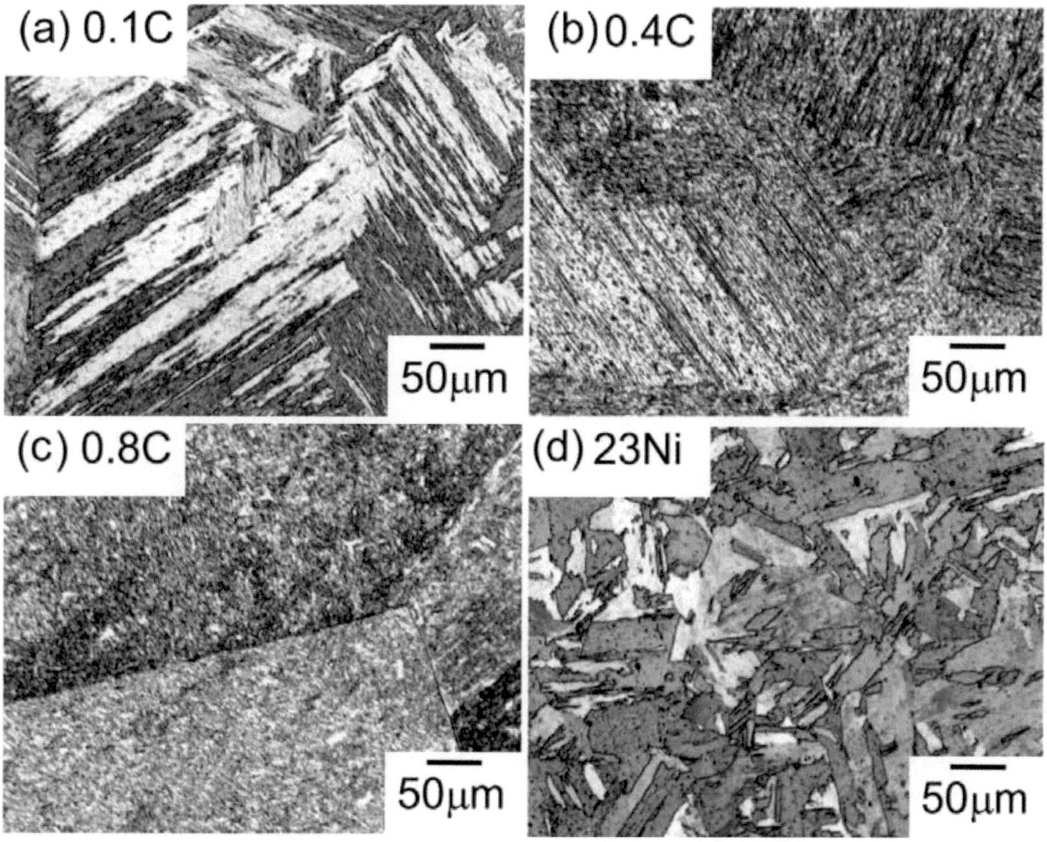

Figure 2. Optical micrographs of the Fe-C lath martensite of (a) 0.1, (b) 0.4, (c) 0.8C and (d) of the Fe-23Ni lath martensite. The block width decreases with an increase in the carbon content. The block width of the Fe-23Ni alloy is much larger than that of the Fe-C alloys.

Both nanohardness H_n and micro Vickers hardness H_v for each specimen are plotted as a function of the carbon content in Fig. 4. The standard deviation of the hardness was plotted as the error bars (an average of more than 20 data with H_n and 5 data with H_v). The hardness obviously increases with increasing carbon content, especially when it is lower than 0.6 wt%. The effect of crystallographic orientation on both the micro Vickers hardness and the nanohardness is thought to be small since the measured hardness value is close when randomly indented at various sites.

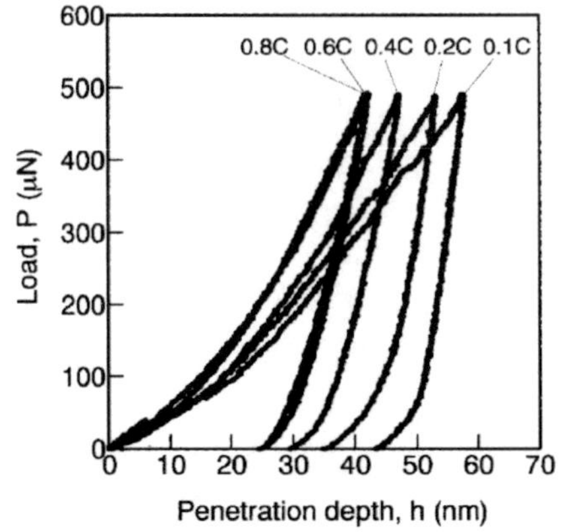

Figure 3. Typical load-displacement curves of five Fe-C martensites. The corresponding carbon content is shown in the figure.

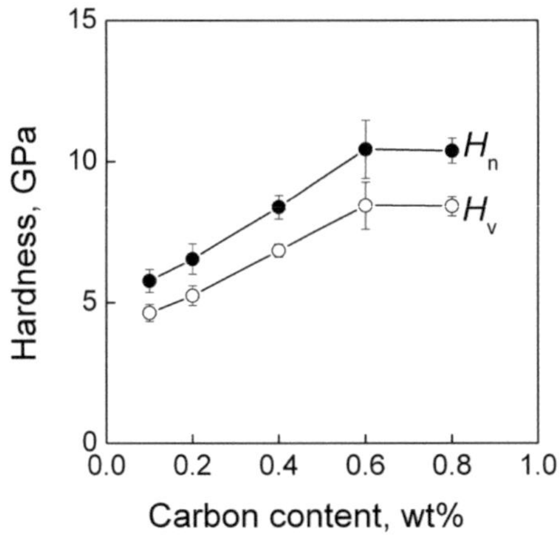

Figure 4. Nanohardness and micro Vickers hardness of the Fe-C martensite plotted as a function of carbon content.

Fig. 5 shows the relationship between the nanohardness H_n and the micro-Vickers hardness H_v for the body-centered-cubic (bcc) single crystals and the Fe–C martensite. For the single crystals that are shown by the open circles, the nanohardness is almost directly proportional to the micro-Vickers hardness. On the other hand, the series of Fe–C specimens shown by the circles are not related to the single crystal. For example, the nanohardness of 0.1C martensite is comparable to that of Mo single crystal, but the Vickers hardness of 0.1C martensite is much higher than that of Mo single crystal. The other Fe-C martensite is similar to that of 0.1C martensite. This trend suggests that the macroscopic strength in the Fe-C martensite is dominated not only by the matrix strength but by other factors on a larger scale such as the high-angle grain boundaries. The data for the Fe-23Ni martensite deviates slightly from those of the single crystals. This means that although the Fe-23Ni has a grain size effect, the contribution of grain boundaries is not very large because the block size is not as small as that of the Fe-C martensite.

The ratio H_n/H_v is used for further consideration of the contributions of the matrix strength to the macroscopic strength. This behavior of grain size effect is express in the Hall-Petch relation as [2]:

$$\frac{H_n}{H_v} = 1 - \frac{k'd^{-1/2}}{H_v} \tag{2}$$

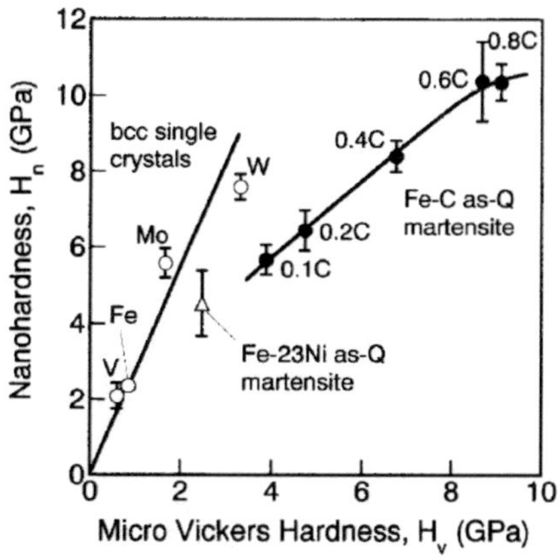

Figure 5. Relationship between nanohardness H_n and micro-Vickers hardness H_v for the bcc single crystals, the Fe–C martensite and the Fe-23Ni lath martensite with a block width of a few 10μm.

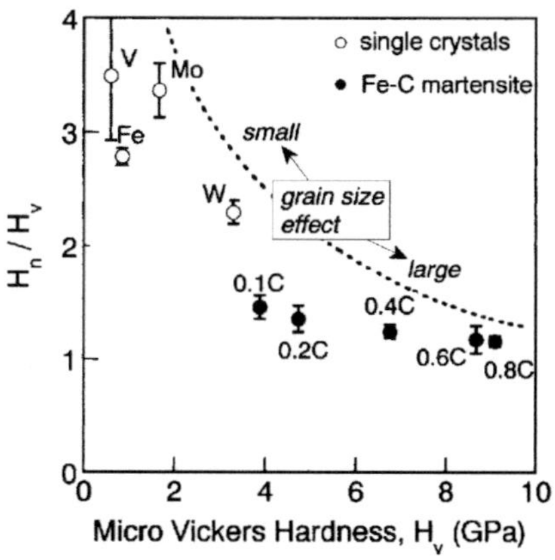

Figure 6. Relationship between micro-Vickers hardness H_v and H_n/H_v for the bcc single crystals and the Fe–C martensite.

When the locking parameter k' is constant and the grain size d approaches zero i.e. the grain size effect is large, the term $k'd^{-1/2}$ is replaced by H_v because $k'd^{-1/2}$ is much larger than H_n in Eq. 2. Accordingly, the ratio H_n/H_v approaches zero. Meanwhile, when d approaches infinity i.e. the grain size effect is small, the term $k'd^{-1/2}$ is zero, and H_v approaches H_n. This result is qualitatively represented by the broken line in Fig. 6. Note that the ration H_n/H_v does not approach 1 because the nanohardness and the Vickers hardness do not coincide due to the difference of indenter geometry and so on. Fig. 6 shows the ratio H_n/H_v for all the specimens plotted as a function of H_v. The much lower ratio for the Fe–C martensite than those for the single crystals suggest that the contribution of the high-angle grain boundaries is large. The ratio H_n/H_v decreases with increasing carbon content, indicating that the grain-boundary effect increases with carbon content, especially for the carbon content range below 0.6 wt%.

3.3. In situ nanoindentation in transmission electron microscope (TEM) with Fe-0.4C tempered martensitic steel

Much deeper investigation was performed in Fe–0.4C tempered martensitic steel through in situ nanoindentation in TEM to study the dislocation–interface interactions in the Fe–C tempered martensitic steel. Two types of boundaries were imaged in the dislocated martensitic structure: a low-angle lath boundary and a coherent, high-angle block boundary.

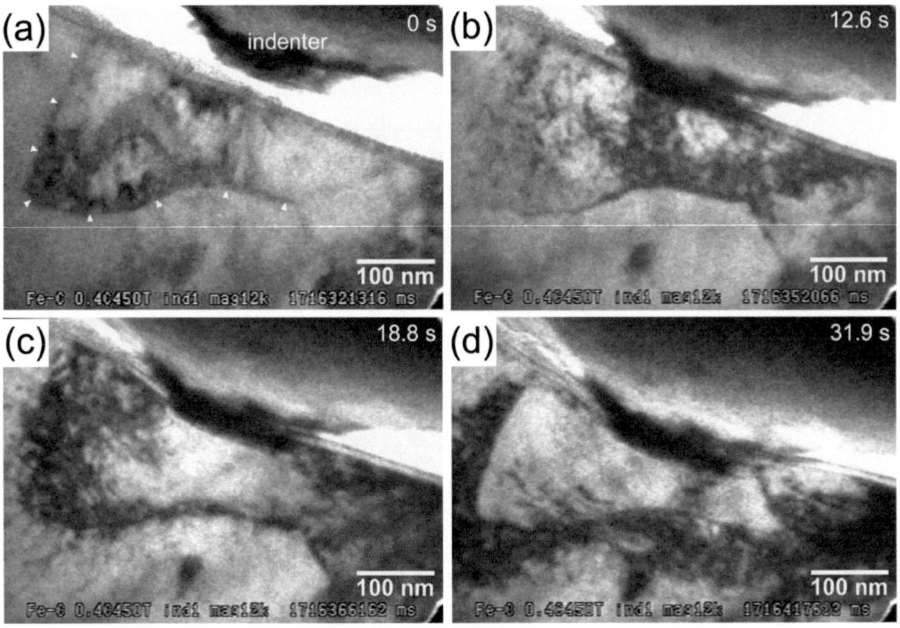

Figure 7. In situ TEM micrographs of the Fe–0.4C martensite, including a low-angle grain boundary: (a) before indentation, (b) at 21 nm penetration depth, showing dislocation emission beneath the indenter, (c) at 46 nm penetration depth, showing dislocation pile-up at the grain boundary, and (d) at 84 nm penetration depth, demonstrating dislocation emission on the far side of the grain boundary.

3.3.1. At low-angle grain boundary

Fig. 7 includes a series of video frames from the in situ TEM nanoindentation of the low-angle grain boundary. The grains on either side of the boundary have the same $[111]_{\alpha'}$ zone axis and the boundary is identified as a lath boundary. Fig. 7(a) shows the diamond indenter prior to indentation, as it approached the surface from the upper-right corner of the figure. The grain boundary is indicated by arrow-heads. As the indenter penetrated into the grain, Fig. 7(b), the deformation was accommodated by the motion of dislocations away from the indenter contact point toward the grain boundary. The low-angle boundary offered some resistance to the dislocation motion, resulting in a pileup at the grain boundary, shown in Fig. 7(c) (a frame taken at an indenter penetration depth of 46 nm). As the indenter penetrated further, a large number of dislocations were emitted on the far side from the indenter tip (left and lower side on the micrograph) of the grain boundary into the adjacent grain. A dense and tangled dislocation structure is seen on the far side of the grain boundary in Fig. 7(d), which was taken at an 84 nm penetration depth. When the indenter was withdrawn, a high density of dislocations was retained near the far side of the boundary.

3.3.2. At high-angle grain boundary (block boundary)

The dislocation interactions with the high-angle grain boundary (block boundary) were dramatically different, as documented in Fig. 8. The misorientation across the grain boundary was determined to be about 56°, with a rotation axis near [0.16, 0.64, 0.76]. Figure 8(a) shows the initial state before indentation, with the high-angle grain boundary indicated by arrowheads. During the early stages of the indentation dislocations can be seen to sweep across the grain from right to left, shown part-way in Fig. 8(b). Note that the pre-existing dislocations in the left area of the indented grain are not seen in the Fig. 8(b) any more, while they are clearly shown in the Fig. 8(a). This could be due to a change of diffraction condition into an invisible condition of the dislocations because of the deformation of the sample. At approximately 60 nm penetration depth, a dense cloud of dislocations reached the high-angle boundary.

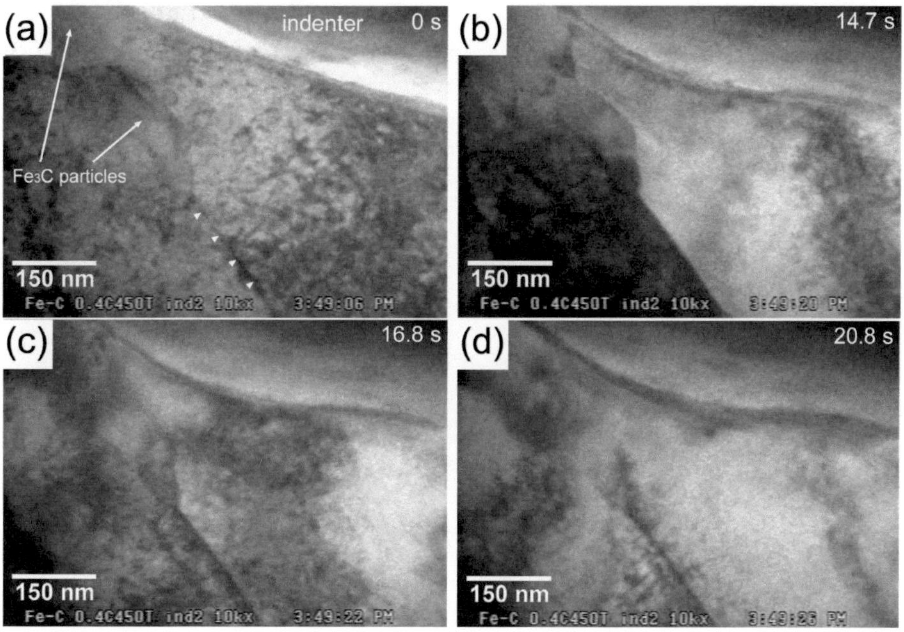

Figure 8. In situ TEM micrographs of the Fe–0.4C martensite, including a high-angle grain boundary: (a) before indentation, (b) at 70 nm penetration depth, showing dislocations sweeping across the grain from right to left, (c) at 83 nm penetration depth, showing disappearance of dislocation at the grain boundary, and (d) at 92 nm penetration depth, demonstrating no dislocation emission on the grain boundary.

However, as shown in Fig. 8(c), on reaching the boundary, the dislocations simply disappeared; there is no indication of a pileup or significant penetration into the adjacent grain. In fact, as illustrated in Fig. 8(d), there is virtually no change in the dislocation configuration on the far side of the boundary. The grain beyond the boundary is essentially unaffected. The behavior of the dislocation "sink" at the high angle grain boundary is illustrated in more detail in Fig. 9, which presents a sequence of images recorded between those in Figs. 8(c) and 8(d). The earliest image, Fig. 9(a), shows a high density of dislocations on both sides of the grain boundary. Many of the dislocations on the right side of the grain boundary are moving from right to left in the indented grain as illustrated in Fig. 8.

The video frames shown in Figs. 9(b)-(d) correspond to increasing penetration depth. As can be seen in these three frames, the dislocations on the right side of the grain boundary vanished at the boundary, while the dislocations on the left side of the grain boundary were essentially unchanged in number, density, or position during the deformation. The stable contrast in these images show that the diffraction condition in the vicinity of the grain boundary was the same throughout the deformation sequence shown, so the behavior observed was not a result of bending in the foil. This sequence demonstrates that the dislocations that approached the boundary were almost completely absorbed by it.

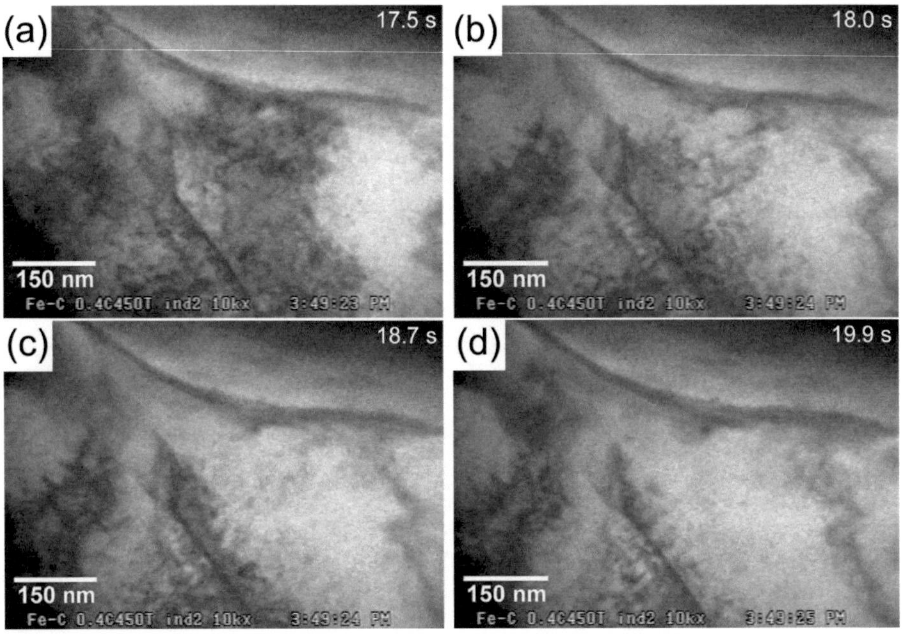

Figure 9. In situ TEM micrographs of the Fe–0.4C martensite showing details of the "sink" of dislocations at the high-angle grain boundary. Frames (a–d) correspond to a sequence between Figs. 8(c) and 8(d). Penetration depth increases from (a) to (d).

3.4. Conclusions

High purity Fe-C as-quenched martensitic steels, four bcc single crystals and an Fe-23Ni lath martensitic steel were investigated using the nanoindentation technique. In the penetration depth range shallower than 60 nm, the indent size is a few times larger than the lath width and smaller than the block size, hence the nanohardness in this scale corresponds to the block matrix strength. The ratio of the nanohardness to the micro Vickers hardness (H_n/H_v) is much smaller for the Fe-C martensite than those of the single crystals, indicating a significant effect of grain size on the macro strength of the Fe-C martensite.

Dislocation–interface interactions in Fe–0.4C tempered martensitic steel were studied through in situ nanoindentation in a TEM. Two types of boundaries were imaged in the dislocated martensitic structure: a low-angle (probable) lath boundary and a coherent, high-angle (probable) block boundary. In the case of a low-angle grain boundary, the dislocations induced by the indenter piled up against the boundary. As the indenter penetrated further, a critical stress appeared to have been reached, and a high density of dislocations was suddenly emitted on the far side of the grain boundary into the adjacent grain. In the case of the high-angle grain boundary, the numerous dislocations that were produced by the indentation were simply absorbed into the boundary, with no indication of pileup or the transmission of strain. This surprising observation is associated with the basis of the crystallography of the block boundary [3].

3.5. Some other related researches

The temper softening behavior of Fe–C binary martensite with various carbon contents [24, 26] was also considered by evaluating the matrix strength through nanoindentation. The tempering temperature dependence of conventional micro-Vickers hardness shows a "hump" around 673 K, while the matrix strength obtained by nanoindentation simply decreases with tempering temperature. This is connected with the microstructure change as discussed in detail in Ref.[27]. Furthermore, the mechanical characterizations using nanoindentation technique were performed for the martensitic steel used as practical dies steel [28] containing carbide-former elements of Cr, Mo, W, and V, which are responsible for secondary hardening by tempering. The nanohardness corresponding to the matrix strength shows obvious secondary hardening, and the hardening-peak temperature coincides with that of the macroscale hardness. The results suggest that the secondary hardening of the dies steel during tempering is attributed not only to the nanoscale strengthening factors such as precipitation hardening by the alloy carbides, but also to some other factors in larger scale. One of the strengthening factors is a decomposition of retained austenite to much harder phases, such as martensite and/or ferrite–cementite constituent.

4. Nanoindentation on Fe-Ni lenticular martensite

4.1. Microstructures of Fe-Ni lenticular martensite

A Fe-33 wt% Ni alloy (shorted for Fe-33Ni) was used in the study. The full austenite phase sample used for comparison is named as RT-austenite sample 'A0'. Transformed lenticular martensite with a straight martensite/austenite (M/A) interface formed by cooling austenite sample 'A0" in liquid nitrogen (77K) is referred to as sample 'M1', which contains the transformed lenticular martensite phase and the retained austenite phase. The substructures of the lenticular martensite are shown by TEM micrographs from Fig. 10(a) ~ (d). As shown in Fig. 10(a), the midrib region is completely twinned with a twinning system of $(112)[\bar{1}\bar{1}1]_M$. In the twinned region (Fig. 10(b)), a small amount of transformation twins exist concurrently with a set of screw dislocations. The Burgers vector ($\vec{b} = a/2[\bar{1}\bar{1}1]_M$) is parallel to the twinning shear direction as indicated by the arrow. The structure in the untwinned region contains several sets of screw dislocations near the twinned region or the curved and tangled dislocations (marked by the circles in Fig. 10(c)) in addition to the straight dislocations (marked by the dashed lines in Fig. 10(c)) near the M/A interface. As shown in Fig. 10(d), the surrounding austenite phase close to the M/A interface also contains a high density of tangled dislocations. These dislocations are probably introduced into the austenite to accommodate the transformation strain and might be inherited during the growth of the lenticular martensite [15]. The dislocation density in the residual austenite decreases with increasing distance away from the M/A interface. In a word, the substructure of the lenticular martensite gradually transited from fully twinned in the midrib to dislocation structure in the untwinned region.

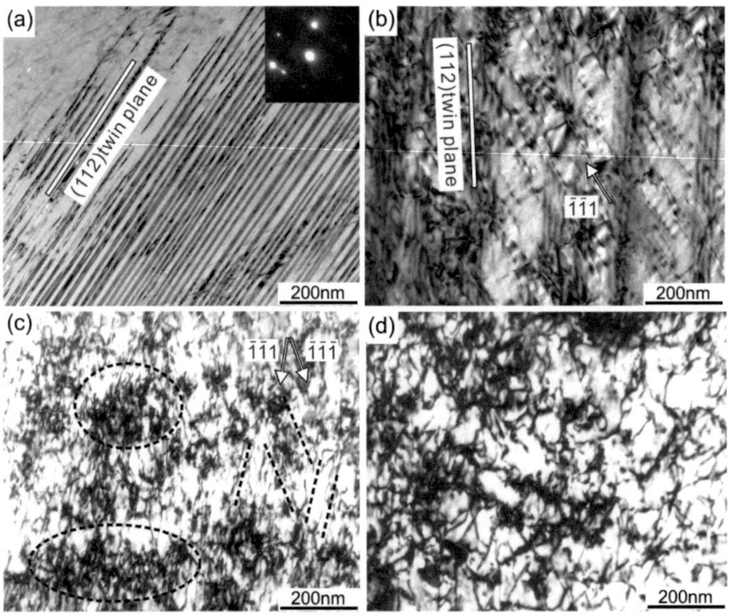

Figure 10. TEM micrographs of the substructure in the lenticular martensite, (a) midrib region, (b) twinned region, (c) untwinned region near the M/A interface and (d) austenite near the M/A interface.

4.2. Results and discussions

Fig. 11(a) is the AFM image after the nanoindentation test. The midrib in the center and the surrounding twinned region of the lenticular martensite plate can be recognized easily by the dark color. The average nanohardness values of Fe-33Ni sample M1, RT-austenite sample A0 and the Fe-23Ni sample are shown in Fig. 11(b). The nanohardness value of the midrib region, 5.22 ± 0.19 GPa, is higher than that of the other regions in the same martensite plate. A drop in nanohardness can be observed from the midrib in the martensite phase to the austenite phase near the M/A interface. The fine twin (the midrib region) shows about a 20% higher resistance to deformation than the dislocated regions (untwined region) in the lenticular martensite (Fe-33Ni alloy) or the lath martensite (Fe-23Ni alloy). It is speculated that the dislocation motion was blocked by the twin boundaries and thus led to strengthening. However, this higher strength in the midrib region cannot be revealed by the conventional micro-Vickers hardness test since the contribution of the fine twinned region to the strength in the lenticular martensite is limited by its small volume fraction. The lath martensite of the Fe-23Ni alloy (dislocated) has an average nanohardness value smaller than that of the midrib region (fully twinned) and larger than that of the untwinned region (dislocated) in the Fe-33Ni lenticular martensite sample M1. The difference in chemical compositions of both alloys might affect their hardness as well, but the influence should be small compared with the difference in the substructure [7, 29]. For both dislocated substructures, the lath martensite had higher dislocation densities (8.5 ~ 12.4 $\times 10^{14}$ m^{-2}) in the Fe-23Ni alloy [8] than that of the untwined region (4.7 ~ 9.1 $\times 10^{14}$ m^{-2}) in the Fe-33Ni [30] lenticular martensite. The nanohardness of the austenite phase near the M/A interface is slightly higher than that of the austenite far from the M/A interface. The dislocation density decreases as the distance from the M/A interface increases and causes a decrease in the nanohardness of the austenite phase.

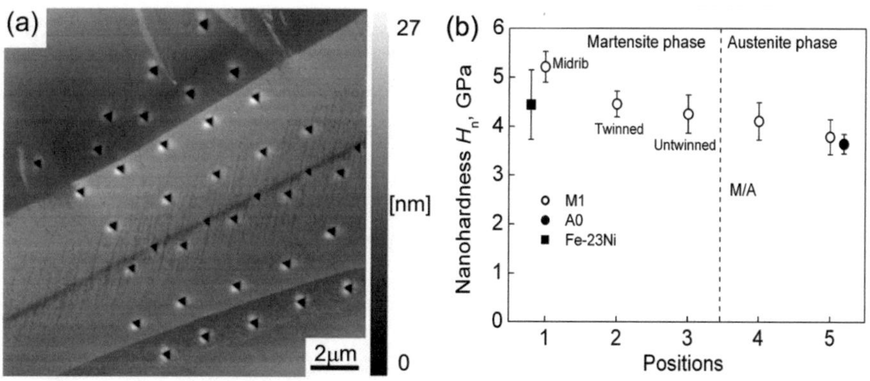

Figure 11. (a) the AFM image after the nanoindentation test and (b) nanohardness values of the transformed martensite sample M1 at various positions, RT-austenite sample A0, and the lath martensite of a Fe-23Ni alloy.

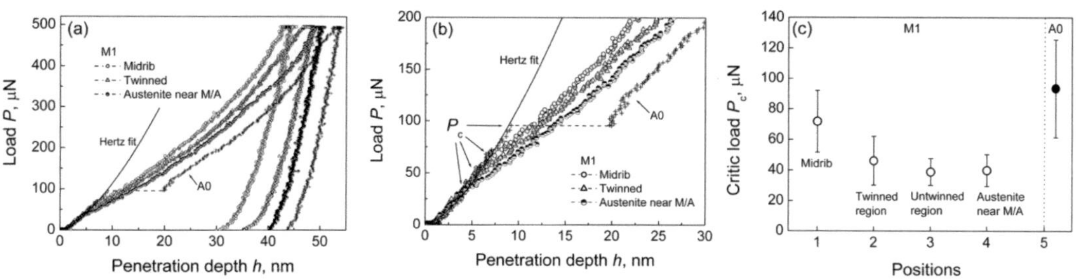

Figure 12. Typical load–depth curves of the transformed martensite sample M1 and RT-austenite sample A0 (a) and a magnified view under ultra-low loads (b); (c) is the average critical load in the various substructures.

The typical load-depth curves of the lenticular martensite sample M1 and RT-austenite sample A0 are shown in Fig. 12(a). According to the earlier reports, the initial stage of loading is of a purely elastic nature and can be described by the Hertz contact theory [31, 32]:

$$P = \frac{4}{3} E_r R^{1/2} h^{3/2}, \tag{3}$$

where P is the applied load, R is the indenter tip radius, h is the displacement/depth, and E_r is the reduced modulus. As shown in Fig. 12(a), the initial portion of the load-depth curve fits the Hertz contact theory well by putting the values of the reduced modulus E_r (173±18 GPa) and the tip radius (R = 200 nm) into the equation (3). When the applied load is higher than about 40 μN, the curves begin to deviate from the Hertz fit, which marks the onset of plastic deformation.

Fig. 12(b) is a magnified view of Fig. 12(a) in the load range below 200 μN, showing the initial stage of the indentation-induced deformation. The load-depth curves of the RT-austenite sample A0 shows a strain burst in the loading curve, which is called a pop-in event. The load where the pop-in event sets in is defined as a critical load P_c. The pop-in behavior has been reported to be associated with a nucleation and propagation of dislocations under an initial elastic strain field or a defect-free region in the metals [24, 33-35]. The obvious pop-in behavior in the sample A0 is consistent with the property where the untransformed RT-austenite phase has a lower dislocation density than that of the transformed martensite phase. The critical stress of the pop-in event can be estimated as 7.3 GPa through the Hertz contact theory where the maximum shear stress τ_{max} [36] beneath the indenter can be given as:

$$\tau_{max} = 0.18 \cdot \left(\frac{E_r}{R} \right)^{\frac{2}{3}} P^{\frac{1}{3}}, \tag{4}$$

which is in an order of an ideal strength. For the transformed martensite sample M1 which contains a high density of defects (twins or dislocations), the load-depth curves do not show a pop-in event but undergo an apparent slope change which marks the onset of plastic deformation. The changing point of the load is also defined as critical load P_c for easy comparison, as indicated by the arrows in Fig. 12(b). The average P_c at different regions of the samples M1 and A0 are shown in Fig. 12(c). In the transformed martensite sample M1, the critical load of the midrib region is higher than that of the other regions but lower than that of the RT-austenite sample A0.

Note that the strain rates vary when the plasticity initiation occurs. A classic model based on the dislocation theory expresses a strain rate $\dot{\gamma}$ as

$$\dot{\gamma} = \rho \cdot b \cdot \overline{\upsilon}, \tag{5}$$

where ρ is the mobile dislocation density, b is the magnitude of the Burgers vector and $\overline{\upsilon}$ is the average dislocation velocity. The difference in the critical load and the strain rate is associated with the different plastic deformation behaviors in the various substructures and will be discussed by the interaction between the pre-existing lattice defects (twin or dislocation structure) and the newly formed dislocations during the indentation process. In the RT-austenite sample A0, the density of pre-existing dislocations is very low, thus the probability of a mobile dislocation existing underneath the indenter is also significantly low. Therefore, the stress required for the nucleation of dislocations in the defect free region is high, which is close to the ideal strength and results in the highest critical load P_c as shown in Fig. 12(c). After the initiation of plastic deformation, the strain rate in the RT-austenite phase is quite high; this is understood by the remarkable strain burst as the obvious pop-in event. Once the plastic deformation is initiated and a dislocation source is activated, many mobile dislocations become present in the stress field producing a high ρ value. Additionally, the $\bar{v}$ could also be high since the density of pre-existing dislocations as obstacles for the dislocation movement is significantly low. After the pop-in event, plastic deformation expands far from the initiation point and the low density of pre-existing dislocations also causes a lower flow stress in a further deformation stage resulting in a relatively low nanohardness in the RT-austenite sample A0. This result indicates that high flow stress is not necessarily obtained in a microstructure with high critical stress for plasticity initiation.

In the regions with high dislocation densities, deformation underneath the indenter could be initiated at much lower loads since the stresses required for the motion of pre-existing dislocations are much lower than the stress required for the nucleation of dislocations [37]. This is an appropriate reason for the much lower critical loads in the dislocated structure of the martensite phase in the sample M1. After the initiation of plastic deformation, moving dislocations interact with other pre-existing dislocations; hence the average dislocation velocity $\bar{v}$ is significantly low. Therefore, the strain rate $\dot{\gamma}$ is also very low and a higher applied stress is necessary for further deformation. This does not result in a pop-in behavior, causing only a change in the slope of loading curve, and inevitably leads to higher nanohardness in the dislocated regions than that of the RT-austenite sample A0.

In the midrib region, the existing twins may act as a stress concentrated site at the subsurface for the nucleation of dislocations, even though there are only a few pre-existing dislocations, and produce a relatively lower critical load than that of the defect-free RT-austenite sample A0. On the other hand, the twin structure could act as an obstacle to the dislocation motion like that of the pre-existing dislocations, and lead to the same deformation behavior as that of the dislocated regions. Furthermore, the very fine twin structure in the midrib region causes a much higher flow stress and higher nanohardness than that of any other regions and phases. The mixed structure of twin and dislocation in the twinned region has a critical load P_c between the midrib and untwinned region.

4.3. Conclusions

Nanoindentation tests were conducted on the lenticular martensite of a Fe–33Ni alloy to reveal the deformation behavior of each component with different lattice defects. A continuous drop in nanohardness from the midrib in the martensite phase to the austenite phase was observed in the transformed martensite sample M1. The nanohardness value of the midrib region was about 20% higher than that of the other regions; however, its contribution to the macroscopic strength was small due to its limited volume fraction. Plasticity initiation and subsequent deformation behavior depended on the microstructures. In the RT-austenite sample A0, the low density of pre-existing dislocation caused a high critical load of plasticity initiation and a low flow stress. In contrast, plasticity initiation occurred rather easily in highly dislocated regions such as the untwinned region and the austenite phase near the M/A interface in the sample M1; however, the subsequent flow stresses were higher than that of the defect-free RT-austenite sample A0. The very fine twin structure in the midrib region in the sample M1 had two roles: one was to assist the plasticity initiation as a stress concentrated site and the other was to act as a strong obstacle to the dislocation motions.

5. Nanoindentation test on surface mechanical attrition treated NiTi shape memory alloy

5.1. Results and discussions

The NiTi samples were subjected to SMAT at room temperature using bearing steel balls with a diameter of 5 mm for 5 and 60 min (hereafter denoted SMAT-5 and SMAT-60, respectively). NiTi samples before SMAT contained a coarse grained austenite (B2) phase at room temperature. After SMAT near equiaxial B2 nanograins were produced on the surface. Variations in grain size (*d*) with respect to the distance from the SMAT surface (*D*) for the SMAT-5 and SMAT-60 samples are shown in Fig. 13(a). For both samples grain size *d* decreased gradually with decreasing distance *D*, whereas the SMAT-5 sample had a larger *d* at the same *D*. Fig. 13(b) shows typical force-indentation depth curves at different *D* for SMAT-60. Significant differences in those curves are clearly visible.

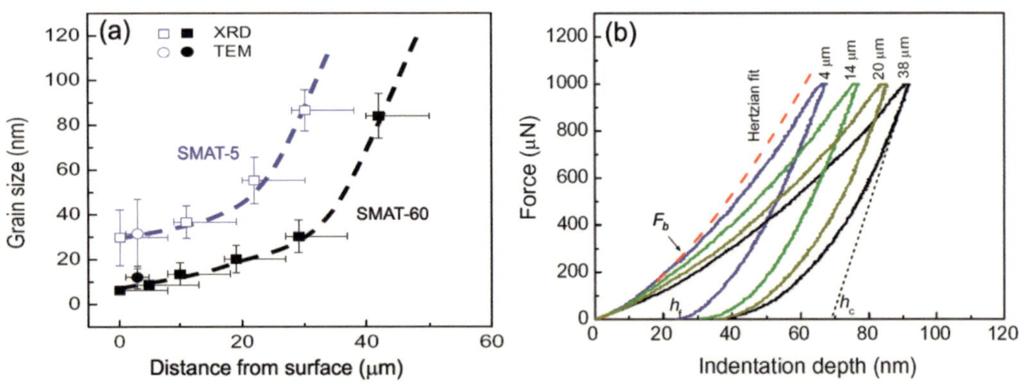

Figure 13. (a) Variation in grain size with distance from the surface for SMAT samples. (b) Typical force–indentation depth curves at different distance *D* for SMAT-60 as indicated. Definition of the bifurcating force F_b, h_c and h_f are indicated.

Fig. 14(a) demonstrates the determined values of Young's moduli *E* as a function of *D* for a large number of measurements in different regions of the two samples. For the SMAT-60 sample a remarkable increase in *E* with decreasing *D* was obtained. *E* increased from about 60 GPa at 40 μm to 85 GPa at 2 μm distance. For the SMAT-5 sample the increase in *E* was less remarkable. E increased from about 60 GPa at 30 μm to about 70 GPa at 2 μm distance. Using the measured grain size, E was plotted against d for the two samples, as shown in Fig. 14(b). An obvious dependence of E on d can be seen in Fig. 14(b). E increased dramatically with decreasing d for d less than 100 nm. For comparison, the Young's modulus of amorphous NiTi was also measured and is in agreement with the reported value (~93 GPa) [38]. The Young's modulus of NiTi, as measured by the rectangular parallelepiped resonance (RPR) method [39], is also included in Fig. 14(b). The maximum Young's modulus obtained for the present NiTi sample (~85 GPa for d ~ 6 nm) is close to that measured by RPR and that of amorphous NiTi, and is higher than that of partially amorphous NiTi (71 GPa) [40].

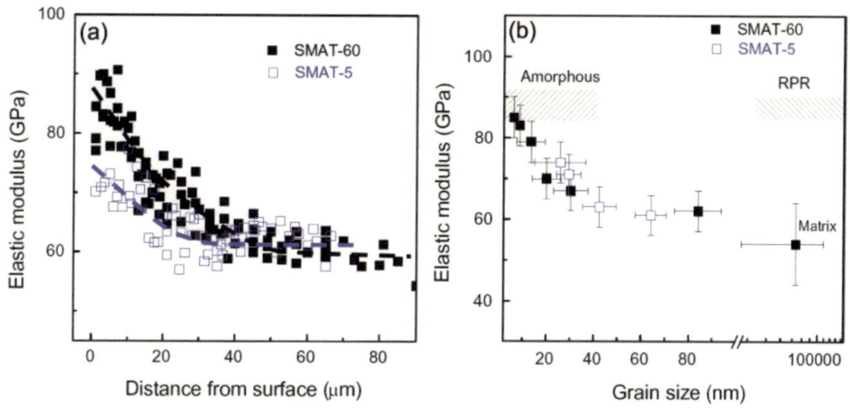

Figure 14. (a) Calculated Young's modulus as a function of distance from the surface. Dashed lines are guides for the eye. (b) Calculated Young's modulus as a function of grain size. The Young's moduli of NiTi determined by the RPR method and that of amorphous NiTi are also indicated.

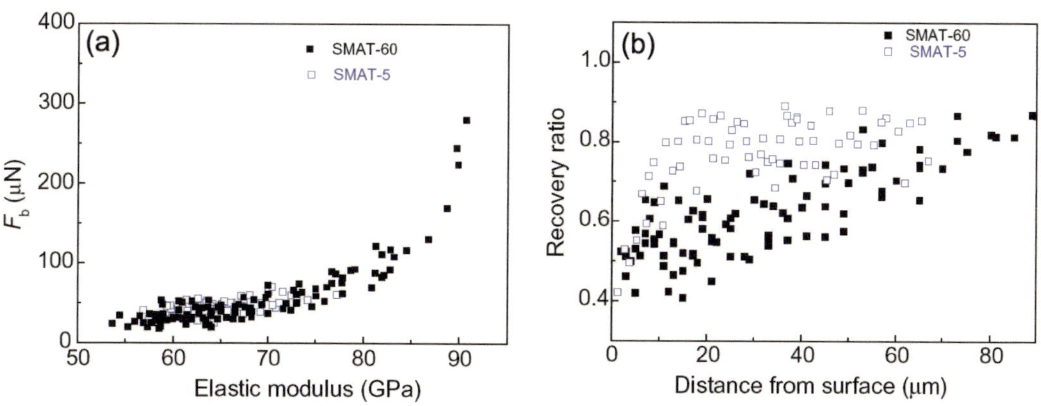

Figure 15. (a) Relationship between the bifurcating force (F_b) and Young's modulus. (b) Indentation recovery ratio as a function of distance from the surface.

The remarkable dependence of E on d was somewhat unexpected, because grain size has usually been believed to have little effect on the Young's modulus of nanocrystalline materials. It should be noted that residual stress may have an effect on the Young's modulus, typically a change of <1% for a compressive stress of 100 MPa [41], which is, however, rather small compared with the large enhancement of Young's modulus (~40%) in the present study. Moreover, it was found that the compressive stress in SMAT samples increased with increasing *D*, with a maximum in the subsurface instead of the top surface [42]. Obviously, the effect of residual stress can be excluded. Actually, no change in Young's modulus has ever been reported in previous SMAT samples [43]. Therefore, the observed dependence of *E* on *d* should be related to the unique properties of NiTi. It is speculated that the transformation stress for the SIM increases with decreasing grain size for nanograins. Thus, the measured dependence of *E* on *d* is related to the grain size effect on the SIM in nanocrystalline NiTi, i.e. suppression of the SIM leads to the increase in *E* with decreasing *d*.

As indicated in Fig. 13, the bifurcating force (F_b) is defined as the point where the indentation loading curve starts to deviate from the purely elastic curve determined by Hertz contact theory. The bifurcating force is related to the transformation stress, i.e. a higher transformation stress leads to a higher bifurcating force, and vice versa. As shown in Fig. 15(a), a higher Young's modulus corresponds to a higher bifurcating force and, hence, a higher transformation stress. This clearly indicates that the increase in Young's modulus is related to suppression of the SIM with the reduction of grain size. Indentation depth recovery is also a strong indication of the SIM effect. The recovery ratio (δ) was introduced as a measure of the recovery due to the SIM and is defined as [44]:

$$\delta = \frac{h_c - h_f}{h_c}, \tag{6}$$

Definitions of h_c and h_f are indicated in Fig. 13(b). Suppression of the SIM can be demonstrated quantitatively by plotting δ against *D*. As shown in Fig. 15(b), δ decreased with decreasing *D*, indicating less recovery due to less SIM for smaller *D* and *d*. Thus, the measured dependence of *E* on *d* is related to the grain size effect on the SIM in nanocrystalline NiTi, i.e. suppression of the SIM leads to the increase in *E* with decreasing *d*. The highest value of *E* for a *d* of ~6 nm was close to the intrinsic value of crystalline and amorphous NiTi, although, as indicated in Fig. 14(b), the SIM may not be completely suppressed for a *d* of ~6 nm. By further extrapolation of the *E*–*d* relationship to a *d* of 0 nm, a value of *E* equal to 90 GPa is derived, which can be considered the intrinsic value for NiTi by complete suppression of the SIM in this study. This value was used for Hertz fitting in Fig. 13, and a good fit of the initial part of the loading curve was obtained.

5.2. Conclusions

In summary, a nanocrystalline surface layer with a graded grain size distribution was produced on NiTi by SMAT. The apparent Young's modulus, as measured by nanoindentation, was found to increase dramatically with decreasing *D* and *d*, and reached ~85 GPa for *d* ~ 6 nm. Such dependence of *E* on *d* can be attributed to the suppression of the SIM in nanocrystalline NiTi. The present study has provided unambiguous experimental evidence indicating that the Young's modulus of NiTi can be underestimated due to the SIM, and by suppression of SIM in nanocrystalline NiTi an enhanced Young's modulus can be obtained, approaching the intrinsic value for NiTi.

5.3. Some other related researches

Furthermore, nanoindentation was conducted in a Fe–28Mn–6Si–5Cr alloy [45] to investigate the stress-induced ε-martensitic transformation behavior and the shape memory effect. The shape memory effect of Fe–Mn-based alloys is derived from the stress-induced ε-martensitic transformation (face-centered cubic (fcc) to hexagonal close packed (hcp)) [46]. Since the pre-strains required for shape memory are 2–6% [47], an understanding of the initial deformation behavior, including the ε-martensitic transformation, is extremely important for any investigation of the shape memory effect. The ε-martensitic transformation has a number of distinctive characteristics, including low critical stress [48], remarkable crystallographic orientation dependence [49], specific surface relief, and reversibility. However, the analyses essential for determining the characteristics of shape memory alloys (SMAs) are difficult due to the existence of grain boundaries, twin boundaries and thermally induced ε-martensite. A nanoindentation test is useful for investigating the initial deformation behavior since the test can be performed on several positions over a limited area and enables the load–displacement behavior of the entire process to be observed. The indentation of austenite has been reported to create high stresses under diamond indenters that can cause stress-induced martensitic transformation in Fe–0.08C–0.5Si–1Al–7Mn (wt.%) [50]. Additionally, the combination of nanoindentation with AFM makes it possible to observe indentation-induced surface relief in the CuAlNi SMA [51] and to evaluate shape memory effects in the NiTi SMA [52]. Hence, the nanoindentation test appears to be suitable for evaluating the ε-martensitic transformation behavior in a specific orientation of a bulk material based on a careful study of the above-mentioned characteristics of the ε-martensitic transformation. Nanoindentation tests and AFM observations were performed in a Fe–28Mn–6Si–5Cr alloy [45] to investigate the stress-induced ε-martensitic transformation behavior and the shape memory effect. The shape memory effect was evaluated from the volume change in an indent mark caused by annealing measured by atomic force microscopy. Using the load–displacement (P–h) data obtained from nanoindentation, the plot of P/h vs. h showed two types of slopes, corresponding to the ε-martensitic transformation and slip deformation, and exhibited a correlation with the shape recovery ratio.

6. Chapter summary

As mentioned above, nanoindentation technique has been used in martensitic structures to fulfill various missions. Such as nanohardness test and Young's modulus measurement which can measure the mechanical properties of a precise site in a microstructure with accuracy within a nanometer. But its capability is not limited to this simple usage. By correlating the microstructure change with the load-depth curve, the underlying mechanism can be revealed. And it is definitely will be used more widely in the future.

Author details

L. Zhang and T. Ohmura
National Institute for Materials Science, 1-2-1 Sengen, Tsukuba, Ibaraki, Japan

K. Tsuzaki
National Institute for Materials Science, 1-2-1 Sengen, Tsukuba, Ibaraki, Japan
Graduate School of Pure and Applied Sciences, University of Tsukuba, Ibaraki, Japan

7. References

[1] A. Gouldstone, N. Chollacoop, M. Dao, J. Li, A. M. Minor, Y.-L. Shen, Indentation across size scales and displines: Recent developments in experimentation and modeling. Acta Mater. 55 (2007) 4015-4039.
[2] T. Ohmura, K. Tsuzaki, S. Matsuoka, Nanohardness measurement of high-purity Fe-C martensite. Scripta Mater. 45 (2001) 889-894.
[3] T. Ohmura, A. M. Minor, E. A. Stach, J. W. Morris Jr., Dislocation–grain boundary interactions in martensitic steel observed through in situ nanoindentation in a transmission electron microscope. J. Mater. Res. 19 (2004) 3626-3632.
[4] L. Zhang, T. Ohmura, A. Shibata, K. Tsuzaki, Characterization of local deformation behavior of Fe-Ni lenticular martensite by nanoindentation. Mater. Sci. Eng. A527 (2010) 1869-1874.
[5] Q. S. Mei, L. Zhang, K. Tsuchiya, H. Gao, T. Ohmura, K. Tsuzaki, Grain size dependence of the elastic modulus in nanostructured NiTi. Scripta Mater. 63 (2010) 977-980.
[6] T. Maki, Microstructure and mechanical behaviour of ferrous martensite. Mater. Sci. Forum 56-58 (1990) 157-168.
[7] M. J. Roberts, W. S. Owen, The strength of martensitic iron-nickel alloys. Trans. ASM 60 (1967) 687-692.
[8] S. Morito, J. Nishikawa, T. Maki, Dislocation density within lath martensite in Fe-C and Fe-Ni alloys. ISIJ Int. 43 (2003) 1475-1477.
[9] A. R. Marder, G.Krauss, The morphology of martensite in iron-carbon alloys. Trans ASM 60 (1967) 651-660.
[10] A. R. Marder, G. Krauss, The Formation of Low-Carbon Martensite in Fe-C Alloys. Trans ASM 62 (1969) 957-964.
[11] J. M. Marder, A. R. Marder, The morphology of iron-nickel massive martensite. Trans ASM 62 (1969) 1-10.
[12] T. Maki, K.Tsuzaki, I. Tamura, The morphology of microstructure composed of lath martensites in steels. Trans ISIJ 20 (1980) 207.
[13] R. A. Grange, C. R. Hribal, L. F. Porter, Hardness of Tempered Martensite in Carbon and Low- Alloy Steels. Metall Trans. A 8 (1977) 1775-1785.
[14] G. R. Speich, W. C. Leslie, Tempering of Steel. Metall Trans 3 (1972) 1043-1054.
[15] A. Shibata, S. Morito, T. Furuhara, T. Maki, Local orientation change inside lenticular martensite plate in Fe–33Ni alloy. Scripta Mater. 53 (2005) 597-602.
[16] A. Shibata, T. Murakami, S. Morito, T. Furuhara, T. Maki, The origin of midrib in lenticular martensite. Mater. Trans. 49 (2008) 1242-1248.
[17] T. Kakeshita, K. Shimizu, T. Maki, I. Tamura, Growth behavior of lenticular and thin plate martensites in ferrous alloys and steels. Scripta Metall. 14 (1980) 1067-1070.
[18] M. Dechamps, L. M. Brown, Structure of twins in Fe-Ni martensite. Acta Metall. 27 (1979) 1281-1291.

[19] K. Otsuka, X. Ren, Physical metallurgy of Ti-Ni-based shape memory alloys. Prog. Mater. Sci. 50 (2005) 511-678.

[20] Y. Liu, H. Xiang, Apparent Modulus of Elasticity of Near-Equiatomic NiTi. J. Alloys Compd. 270 (1998) 154-159.

[21] C. P. Frick, A. M. Ortega, J. Tyber, K. Gall, H. J. Maier, Multiscale Structure and Properties of Cast and Deformation Processed Polycrystalline NiTi Shape-Memory Alloys. Metall. Mater. Trans. A 35 (2004) 2013-2025.

[22] L. M. Qian, X. D. Xian, Q. P. Sun, T. X. Yu, Anomalous relationship between hardness and wear properties of a superelastic nickel-titanium alloy. Appl. Phys. Lett. 84 (2004) 1076-1078.

[23] T. Waitz, T. Antretter, F. G. Fischer, N. K. Simha, H. P. Karnthaler, Size Effects on the Martensitic Phase Transformation of NiTi Nanograins. J. Mech. Phys. Solids 55 (2007) 419-444.

[24] T. Ohmura, K. Tsuzaki, S. Matsuoka, Evaluation of the matrix strength of Fe-0.4 wt% C tempered martensite using nanoindentation techniques. Philos. Mag. A 82 (2002) 1903-1910.

[25] M. Itokazu, Y. Murakami, Elastic-plastic analysis of triangular pyramidal indentation. Trans. Jpn. Soc. Mech. Eng. A59 (1993) 2560-2568.

[26] T. Ohmura, T. Hara, K. Tsuzaki., Relationship between nanohardness and microstructures in high-purity Fe-C as-quenched and quench-tempered martensite. J. Mater. Res. 18 (2003) 1465-1470.

[27] T. Ohmura, T. Hara, K. Tsuzaki, Evaluation of temper softening behavior of Fe-C binary martensitic steels by nanoindentation. Scripta Mater. 49 (2003) 1157-1162.

[28] T. Ohmura, T. Hara, K. Tsuzaki, H. Nakatsu, Y. Tamura, Mechanical characterization of secondary-hardening martensitic steel using nanoindentation. J. Mater. Res. 19 (2004) 79-84.

[29] J. M. Chilton, P. M. Kelly, The strength of ferrous martensite. Acta Metall. 16 (1968) 637-656.

[30] A. Shibata, S. Morito, T. Furuhara, T. Maki, Substructures of lenticular martensites with different martensite start temperatures in ferrous alloys. Acta Mater. 57 (2009) 483-492.

[31] B. Yang, H. Vehoff, Dependence of nanohardness upon indentation size and grain size - a local examination of the interaction between dislocations and grain boundaries. Acta Mater. 55 (2007) 849.

[32] K. Durst, B. Backes, O. Franke, M. Göken, Indentation size effect in metallic material: Modeling strength from pop-in to macroscopic hardness using geometrically necessary dislocations. Acta Mater. 54 (2006) 2547-2555.

[33] A. Gouldstone, H.-J. Koh, K.-Y. Zeng, A. E. Giannakopoulos, S. Suresh, Discrete and continuous deformation during nanoindentation of thin films. Acta Mater. 48 (2000) 2277-2295.

[34] A. M. Minor, J. W. Morris Jr., E. A. Stach, Quantitative in situ nanoindentation in an electron microscope. Appl. Phys. Lett. 79 (2001) 1625-1627.

[35] L. Zhang, T. Ohmura, K. Seikido, K. Nakajima, T. Hara, K. Tsuzaki, Direct observation of plastic deformation in iron–3% silicon single crystal by in situ nanoindentation in transmission electron microscopy. Scripta Mater. 64 (2011) 919-922.

[36] T. Ohmura, K. Tsuzaki, F. Yin, Nanoindentation-induced deformation behavior in the vicinity of single grain boundary of interstitial-free steel. Mater. Trans. 46 (2005) 2026-2029.

[37] D. Lorenz, A. Zeckzer, U. Hilpert, P. Grau, H. Johansen, H. S. Leipner, Pop-in effect as homogeneous nucleation of dislocations during nanoindentation. Phys. Rev. B 67 (2003) 172101.

[38] A. Gyobu, Y. Kawamura, T. Saburi, H. Horikawa, in: M. A. Imam, R. Denale, S. Hanada, Z. Zhong, D. N. Lee (Eds.), PRICM 3, TMS, Warrendale, PA, 1998, p. 2719.
[39] X. Ren, N. Miura, J. Zhang, K. Otsuka, K. Tanaka, M. Koiwa, A comparative study of elastic constants of Ti–Ni-based alloys prior to martensitic transformation. Mater. Sci. Eng. A 312 (2001) 196-206.
[40] K. Tsuchiya, Y. Hada, T. Koyano, K. Nakajima, M. Ohnuma, T. Koike, Y. Todaka, M. Umemoto, Production of TiNi amorphous/nanocrystalline wires with high strength and elastic modulus by severe cold drawing. Scripta Mater. 60 (2009) 749-752.
[41] T. Y. Tsui, W. C. Oliver, G. M. Pharr, Influences of stress on the measurement of mechanical properties using nanoindentation. 1. Experimental studies in an aluminum alloy. J. Mater. Res. 11 (1996) 752-759.
[42] A. L. Ortiz, J. Tian, L. L. Shaw, P. K. Liaw, Experimental study of the microstructure and stress state of shot peened and surface mechanical attrition treated nickel alloys. Scripta Mater. 62 (2010) 129-132.
[43] N. R. Tao, W. P. Tong, Z. B. Wang, W. Wang, M. L. Sui, J. Lu, K. Lu, Mechanical and wear properties of nanostructured surface layer in iron induced by surface mechanical attrition treatment. J. Mater. Sci. Tech. (2003) 563-566.
[44] W. Ni, Y.-T. Cheng, D. S. Grummon, Recovery of microindents in a nickel– titanium shape-memory alloy: a self-healing effect. Appl. Phys. Lett. 80 (2002) 3310-3312.
[45] K. Sekido, T. Ohmura, T. Sawaguchi, M. Koyama, H. W. Park, K. Tsuzaki, Nanoindentation/atomic force microscopy analyses of e-martensitic transformation and shape memory effect in Fe-28Mn-6Si-5Cr alloy. Scripta Mater. 65 (2011) 942-945.
[46] K. Enami, A. Nagasawa, S. Nenno, Reversible Shape Memory Effect in Fe-base Alloys. Scripta Metall. 9 (1975) 941-948.
[47] M. Koyama, T. Sawaguchi, K. Tsuzaki, Si content dependence on shape memory and tensile properties in Fe–Mn–Si–C alloys. Mater. Sci. Eng. A 528 (2011) 2882-2888.
[48] A. Sato, H. Kubo, T. Maruyama, Mechanical Properties of Fe-Mn-Si Based SMA and the Application. Mater. Trans. 47 (2006) 571-579.
[49] A. Sato, E. Chishima, Y. Yamaji, T. Mori, Orientation and composition dependencias of shape memory effect in Fe-Mn-Si alloys. Acta Metall. 32 (1984) 539-547.
[50] T. H. Ahn, C. S. Oh, D. H. Kim, K. H. Oh, H. Bei, E. P. George, H. N. Han, Investigation of strain-induced martensitic transformation in metastable austenite using nanoindentation. Scripta Mater. 63 (2010) 540-543.
[51] W. C. Crone, H. Brock, A. Creuziger, Nanoindentation and microindentation of CuAlNi shape memory alloy. Exp. Mech. 47 (2007) 133-142.
[52] C. P. Frick, T. W. Lang, K. Spark, K. Gall, Stress-Induced Martensitic Transformations and Shape Memory at Nanometer Scales. Acta Mater. 54 (2006) 2223-2234.

Section 3

Nanoindentation on Thin Layers and Films

Chapter 6

Indentation and Fracture of Hybrid Sol-Gel Silica Films

Bruno A. Latella, Michael V. Swain and Michel Ignat

Additional information is available at the end of the chapter

http://dx.doi.org/10.5772/48140

1. Introduction

Organic-inorganic hybrid thin films fabricated using sol-gel processing have many compelling properties that render them quite attractive for many applications, including optics, electronics, sensors and corrosion and scratch resistant films (Haas & Wolter, 1999; Sanchez et al., 2005). Organic-inorganic hybrid network materials have received much interest as transparent functional coatings on polymer substrates (Haas & Wolter, 1999; Haas et al., 1999a) and barrier coatings on metals (Metroke et al., 2001). Compared to glass, polymers such as polycarbonate (PC) and glycol bis(allyl carbonate) (CR-39) exhibit several advantageous physical and mechanical properties, such as high impact resistance and reduced weight, but also have the significant disadvantage of higher refractive index resulting in greater surface reflections as well as a much lower tolerance to abrasion. These drawbacks have limited their exploitation as a replacement to glass, especially for ophthalmic lenses where reflections and scratches on lenses can significantly obscure vision. The incorporation of a film or coating on glass or polymer can have immense benefits as is the case in the eyewear industry where several layers are deposited on polymer substrates to overcome substrate limitations (Samson, 1996; Schottner, 2001). By controlling the chemistry of the organic component incorporated in hybrid films, the physical and mechanical properties can be readily adjusted to realise specific attributes such as scratch resistance. Yet a vital reliability issue for film-on-substrate systems is the intrinsic mechanical properties of the film and adhesion to the substrate (Ignat et al., 1999).

In order to achieve good scratch resistance, two properties need to be optimised: *adhesion* of the film to the substrate and film *hardness*. In hybrid films, hardness is provided by the inorganic ceramic phase or from nano-particle inclusions. Not as much attention has been paid to adhesion behaviour of these hybrid film-substrate systems although enhancements in adhesion may be achieved using organic materials, which are softer and generally more flexible as compared to the inorganic, which are typically harder and more brittle. Hence, hybrid coatings are considered extremely versatile given the combination of these two very different material characteristics for films to be tailored to achieve a range of functional responses. Accordingly, in this chapter nanoindentation and tensile testing are surveyed as tools to characterise film properties, fracture behaviour and adhesion to the underlying substrate of a variety of model hybrid films. This begins with an overview of hybrid sol-gel film processing. The key principles of nanoindentation focussing on spherical indentation to examine elastic-plastic response and creep behaviour are then outlined. Finally, tensile testing and the mechanics for ascertaining film fracture properties and film-substrate adhesion are described along with specific examples to illustrate the combined power of the techniques.

2. Hybrid sol-gel films – Overview

A hybrid material is any organic-inorganic system in which at least one of the components, organic or inorganic, is present with a size scaling from tenths to tens of nanometres. Components used to make hybrids can be molecules, oligomers or polymers, aggregates and even particles. Therefore they can be considered as nanocomposites at the molecular scale. Schmidt (Schmidt, 1985) and Wilkes (Wen & Wilkes, 1996) have been widely credited with pioneering the research into organic-inorganic hybrids using the sol-gel process. They both showed that an organic polymer can be chemically bonded to an inorganic oxide network to form a new type of polymer. Schmidt named his hybrid material "ORMOSILs" (for ORganically MOdified SILicates) or "ORMOCERs" (for ORganically MOdified CERamics) while Wilkes named his materials "CERAMERs" (for CERAmic polyMERS).

Hybrid materials can be classified by their chemical composition or by the nature of the chemical interactions (Sanchez et al., 2005). Reactive monomers linked through covalent chemical bonds to the inorganic network react in the wet film through organic cross-linking reactions. Depending on the chemical nature of the reactive species (vinyl, epoxy, acrylic, etc.), various organic network types can be formed. They can be classified by four compositional parameters (Haas et al., 1999b): (i) Type I: nonorganically modified Si alkoxides; (ii) Type II: heterometal alkoxides; (iii) Type III: organically modified reactive Si alkoxides; and (iv) Type IV: functional organically modified Si alkoxides.

The adaptation of materials for special applications is mainly determined through the use of these four structural elements and the conditions for forming inorganic and organic networks (Mackenzie & Bescher, 1998; Mackenzie & Bescher, 2003). For example, the amount of inorganic structures and the extent of organic cross-linking can have a dramatic influence on the mechanical properties (Mackenzie & Bescher, 2000) as will be shown in section 5. A high inorganic content leads to stiff but brittle materials. Hardness combined with elasticity is realised by using inorganic structures with a certain amount of organic cross-linking (Mammeri et al., 2005).

Sol-gel technology is used to produce hybrid coatings because: (i) it allows the formation of hybrid structure at temperatures below 150 °C; (ii) it is attractive for coating polymers which have melting points between 150°C and 300°C; (iii) coatings can easily be produced by dip or spin coating; (iv) the technology is simple to implement on a large scale and it is cheap; and (v) ceramic, metal and polymer substrates can be easily coated. For a comprehensive understanding of sol-gel technology and hybrid film processing see refs (Brinker & Scherrer, 1990; Haas et al., 1999b; Letailleur et al., 2011).

The general processing scheme for hybrids is shown in Figure 1. The reaction is divided into hydrolysis and polycondensation (Brinker & Scherrer, 1990). The hydrolysis reaction induces the substitution of OR groups linked to silicon by silanol Si-OH groups:

$$\equiv Si\text{–}OR + H_2O \rightarrow \equiv Si\text{-}OH + ROH \tag{1}$$

These chemical species may react together to form Si-O-Si (siloxane) bonds which lead to the silica network formation. The polycondensation equations are:

$$\equiv Si\text{–}OH + \equiv Si\text{–}OH \rightarrow \equiv Si\text{–}O\text{–}Si\equiv + H_2O \tag{2a}$$

$$\equiv Si\text{–}OH + \equiv Si\text{–}OR \rightarrow \equiv Si\text{–}O\text{–}Si\equiv + ROH \tag{2b}$$

Polycondensation leads to the formation of a sol which can be deposited on a substrate using spin, spray, flow or dip coating. When the sol is applied on a substrate, the wet film can be further cross-linked thermally or by using UV/IR radiation to evaporate the water and alcohol remaining in the pores and increase the bonding to the substrate.

3. Nanoindentation

Nanoindentation is an exceptionally versatile technique and is ideal for quantifying mechanical properties of materials at the sub-micron scale (Oliver & Pharr, 1992; Fischer-Cripps, 2002). The growing need to study the mechanical properties of small volumes, thin films and surface treated materials has seen dramatic developments in sub-micron indentation testing and instrumentation capable of loads down to tens of micro-newtons to produce nanometre size indentations. The two basic properties readily obtained are hardness (H) and Young's modulus (E).

3.1. Spherical versus sharp indentation

Indenters can generally be classed into two categories: *sharp* (pointed) or *blunt* (spherical). The fundamental difference between the indentation of a pointed indenter and a spherical indenter in a material is that the pointed indenter induces an immediate plastic response at the point of first contact with the material while the spherical indenter induces an elastic-plastic response. Examples of these differences on the resultant load-displacement curves of a silica glass (E = 70 GPa, ν= 0.17) using four typical indenter types with a maximum load of 1 mN are shown in Figure 2. In these tests only the spherical indenter displayed a completely reversible elastic contact.

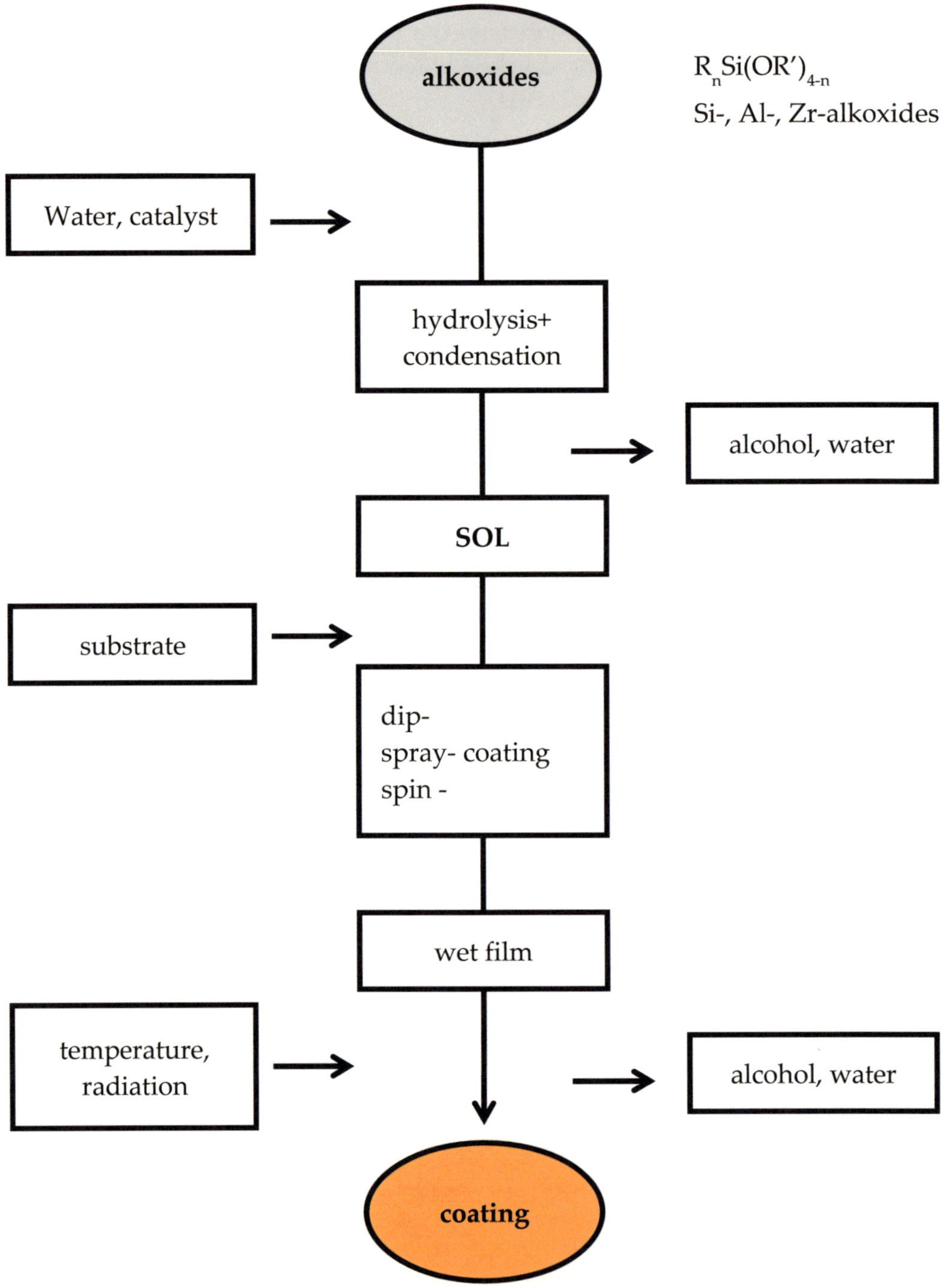

Figure 1. Ormocer® processing for coatings [Redrawn from ref (Haas et al., 1999b) with permission from Elsevier].

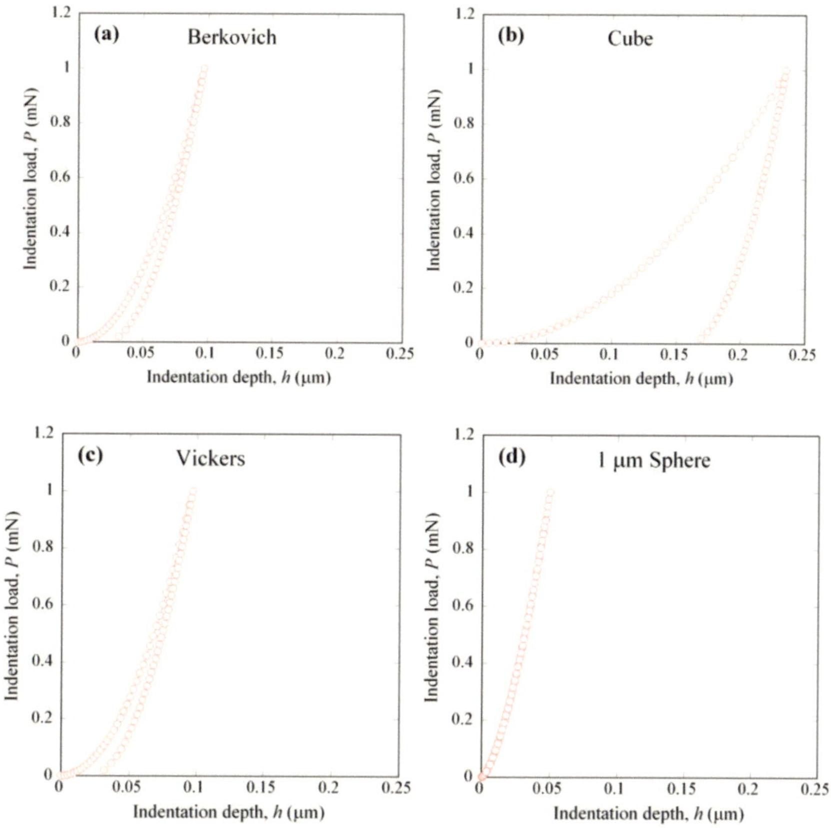

Figure 2. Load versus depth profiles for silica glass ($E = 70$ GPa, $\upsilon = 0.17$) at 1 mN maximum load for diamond indenter types: (a) Berkovich, (b) cube-corner, (c) Vickers and (d) 1 μm sphere.

3.1.1. Pointed indenters

Pointed indenters used in nanoindentation are made of diamond and are either Berkovich, Vickers, Knoop or cube-corner geometry (Fischer-Cripps, 2002). The radius of the indenter tip is significantly smaller than that used for microhardness testing. The Berkovich indenter is often favoured as its three-face pyramid geometry is much easier to grind to a sharp point compared to the four-face pyramid geometry of Vickers and Knoop. The cube-corner indenter is now used widely for initiating cracks and therefore facilitates fracture toughness measurements (Fischer-Cripps, 2002; Volinsky et al., 2003). Stresses beneath pointed indenter tips are very high and in theory are infinite at the point of contact with an elastic body. In reality the tip is always a little blunt as no indenter is ideally sharp. Berkovich indenters can be used as a spherical tip for ultra-low contact loading. For pointed indenters the distribution of stress in a homogeneous material remains constant regardless of the penetration depth of the indenter. The mean contact pressure or indentation hardness and the average strain are constant and are the reasons why it is good for hardness determinations. The average or equivalent strain is only dependent on the indenter angle.

3.1.2. Spherical indenters

With spherical indenters the stresses are symmetrical and, unlike pointer indenters have no preferred direction; this means that the orientation of the indenter plays no role in the determination of the properties of crystals or anisotropic materials. The primary use of a spherical indenter is to reveal information on the transition from elastic to elastic-plastic or elastic-brittle behaviour in materials. During the initial stage of penetration into the surface the contact zone is deformed elastically but at larger loads a transition to elastic-plastic occurs and the average strain increases with the depth of penetration as the depth of contact area grows faster than the indenter radius. Therefore, it is possible to construct diagrams of indentation stress versus strain (He et al., 2008). As opposed to pointed indenters the strain increases with increasing depth of penetration for spherical indenters. For thin, soft and compliant (low *H* and *E)* coatings spherical indentation is useful for characterising film behaviour particularly the elastic to plastic transition and viscoelastic properties (Oyen, 2006; Latella et al., 2008a) and for hard coatings the evolution in damage (Haq et al., 2010). For these reasons the discussion and mechanics that follows is restricted to spherical indentation.

3.2. Measurement of hardness and elastic modulus

Since the pioneering work of Hertz (Hertz, 1896) the nature of contact damage that forms in a brittle solid loaded with a sphere has been extensively studied (Lawn, 1998). For spherical indentation of a rigid sphere into a specimen at low loads the Hertz equation for elastic loading is:

$$P = \frac{3}{4} E_{eff} R^{1/2} h_e^{3/2} \tag{3}$$

where R is the sphere radius, h_e the elastic penetration depth and E_{eff} is the effective elastic modulus given by:

$$\frac{1}{E_{eff}} = \frac{1-\nu_m^2}{E_m} + \frac{1-\nu_i^2}{E_i} \tag{4}$$

where E is Young's modulus, ν is Poisson's ratio and the subscript i denotes the indenter material and m the sample. For higher loading elastic and plastic deformations occur within the specimen material. At full load the depth of penetration of the sphere below the original specimen surface is h_t. During unloading the response is elastic and at complete unload a residual impression depth, h_r is left. If the load is then re-applied, the loading is again elastic through the distance $h_e = h_t - h_r$, according to equation 3.

There are two possible approaches with spherical indentation, the continuous load-unload cycle (Oliver & Pharr, 1992) or the load-partial unload cycle (Field & Swain, 1993; Field & Swain, 1995). The continuous load–unload sequence for spherical indenters is essentially the same as the continuous load-unload sequence used for pointed indenters. However, the continuous load-unload data for spherical indenters is usually more difficult to analyse than it is for pointed indenters due to the changing strain and the need to determine and separate the elastic and plastic components of the load-displacement data. This difficulty is overcome by the load-partial unload technique which leads to a relatively simple expression for materials exhibiting permanent plastic deformation during indentation.

The continuous load-unload cycle (Oliver & Pharr, 1992) uses multiple points from the maximum load to determine the slope of the initial unloading portion of a single-cycle load-displacement curve. In contrast for the load partial-unload technique (Field & Swain, 1993; Field & Swain, 1995) typically a 50% unload from the maximum load for multiple increments of loading is used. From the tests the effective Young's modulus can be determined:

$$E = \frac{3P}{4ah_e} \tag{5}$$

where a is the contact radius $a = \sqrt{2Rh_p - h_p^2}$

with R the sphere radius and h_p the plastic penetration depth or depth of the circle of contact, h_e is the elastic penetration depth ($h_e = h_t - h_r$) with h_t the maximum penetration depth at full load P and h_r the residual depth of the impression upon unloading. The depth of the residual impression is obtained from the measurement of load P_s and penetration h_s at partial unload from the higher load P_t and forming the ratio of the elastic displacements:

$$h_r = \frac{h_s(P_t/P_s)^{2/3} - h_t}{(P_t/P_s)^{2/3} - 1} \tag{6}$$

Similarly the hardness, H, or contact pressure is given as:

$$H = \frac{P}{A} \tag{7}$$

where $A = \pi a^2 = 2\pi R h_p - \pi h_p^2$.

The strain increases with indentation impression depth, and an appropriate equivalent expression is:

$$\epsilon = 0.2\, a/R \tag{8}$$

Hence plotting H versus a/R is the contact stress-strain curve. The contact area and contact pressure are calculated for every partial unloading step and provides a measure of hardness and modulus as a function of depth of penetration of materials whose properties vary with penetration.

3.3. Indentation creep

The discussion above has assumed time-independent material behaviour but this is not always the case for materials such as polymers which exhibit time-dependent deformation under loading. The time-dependent deformation can be described in terms of creep and stress relaxation. Creep is the time-dependent deformation that occurs under constant stress while stress relaxation is the stress response under constant strain. The discussion in this chapter will be restricted to creep.

3.3.1. *Methodology*

As with conventional creep testing of structural materials, nanoindentation creep testing provides an accurate measure of indentation depth changes as a function of time (Lucas & Oliver, 1999; Oyen, 2006). The method is best performed using fast loading (high strain rate) to the desired load, held at the load for a predetermined time and then unloaded. If a slow step loading is performed then differences in the creep deformation and creep parameters are expected purely due to the effect of strain rate.

3.3.2. *Models*

It is well known that the organic component in hybrid films exhibits polymeric-like behaviour. Hence the creep in these films can be readily modelled using a combination of springs and dashpots (Bland, 1960; Fischer-Cripps, 2004). The starting point is that the elastic deformation in a material can be described by a *spring* that responds to an applied stress (Hooke's law):

$$\sigma = E\epsilon \tag{9}$$

With the added influence of time dependency the *dashpot* represents a Newtonian viscous substance as follows:

$$\sigma = \eta\left(\frac{d\epsilon}{dt}\right) \tag{10}$$

Therefore by connecting springs and dashpots together in series and parallel combinations various models can be devised (Bland, 1960). The phenomenological spring dashpot models used widely for analysing indentation creep are illustrated in Figure 3.

Lee and Radok (Lee & Radok, 1960) addressed the problem for indentation of a smooth rigid sphere on a semi-infinite viscoelastic plane to determine the relation between indentation force and displacement. The solution is based on viscoelastic extensions of Hertzian contact (see equation 3) by combining the elastic and dissipative components (Bland, 1960; Kumar & Narasimhan, 2004):

$$h^{3/2}(t) = \frac{a^3}{R^{3/2}} = \frac{3}{4}\frac{P}{R^{1/2}}\left[\frac{1}{E_s} + \frac{t}{\eta_s} + \psi(t)\right] \tag{11}$$

where E_s, η_s and $\psi(t)$ are the instantaneous elastic modulus, the long term viscous flow constant and the creep response function of the form $\psi(t) = 1\text{-}e^{-Et/\eta}$.

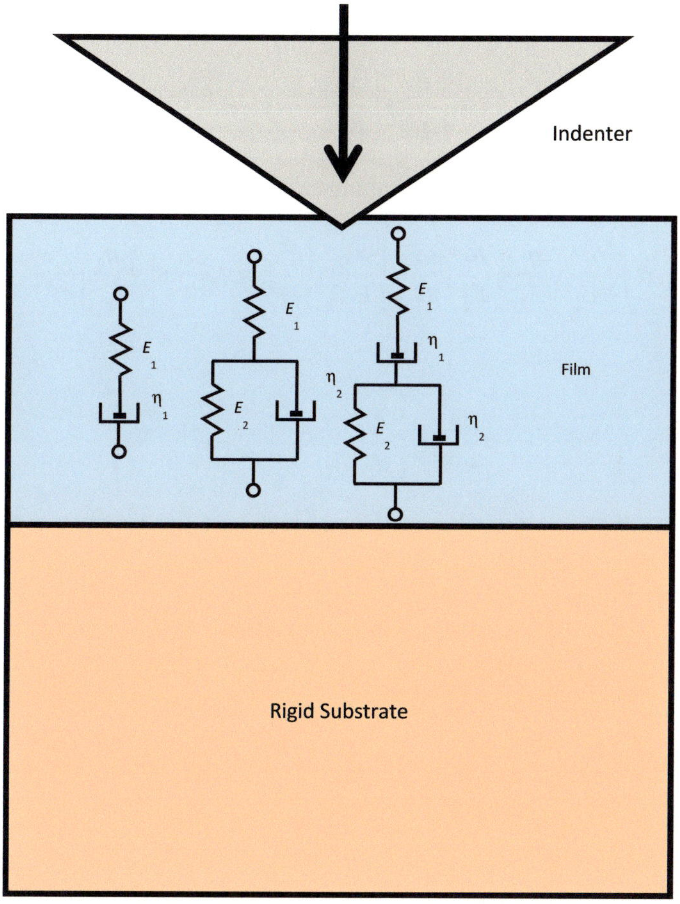

Figure 3. Schematic illustration of spring-dashpot models for indentation on film-substrate system. From left to right: Maxwell, three-element and four-element (Bürger) models.

For the Maxwell model it can be shown that:

$$\frac{d\varepsilon}{dt} = \frac{1}{E_1}\frac{d\sigma}{dt} + \frac{\sigma}{\eta_1} \tag{12}$$

where ε is the strain and σ is the stress.

So that the time-dependent depth of penetration for a spherical indenter in this case is:

$$h^{3/2}(t) = \frac{3}{4}\frac{P_{\max}}{R^{1/2}}\left[\frac{1}{E_1} + \frac{t}{\eta_1}\right] \tag{13}$$

Likewise, for the standard linear solid (three-element) model the constitutive relation is:

$$\left(1 + \frac{E_1}{E_2}\right)\sigma + \left(\frac{\eta_1}{E_2}\right)\frac{d\sigma}{dt} = E_1\varepsilon + \eta\frac{d\varepsilon}{dt} \tag{14}$$

and:

$$h^{3/2}(t) = \frac{3}{4}\frac{P_{max}}{R^{1/2}}\left[\frac{1}{E_1} + \frac{1}{E_2}(1 - e^{-E_2 t/\eta_1})\right] \quad (15)$$

The constitutive equation for the four-element model is:

$$\sigma + \left[\frac{\eta_1}{E_1} + \frac{\eta_1}{E_2} + \frac{\eta_2}{E_2}\right]\frac{d\sigma}{dt} + \left[\frac{\eta_1\eta_2}{E_1E_2}\right]\frac{d^2\sigma}{dt^2} = \eta_1\frac{d\varepsilon}{dt} + \left[\frac{\eta_1\eta_2}{E_2}\right]\frac{d^2\varepsilon}{dt^2} \quad (16)$$

and:

$$h^{3/2}(t) = \frac{3}{4}\frac{P_{max}}{R^{1/2}}\left[\frac{1}{E_1} + \frac{t}{\eta_1} + \frac{1}{E_2}(1 - e^{-E_2 t/\eta_2})\right] \quad (17)$$

Equations 13, 15 and 17 can then be used to obtain best fits to the experimental data by systematically adjusting the fitting parameters (E_1, E_2, η_1, η_2) using an iterative procedure with the Levenberg–Marquardt algorithm. The starting values for the fitting parameters are based on nanoindentation results and a refinement of estimates for the other parameters to achieve fits with correlation coefficient $R^2 > 0.95$. Similarly creep in thin films has been analysed using logarithmic relations (Berthoud et al., 1999; Chudoba & Richter, 2001; Beake, 2006) such as:

$$h(t) = A + B\ln\left[(Ct) + 1\right] \quad (18)$$

where *A*, *B* and *C* are fitting constants and t is the time. This equation does not give *E* and η but the coefficient *B* is defined as an extent term and *C* as a rate term for deformation (Beake, 2006).

4. Microtensile testing

4.1. Background

Characterising the cracking evolution, debonding behaviour and adhesion performance of thin films subject to external applied stresses is an important aspect in materials selection for specific applications. As a complement to nanoindentation testing, micro-mechanical tensile testing is valuable in elucidating the critical conditions for cracking and debonding of thin brittle films on ductile substrates (Ignat, 1996; Ignat et al., 1999). These types of experiments have been shown to offer insights into evaluating interfacial adhesion of thin films and multilayered structures (Agrawal & Raj, 1989; Filiaggi et al., 1996; Scafidi & Ignat, 1998; Wang et al., 1998; Harry et al., 2000; Latella et al., 2007a; Roest et al., 2011).

In this type of test a film is deposited on a tensile coupon, which can then be pulled in a universal testing machine or a specialized device and the surface can be viewed with an optical microscope or in a scanning electron microscope. Brittle coatings produce parallel cracks on ductile substrates when uniaxially stressed perpendicular to the tensile axis – see section 5 for examples. These cracks generally extend through the thickness of the coating and along the width of the sample and increase in number with additional elongation, leading to a decrease in the crack spacing. For some systems, cracks may also be accompanied by localized delamination of the coating from the substrate. Eventually, delamination of the coating signals the end of the lifetime of the coated system. For more compliant films, cracking can be irregular and film debonding reduced substantially.

Tensile testing is advantageous in that the stress field is uniform along the gauge length of the sample and relatively small specimens can be used. Similarly, using optical or scanning electron microscopy (SEM) to view the damage *in-situ* during loading reveals fracture and film failure mechanisms (Ignat et al., 1999; Latella et al., 2004; Latella et al., 2007b). The only prerequisite for this type of test is that for analysis of the coating behaviour, the residual stress, and Young's modulus of the coating are required by other means, such as from substrate curvature measurements and nanoindentation, respectively.

4.2. Mechanics

It is recognised that cracking of a film and its detachment from an underlying substrate are controlled by the intensity of the stored elastic energy. For a thin film subjected to an in plane isotropic stress, the elastic stored energy is:

$$U = \frac{1-\nu_f}{E_f}{\sigma_f}^2 t \qquad (19)$$

where σ_f is the normal stress in the film, ν_f, E_f and t are the Poisson's ratio, Young's modulus and thickness of the film, respectively. Hence a film under tension will crack when U equals the films cracking energy and for a film under compression will delaminate when U equals the interfacial cracking energy. Accordingly the mechanical stability of the film depends on its strength and fracture toughness and adhesion behaviour. Micromechanical tensile testing is useful because these key material parameters can be readily studied.

For a film-substrate system that is strained in tension the requirement is to determine the instant of first cracking in the film, which corresponds to a strain ε_c. Using Young's modulus of the film (E_f) the critical stress, σ_c, for cracking or film strength is calculated as follows:

$$\sigma_c = \epsilon_c E_f + \sigma_r \qquad (20)$$

where σ_r is the residual stress in the film.

The fracture energy of the coating is obtained from (Hu & Evans, 1989):

$$\gamma_f = \frac{\sigma_c^2 t}{E_f}\left(\pi g(\alpha) + \frac{\sigma_c}{\sqrt{3}\sigma_y}\right) \qquad (21)$$

where γ_f is the fracture energy, t is the thickness of the film, σ_y is the yield stress of the substrate and α is Dundar's parameter $\alpha = (E_f - E_s)/(E_f + E_s)$, where E_s is Young's modulus of the substrate and $g(\alpha)$ is obtained from (Beuth & Klingbeil, 1996).

Adhesion of the film to the substrate is determined by the measurement of the interfacial fracture energy. The instant of first debonding of the film during tensile loading corresponds to a strain ε_d. The apparent interfacial fracture energy is given by (Hu & Evans, 1989):

$$\gamma_i = \frac{E_f}{2} t \epsilon_d^2 \tag{22}$$

5. Experimental studies

5.1. Case study 1 – Different length and functionality of organic

Sol-gel coating solutions were prepared by adding a 0.01 M solution of nitric acid (HNO_3) to equimolar mixtures of tetraethylorthosilicate (TEOS) and selected alkyltriethoxysilanes in dry ethanol with equivalent SiO_2 concentrations of 5 wt%, specifically, methyltrimethoxysilane (MTMS), vinyltrimethoxysilane (VTMS) and 3-glycidoxypropyltrimethoxysilane (GTMS). GTMS is a low cost and readily available commercial compound and is of major interest as it is widely used for coatings in optical and anti-corrosion applications. The molecule has a long organic chain composed of seven carbons and an epoxy ring polymerisable at its end group.

A solution of 100% TEOS was also prepared as the control. A water-to-alkoxide ratio of 10 was used in all cases and the solutions were aged at room temperature for 24 h before use. The chemical structures of the organic constituents are given in (Atanacio et al., 2005). Thin film coatings were deposited on silicon wafers (25.4 mm diameter; thickness, 0.5 mm; single sided polished) and polished stainless steel coupons by spin coating at 5000 rpm for two minutes. The coated specimens were then allowed to dry for 24 h at 60°C. The coatings produced were transparent and amorphous in nature and given the following designations: (i) TEOS (thickness, t = 270 nm), (ii) MTMS (t = 280 nm), (iii) VTMS (t = 250 nm) and (iv) GTMS (t = 620 nm).

Figure 4 shows full cycle spherical indentation load-displacement curves for the films with a 30 s dwell at 1 mN maximum load. The key features to note from the load-displacement curves are differences in the maximum penetration depth, the increase in penetration for the 30 s dwell at peak load and the recovery behaviour of the films during the unloading cycle. The TEOS film initially displays elastic behaviour, which is then followed by small deviations from the ideal elastic behaviour based on computation of simulated load-depth curves. The MTMS and VTMS films show similar trends, although with a much greater degree of compliance, with their response curves displaced to the right. The GTMS film, on loading, displays a dramatic increase in penetration far exceeding those of the other films, and even more striking on unloading, is the dramatic recovery from 0.4 mN to complete unload of approximately 340 nm, not evident in the other films and indicative of polymer-like behaviour. This is most likely due to viscoelastic flow and relaxation processes, as there is little permanent deformation with the creep and recovery being almost reversible. The TEOS film shows the least amount of deformation, attributable to the predominantly silica comprised network providing rigidity and hence less molecular movement under constant load. The MTMS and VTMS are intermediate and the GTMS film shows the greatest deformation. A study of silica nano-particle filled hybrid films on glass showed similar mechanical responses (Malzbender et al., 2002).

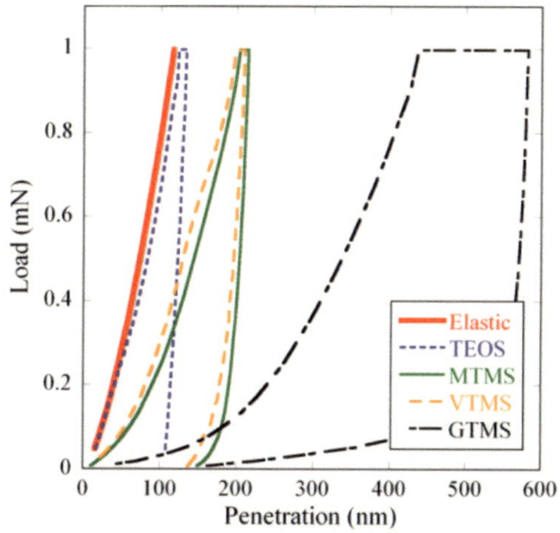

Figure 4. Load-displacement response for spherical indentation of the four coatings on silicon. The bold red curve denoted Elastic is the calculated Hertzian elastic response for the TEOS film [Redrawn from ref (Latella et al., 2003)].

The derived indentation stress-strain curves of the films from the load-partial unload method are shown in Figure 5. Again the results show the increasing deviation in mechanical response of the films from nominally elastic-brittle for TEOS (E = 18.8 GPa; H = 0.6 GPa), intermediate for MTMS (E = 5.6 GPa; H = 0.3 GPa) and VTMS (E = 4.8 GPa; H = 0.3 GPa) films to elastic-plastic for GTMS (E = 0.9 GPa; H = 0.15 GPa). The curves indicate that the addition of organics leads to a decrease in Young's modulus and a greater tendency for energy absorbing behaviour particularly in GTMS to minimise damage under contact loading. The transition is analogous to that observed in porous hydroxyapatites (He et al., 2008).

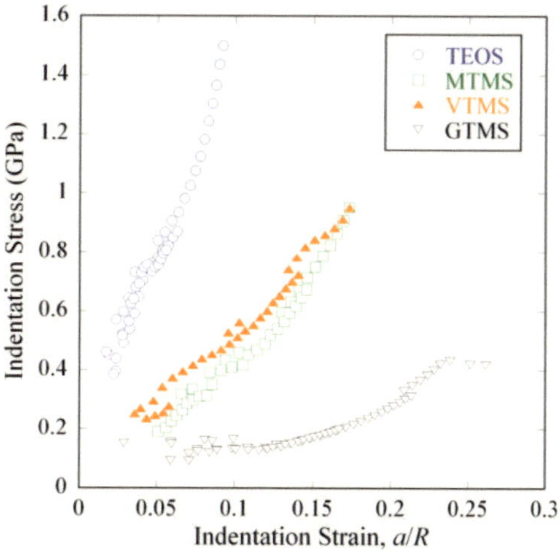

Figure 5. Indentation stress-strain behaviour of the four films.

The results concerning the influence of chain length and functionality of the organic precursors introduced in the inorganic network on the mechanical properties are linked to the structure and the network. By introducing different organic chain lengths dramatic modifications in the connectivity of the network are expected. For pure inorganic silica coatings the structure is dense but with the addition of a small chained organic component the short-range network is significantly modified. For example when MTMS (1 carbon chain length) is introduced, some silica domains may be formed but the structure is not dramatically modified suggesting that the silica domains are still closely grouped. However, the modification that occurs leads to a discernible difference in the mechanical properties compared to the pure inorganic coating (TEOS). By comparison when GTMS (7 carbon chain length) is introduced the mechanical properties are reduced further. In this case, it is thought that the longer GTMS chain creates larger gaps between the silica-rich domains, which are much further apart. This result in the connectivity of the network to be significantly lowered compared to a pure inorganic network.

To complement the indentation testing results, similar composition films deposited on stainless steel dogbones were uniaxially loaded in tension at a rate of 0.003 mm/s using a high-stiffness mechanical testing device (Ignat et al., 1999) positioned directly under the objective lens of an optical microscope (Zeiss Axioplan) at a fixed magnification. This allowed direct observation of crack initiation and evolution and debonding of the thin films on the steel specimens (see section 4). The applied load and the imposed displacement were recorded during the tests and optical images were captured at designated points as shown in the example in Figure 6 for the base TEOS film. Higher magnification views of the four films were obtained on carbon coated samples using SEM (JEOL 6300).

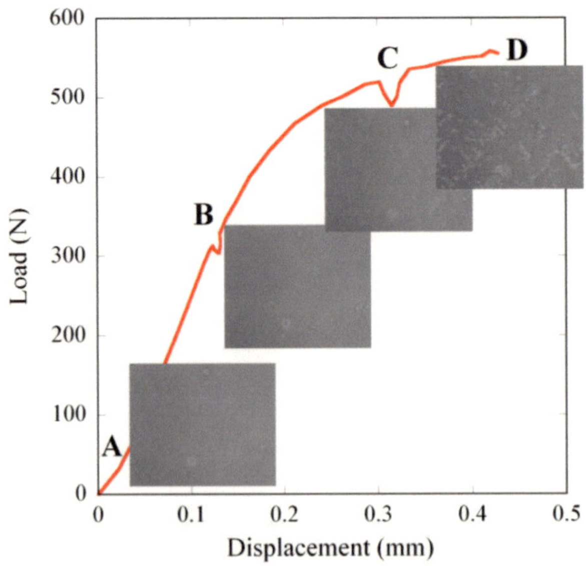

Figure 6. Load-displacement curve from tensile test of the TEOS coated stainless steel. Inset images at points A (0 N), B (320 N), C (520 N) and D (560 N) correspond to the specimen surface during loading (field of view in each image is 400 μm) [After (Latella et al., 2003)].

Figure 7 shows SEM images of the four thin film coatings after tensile testing. The baseline TEOS film (Fig. 7(a)) displays characteristic brittle behaviour with cracks at about a 90° angle to the loading direction (transverse cracking) and normal to the interface. Delamination of the coating is obvious and buckling occurs readily in these areas. The MTMS film (Fig. 7(b)) also showed cracking but it was more irregular and there was a great deal of debonding. The VTMS film (Fig. 7(c)) exhibited less debonding than MTMS but it had transverse cracking with a similar inter-cracking distance. By contrast the longer-chained GTMS film (Fig. 7(d)) resulted in excellent substrate–film bonding with little cracking and decohesion of the film. The same type of cracking and debonding is observed in related tensile testing of these films on copper substrates – see ref (Atanacio et al., 2005) and later in section 5.2.

Clearly the size of the organic chain has a dramatic influence on the mechanical response of the films in contrast to the relatively brittle baseline TEOS film (Schmidt 1985). The reduction in Young's modulus and the greater resistance to cracking and debonding of the films with increasing organic-modifier showed that the films can be tailored by simple manipulation of the sol-gel chemistry. The larger chained organic component results in a structure which is not unlike a polymer, as is the case in the GTMS film, which minimises film cracking. The mechanical property results suggest that there is an important rearrangement of the organic modifier links in these hybrids that controls the deformation, hence the transition to semi-brittle or viscoelastic response (Latella et al., 2003; Atanacio et al., 2005). Similar behaviour has been observed in bulk samples made from TEOS and polydimethylsiloxane (PDMS) (Mackenzie, 1994). Mackenzie demonstrated that the mechanical properties could vary from being hard and brittle to rubbery and soft, depending on the ratio of organic to inorganic constituents. Samples were found to retain a rubbery nature even when the inorganic constituent was in excess of 70 wt%. However, when the PDMS content was less than 10 wt%, the sample became brittle. Mackenzie proposed that small concentrations of PDMS react in solution to form gels with a porous three dimensional network still dominated by the Si-O-Si linkages. However, as the PDMS content increases, the structure is characterised more by silicon clusters linked with flexible chains of PDMS.

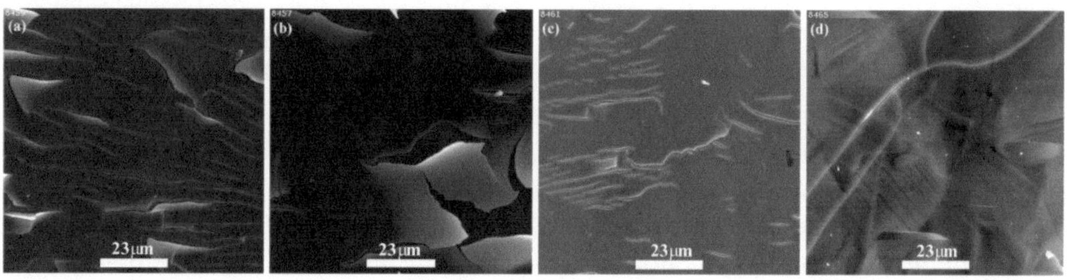

Figure 7. SEM images of the films: (a) TEOS, (b) MTMS, (c) VTMS and (d) GTMS on stainless steel substrates after tensile loading [from ref (Latella et al., 2003)]. Heterogeneous cracking with marked debonding is obvious in (a) and (b). Tensile axis is vertical in these images.

5.2. Case study 2 – Different amount of organic

In this study the effect of increasing amount of GTMS was examined for films deposited on copper sheet. The Cu sheet was cut into samples of about 40 mm in length and 20 mm width and as tensile dogbones (12 mm gauge length, 3 mm width at the gauge and 1 mm thickness). The samples were polished to a 1 μm finish then cleaned in soap solution and ethanol and then dried. Sol-gel solutions were prepared by adding a 0.01 M solution of HNO_3 to (i) TEOS and (ii) a mixture of 25%, 50% or 75% of the organic GTMS and TEOS (75%, 50% and 25%) in ethanol. Each solution contained an equivalent SiO_2 concentration of 5 wt% and a water-to-alkoxide mole ratio of 10. The solutions were spin coated on the copper samples and then dried at 60°C for 24 h in a clean room environment. Thickness of the coatings was determined using spectroscopic ellipsometry (Sopra GES). The coatings produced were transparent and amorphous in nature and given the following designations based on the precursors used: (i) TEOS (thickness, t = 190 nm), (ii) 25% GTMS (t = 290 nm), (iii) 50% GTMS (t = 450 nm) and (iv) 75% GTMS (t = 600 nm).

The indentation load-displacement curves (P_{max} = 1 mN) are shown in Figure 8 for the TEOS and the 25%, 50% and 75% GTMS films deposited on the copper substrates using a nominal 1 μm spherical indenter. A 10 s hold at maximum load was used to provide a qualitative assessment of creep. The load-displacement curve of the TEOS coating is typical of an elastic-brittle material, showing initially elastic loading then elastic-plastic behaviour up to maximum load. The GTMS films show an increasing tendency, with higher organic, for greater penetration on loading indicative of soft and compliant coatings. It is important to note that the TEOS film is thin and Cu is much softer than Si (*cf.* with Figure 4) so plastic deformation of the substrate is more prevalent. Also with the increased % GTMS the films are progressively thicker and softer so now most of the deformation is in the film rather than in the substrate. At peak load, there was detectable creep, particularly for the 75% GTMS, and then on unloading there was recovery back to a low residual penetration, symptomatic of viscoelastic behaviour. A better approach here would be to hold at low load to quantify the recovery with time to give a clearer indication of viscoelastic response (He & Swain, 2009).

The hardness, H, and Young's modulus, E, of the three GTMS films, determined using the load partial unload technique (see section 3.2), as a function of percentage organic is given in Figure 9. The hatched boxes at the left are for the baseline silica film (100% TEOS): H = 2.15 GPa and E = 55 GPa. Compared to the TEOS film there was a large drop in both hardness and Young's modulus of the GTMS films, which decreased with increasing organic, confirming the observations of the load-displacement curves in Figure 8. Clearly, with the introduction of the long-chained organic there is a prominent drop in the mechanical properties and evidence of a change from elastic-brittle to viscoelastic behaviour in the sol-gel matrix due to the influence of organic species and its modifying ability on the inorganic network structure (Metroke et al., 2001; Atanacio et al., 2005). The H and E values are comparable to that for bulk GTMS hybrids (Innocenzi et al., 2001) and comparisons with a myriad of hybrid coatings can be found in (Mammeri et al., 2005).

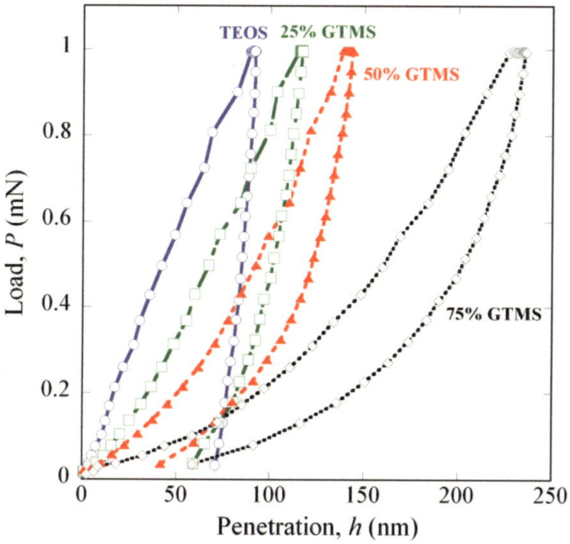

Figure 8. Indentation load-displacement curves for TEOS and GTMS on Cu substrates. $P = 1$ mN with 10 s dwell at maximum force [Redrawn from ref (Latella et al., 2008a)].

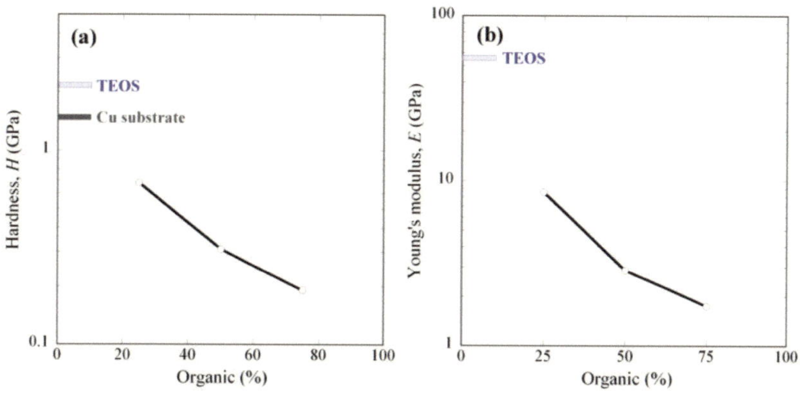

Figure 9. Plots of (a) Hardness and (b) Young's modulus of the GTMS coatings versus percentage organic addition. Hatched boxes at the right correspond to the properties of the TEOS coating and the Cu substrate (not shown is $E_{Cu} \approx 120$ GPa) [From ref (Latella, 2008b)].

Figure 10(a) shows the creep data for the TEOS and the three GTMS films. Creep penetration as a function of time, taken from five indents at each hold time, for the various coatings was examined using step loading (Oyen, 2005) to $P_{max} = 0.5$ mN for a 90 s hold with the 1 μm spherical indenter. Examples of the best fit curves of various spring-dashpot models (see section 3.3.2) for the 50% and 75% GTMS films are shown in Figure 10(b). The results from the fitting of the creep curves using the three- and four-element mechanical models for all coatings are presented in Table 1. Noting that the *E* values quoted were corrected from the best-fit parameters (*E'*): $E'(1-\nu_m^2)$. Figure 10(a) shows clearly the effect of the organic addition resulted in films with increased creep behaviour compared to the baseline inorganic TEOS film. Furthermore the initial penetration and creep deformation escalated substantially with the increasing level of the GTMS addition. Comparing the E_1 and η_1 values for the spring-dashpot models of the materials, the trend is for Young's modulus and viscosity to decrease with increasing level of organic in accord with the trends observed in the mechanical property results. There was little difference in the E_1 values obtained using the three-element and four-element models although there is some discrepancy with the values obtained using the indentation load-partial unload method of analysis (Latella et al., 2008a). Clearly a three-element model is sufficient for extracting the key material parameters of these films even though slightly better fits using four adjustable parameters are obtained in some instances based on the R^2 values. Irrespective of model the standard error for each parameter ranged from 0.2% to a maximum of 5%.

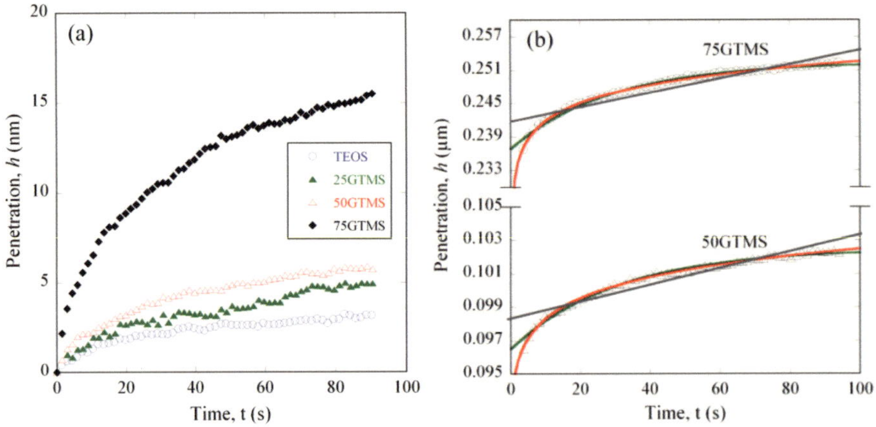

Figure 10. (a) Creep curves corrected for initial penetration for GTMS films. TEOS data is shown in both for comparison. Step loading to $P = 0.5$ mN for 90 s. (b) Examples of fits to raw creep data of the 50% GTMS and 75% GTMS. Solid lines are fits from Maxwell (grey), three-element (green) and logarithmic equations (red) [Redrawn from ref (Latella et al., 2008a)].

	TEOS	25% GTMS	50% GTMS	75% GTMS
Three-element model				
E_1 [GPa]	35.3	21.8	11.7	3.06
E_2 [GPa]	354	174	132	32
η_1 [GPa s]	8814	10178	4045	945
R^2	0.97	0.95	0.99	0.99
Four-element model				
E_1 [GPa]	36.0	22.8	11.8	3.10
E_2 [GPa]	436	323	186	44
η_1 [GPa s]	85974	27139	30715	8035
η_2 [GPa s]	4958	2267	2632	624
R^2	0.98	0.98	0.99	0.99

Table 1. Creep fit parameters for the TEOS and GTMS films.

SEM images of cracking behaviour from microtensile tests of the TEOS and 50% GTMS films on the copper are shown in Figures 11 and 12. The tests have been made at a common imposed total strain of ≈15%. For the TEOS film (Fig. 11(a)) regularly spaced parallel cracks in the coating layer perpendicular to the tensile axis are evident and normal to the interface (loading direction is horizontal) that appear throughout the entire gauge length of the specimen. These cracks multiply in number with increasing elongation leading to a decrease in the crack spacing to a saturation level with no further cracking i.e. intercracking distance of ≈ 5 μm. Fig. 11(b) shows a higher magnification image of a region in the coating that is heavily delaminated from the Cu substrate. These localised debonded zones vary in size with some buckled and fractured fragments evident between parallel cracks. The damage in the 50% GTMS film is shown in Figure 12. In stark contrast to the TEOS film, short cracks scattered throughout the coating are apparent (Fig. 12(a)). It shows excellent coating–substrate adhesive bonding (Fig. 12(b)). Some areas of the coating are free from cracking and there is negligible debonding, which can be attributed to the viscoelastic behaviour of the GTMS film. The difference in the cracking behaviour of the coatings is consistent with the nanoindentation tests, confirming a brittle to viscoelastic change in mechanical response due to the addition of the long-chained GTMS species.

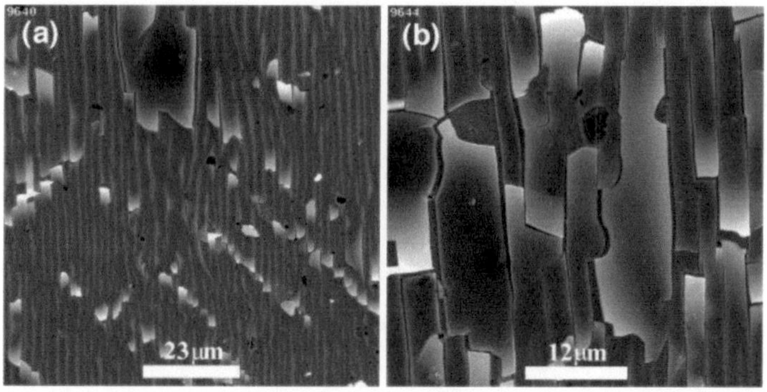

Figure 11. SEM images of cracking and debonding in the TEOS sol-gel film on Cu after tensile testing (15% total strain) showing (a) overall cracking and (b) small cracks.

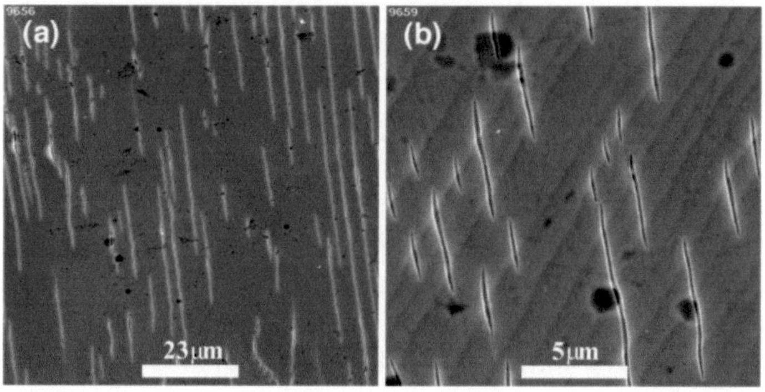

Figure 12. SEM images of cracking in the 50% GTMS sol-gel film on Cu after tensile testing (15% total strain) showing (a) overall cracking and (b) small cracks. Note the absence of debonding in GTMS compared to TEOS (Fig. 11). Tensile axis is horizontal in all images.

From the tensile tests the first signs of coating separation from the substrate observed as buckling in the optical microscope, irrespective of the deadhesion size or extent, was the criterion used for ascertaining the strain for debonding. The critical strains for debonding, ε_d, of the TEOS and the 50% GTMS films from the Cu substrates observed in the tensile tests are presented in Table 2 along with the apparent interfacial fracture energy (see equation 22). Because of the viscoelastic nature of the 50% GTMS film, the irregular nature of cracking and the absence of debonding, the interface fracture energy is a lower-bound estimate given that the calculation assumes linear elastic behaviour. The tensile strain required for debonding clearly shows that higher strains are required to generate debonding in the 50% GTMS film compared to TEOS. The interfacial energy for TEOS of 22 Jm^{-2} appears reasonable given that the fracture energy of soda-lime glass is $\gamma \approx 10$ Jm^{-2}. In the absence of debonding in the 50% GTMS film at a strain of 15% the interfacial fracture energy is clearly much greater than 14.5 Jm^{-2}.

	TEOS (0% GTMS)	50% GTMS
Critical applied strain for film debonding, ε_d [%]	6.5	>15
Apparent interfacial fracture energy, γ_i [Jm^{-2}]	22	>> 14.5

Table 2. Debonding parameters from tensile testing of the TEOS and 50% GTMS films on Cu substrates.

5.3. Case study 3 – Similar chain length and polymerisation

5.3.1. Similar chain length

To study the influence of the nature of the organic substituent on the mechanical properties of the hybrid film, GTMS was substituted by different organotrialkoxysilanes with similar chain lengths. All solutions were prepared at pH=2 with an equivalent SiO_2 concentration of 5 wt% and aged for 24 h at room temperature. A mixture of 50% TEOS and 50% organic: n-[3-(trimethoxysilyl)propyl]ethylene diamine (designated TMOSPEDA), n-octyltrimethoxysilane (designated OTES) were prepared. The GTMS film was prepared using THF (to avoid ring opening – see section 5.3.2) whereas ethanol was used as the solvent for TMOSPEDA and OTES. Various parameters in the preparation of the films could not be easily controlled. For example TMOSPEDA hydrolyses very quickly so the rate of hydrolysis between the samples was not the same for a given ageing time. The morphology and the roughness of the spin-coated sol-gel films on Si wafers were quite different as shown in Figure 13. The GTMS film is smooth and featureless whereas TMOSPEDA is striated and OTES results in a non-uniform coating with many pinholes.

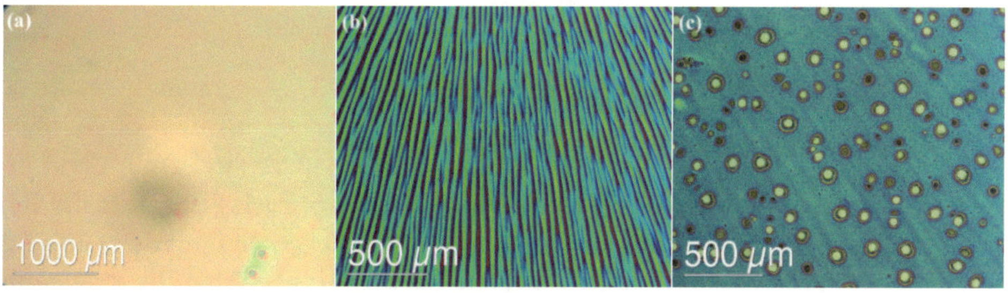

Figure 13. Optical micrographs of spin coated surfaces (a) GTMS (thickness, t = 590 nm), (b) TMOSPEDA (t = 270 nm) and (c) OTES (t = 320 nm).

Figure 14 show Young's modulus for the three sol-gel films with different end group precursors from indentations, away from pinholes and striations, on the solid phase of the films where applicable. The GTMS film has the higher modulus followed by TMOSPEDA and then OTES. The same trend in hardness is also observed. The data indicates better network connectivity in the GTMS structure compared to TMOSPEDA and OTES.

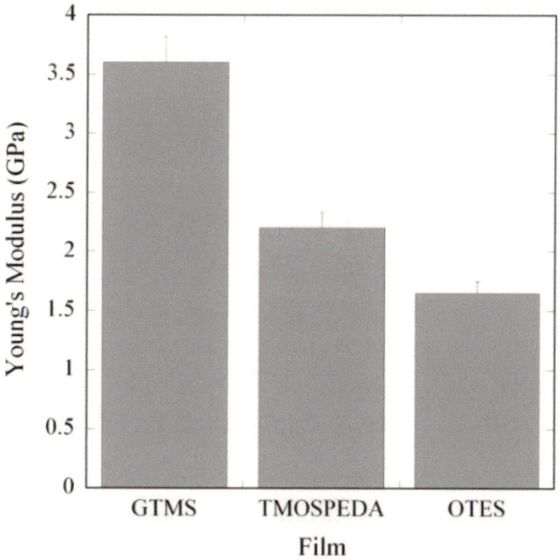

Figure 14. Young's modulus of the hybrid sol-gel films with similar chain lengths deposited on Si wafers.

Although not shown the tensile testing experiments agree with the nanoindentation results with the film strength following the same trend: GTMS (σ_c = 45 MPa), TMOSPEDA (σ_c = 40 MPa) and OTES (σ_c = 22 MPa) but interface fracture energy for the lower modulus and lower strength films is improved. Likewise the film damage studies indicated less delamination failures of these softer films, which were expected as they are more compliant, deformable and exhibit viscoelastic tendencies which provides greater resilience to fracture. Hence these types of soft films yield larger cohesive zones under externally applied stresses inhibiting catastrophic interfacial cracking and debonding as opposed to more brittle coatings (*cf.* Figs. 11 and 12).

5.3.2. *Polymerisation*

The previous case studies have illustrated that GTMS is commonly used as an organometallic precursor in organic-inorganic hybrid coatings because of its ability to undergo both hydrolysis-polycondensation (through the trialkoxysilyl group) and organic polymerisation by ring opening of the terminal epoxy group. Such features make GTMS a very attractive compound for the fabrication of hybrid co-polymers where the organic chains and the inorganic tridimensional network are interpenetrated, by provoking the hydrolysis and the ring opening polymerisation either simultaneously or in a controlled two-step process. To investigate the effects of ring opening in GTMS, samples were prepared using the same sol-gel chemistry as described above but here the solvents ethanol and THF were compared. The coatings produced ranged from 600 to 700 nm in thickness. In ethanol, it was determined by ^{13}C NMR that 10% of the ring was opened after 1 day at room temperature and 60% after 3 days at 60°C. The corresponding samples in THF after 1 day at room temperature showed no consequent ring opening and 10% opening after 3 days at 60°C.

Figure 15 compares the Young's modulus and hardness of the coatings resulting from GTMS sols prepared in either ethanol or THF and aged for different times. This allowed an estimation of the effect of the ring cleavage, without polymerisation, on the mechanical properties of the hybrid film. One consideration was that the presence of alcohol or ether groups at the end of the organic chain could allow some connectivity with the inorganic network through sol-gel type reactions with silanols or strong hydrogen bonding, thus improving the overall strength of the film (Metroke et al., 2002). The results showed that samples aged under the same conditions presented similar mechanical responses. More interestingly, a pronounced difference was observed between samples aged for 1 day at room temperature and those aged 3 days at 60°C. While the expected result was an increase of the Young's modulus with an increase of the silica network condensation, the opposite result was observed.

The presence of long organic chains in the precursor sol appears to reduce the long range connectivity of the inorganic network. This nano-segregation affects the overall cohesion of the hybrid coating and its mechanical strength drops with prolonged ageing. This emphasises the importance of cross-polymerisation of the organic groups in the film to maintain a strong interpenetrated network. This is further reinforced by the tensile testing results as illustrated in Figure 16. The 0% ring opened film shows extensive damage and delamination (Fig. 16(a)). By contrast the adhesion behaviour of the 10% ring opened film (Fig. 16(b)) is better with small cracks and debonded regions – typical of the standard GTMS film behaviour (see sections 5.1 and 5.2). In the 60% ring opened film (3 day aged at 60°C), there is slightly more debonding than the 10% ring opened film but the cracking is rather more irregular similar to a tearing appearance.

Opening of the epoxy ring in GTMS during the sol-gel process does not influence significantly the mechanical properties, for the same ageing time. Although when GTMS is allowed to reach higher hydrolysis-polycondensation states, i.e. longer ageing times, the mechanical properties of the coatings are degraded significantly and the surface roughness is increased. Figure 17 shows the strength, film toughness and interfacial fracture energy for the coatings. Again the same trends are apparent with nano-segregation between the organic and inorganic part of the hybrid presumably responsible for the decrease in strength, toughness and adhesion of the film with ageing time.

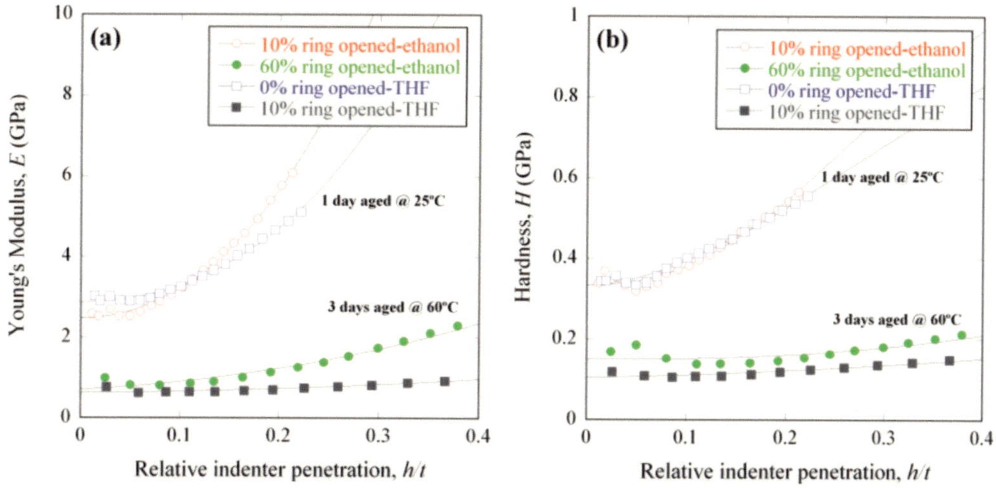

Figure 15. Plots of (a) Young's modulus and (b) hardness as a function of relative indenter penetration (depth to thickness, *h/t*) of the GTMS films at specific ageing time and temperature. Indicated is the amount of ring opening of the structure – for the THF based sols there is 0% and 10% ring opening but for ethanol based sols there is 10% and 60% ring opening. Solid lines are fits to the data using the method in ref (Jung et al., 2004).

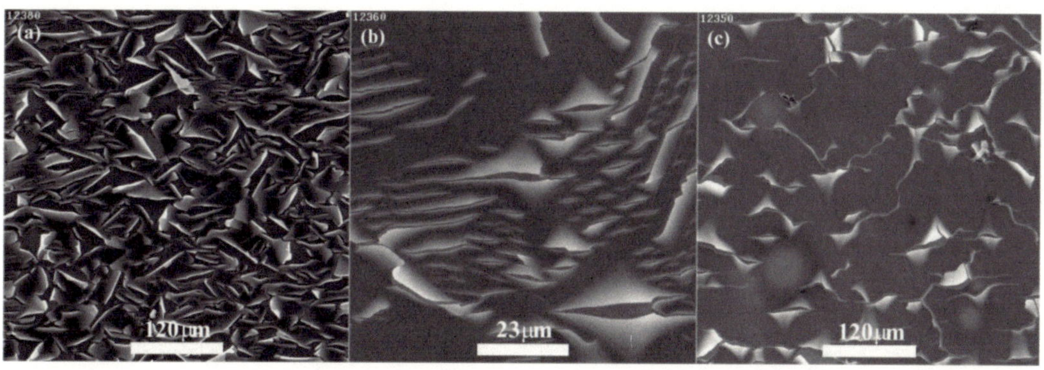

Figure 16. SEM images of damage in GTMS films on stainless steel after tensile testing: (a) 0% ring opening (THF-1 day aged at 25°C), (b) 10% ring opening (ethanol-1 day aged at 25°C) and (c) 60% ring opening (ethanol-3 days aged at 60ºC). Tensile axis is vertical in these images.

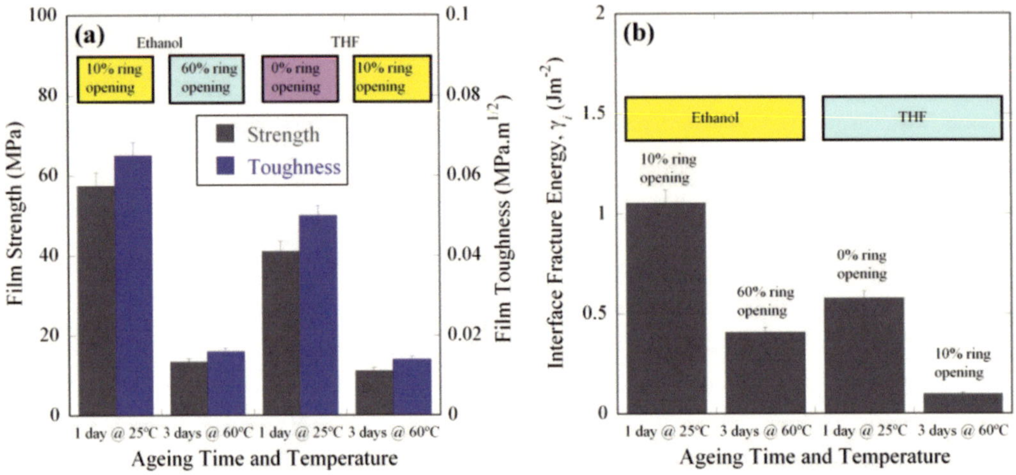

Figure 17. (a) Strength and fracture toughness and (b) interfacial fracture energy of the GTMS thin film coatings at specific ageing time and temperature. Also indicated is the amount of ring opening of the structure – for the THF based sols there is 0% and 10% ring opening but for ethanol based sols there is 10% and 60% ring opening.

6. Conclusion

The approach described in this chapter using both instrumented nanoindentation and micromechanical tensile testing provides significant insights into the effects of organic substituent (type/quantity) in these hybrid thin film systems regarding their mechanical and adhesion behaviour. The advantages of using nanoindentation for thin film characterisation are widely known and the work surveyed here has shown that its flexibility can be used to extract intrinsic film properties and also qualitatively provide insights on film attributes from load-displacement responses. Tensile testing is a practical complementary tool as it provides qualitative and quantitative appreciation of film fracture and damage evolution under controlled strains. Likewise, intrinsic film properties and interfacial adhesion energies can be extracted from *in-situ* experiments. The application of these techniques has been demonstrated on model sol-gel hybrid films with the following key findings:

1. Film properties and adhesion behaviour are dramatically affected by the nature of the organometallic precursors. Shorter chain length gives rise to higher Young's moduli. Smaller differences are observed between precursors with similar chain length but different functionality.
2. GTMS films on a variety of substrates exhibit excellent adhesion and minimal damage under external loading. The modifying ability of the long-chained GTMS molecule affects the network structure to such an extent resulting in viscoelastic flow and relaxation processes to occur under contact and external loading similar to those commonly seen in polymeric materials.
3. The balance between mechanical rigidity and adhesion is dependent on the proportion of $Si(OR)_4$ used in the hybrid films.

4. The epoxy ring opening in GTMS films did not influence the mechanical behaviour of the film to any great extent suggesting that polymerisation did not proceed.
5. Ageing of the hybrid sols for extended times resulted in a dramatic drop in mechanical properties.

Author details

Bruno A. Latella
Commonwealth Science and Industrial Research Organisation, WA, Australia

Michael V. Swain
Biomaterials Science, Faculty of Dentistry, University of Sydney, NSW, Australia

Michel Ignat
Physics Department, School of Engineering, University of Chile, Beauchef, Santiago, Chile

Acknowledgement

The authors wish to thank Australian Nuclear Science and Technology Organisation (ANSTO) colleagues that contributed to the work reviewed in this chapter: A. Atanacio, C.J. Barbé, J.R. Bartlett, G. Calleja, D.J. Cassidy, G. Triani and C. Tartivel. The research contribution presented in this study was undertaken and supported by the Materials Division, ANSTO.

7. References

Agrawal, D.C. & Raj, R. (1989). Measurement of the ultimate shear-strength of a metal ceramic interface. *Acta Metallurgica,* 37(4) pp. 1265-1270.

Atanacio, A.J., Latella, B.A., Barbé, C.J. & Swain, M.V. (2005). Mechanical properties and adhesion characteristics of hybrid sol-gel thin films. *Surface and Coatings Technology,* 192(2-3) pp. 354-364.

Beake, B. (2006). Modelling indentation creep of polymers: a phenomenological approach. *Journal of Physics D: Applied Physics,* 39 pp. 4478-4485.

Berthoud, P., Sell, C.G. & Hiver, J-M. (1999). Elastic-plastic indentation creep of glassy poly(methyl methacrylate) and polystyrene: characterisation using uniaxial compression and indentation. *Journal of Physics D: Applied Physics,* 32 pp. 2923-2932.

Beuth, J.L. & Klingbeil, N.W. (1996). Cracking of thin films bonded to elastic-plastic substrates. *Journal of Mechanics and Physics of Solids,* 44(9) pp. 1411-1428.

Bland, D.R. (1960). *Theory of Linear Viscoelasticity*. Pergamon, Oxford.

Brinker, C.J. & Scherrer, G.W. (1990). *Sol-Gel Science: The Physics and Chemistry of Sol-Gel Processing*. Academic Press, San Diego.

Chudoba, T. & Richter, F. (2001). Investigation of creep behaviour under load during indentation experiments and its influence on hardness and modulus results. *Surface and Coatings Technology,* 148(2-3) pp. 191-198.

Field, J.S. & Swain, M.V. (1993). A simple predictive model for spherical indentation. *Journal of Materials Research,* 8(2) pp. 297-306.

Field, J.S. & Swain, M.V. (1995). Determining the mechanical properties of small volumes of material from submicrometer spherical indentations. *Journal of Materials Research,* 10(1) pp. 101-112.

Filiaggi, M.J., Pilliar, R.M. & Abdulla, D. (1996). Evaluating sol-gel ceramic thin films for metal implant applications. II. Adhesion and fatigue properties of zirconia films on Ti-6Al-4V. *Journal of Biomedical Materials Research,* 33(4) pp. 239-256.

Fischer-Cripps, A.C. (2002). *Introduction to Nanoindentation.* Springer, New York.

Fischer-Cripps, A.C. (2004). A Simple phenomenological approach to nanoindentation creep. *Materials Science and Engineering A,* 385 pp. 74-82.

Haas, K.H. & Wolter, H. (1999). Synthesis, properties and applications of inorganic-organic copolymers (ORMOCER®s). *Current Opinion in Solid State and Materials Science,* 4 pp. 571-580.

Haas, K.H., Amberg-Schwab, S. & Rose, K. (1999a). Functionalised coating materials based on inorganic-organic polymers. *Thin Solid Films,* 351(1-2) pp. 198-203.

Haas, K.H., Amberg-Schwab, S., Rose, K. & Schottner, G. (1999b). Functionalized coatings based on inorganic-organic polymers (ORMOCER®s) and their combination with vapor deposited inorganic thin films. *Surface & Coatings Technology,* 111(1) pp. 72-79.

Haq, A., Munroe, P.R., Hoffman, M., Martin, P.J. & Bendavid, A. (2010). Effect of coating thickness on the deformation behaviour of diamond-like carbon-silicon system. *Thin Solid Films,* 518(8) pp. 2021-2028.

Harry, E., Ignat, M., Pauleau, Y., Rouzard, A. & Juliet, P. (2000). Mechanical behaviour of hard PVD multilayered coatings. *Surface and Coatings Technology,* 125(1-3) pp. 185-189.

He, L.H., Standard, O.C., Huang, T.T.Y., Latella, B.A. & Swain, M.V. (2008). Mechanical behaviour of porous hydroxyapatite. *Acta Biomaterialia,* 4(3) pp. 577-586.

He, L.H. & Swain, M.V. (2009). Nanoindentation creep behavior of human enamel. *Journal of Biomedical Materials Research Part A,* 91(2) pp. 352-359.

Hertz, H. (1896). *Hertz's Miscellaneous Papers.* Chs. 5,6. Macmillan, London.

Hu, M.S. & Evans, A.G. (1989). The cracking and decohesion of thin films on ductile substrates. *Acta Metallurgica,* 37(3) pp. 917-925.

Ignat, M. (1996). Mechanical response of multilayers submitted to in-situ experiments. *Key Engineering Materials,* 116-117 pp. 279-290.

Ignat, M., Marieb, T., Fujimoto, H. & Flinn, P.A. (1999). Mechanical behaviour of submicron multilayers submitted to microtensile experiments. *Thin Solid Films,* 353(1-2) pp. 201-207.

Innocenzi, P., Esposto, M. & Maddalena, A. (2001). Mechanical properties of 3-glycidoxypropyltrimethoxysilane based hybrid organic-inorganic materials. *Journal of Sol-Gel Science and Technology,* 20(3) pp. 293-301.

Jung, Y.-G., Lawn, B.R., Martyniuk, M., Huang, H. & Hu, X.Z. (2004). Evaluation of elastic modulus and hardness of thin films by nanoindentation. *Journal of Materials Research,* 19(10) pp. 3076-3080.

Kumar, M.V.R. & Narasimhan, R. (2004). Analysis of spherical indentation of linear viscoelastic materials. *Current Science*, 87(8) pp. 1088-1095.

Latella, B.A., Ignat, M., Barbé, C.J., Cassidy, D.J. & Bartlett, J.R. (2003). Adhesion behaviour of organically-modified silicate coatings on stainless steel. *Journal of Sol-Gel Science and Technology*, 26(1-3) pp. 765-770.

Latella, B.A., Ignat, M., Barbé, C.J., Cassidy, D.J. & Li, H. (2004). Cracking and decohesion of sol-gel hybrid coatings on metallic substrates. *Journal of Sol-Gel Science and Technology*, 31(1-3) pp. 143-149.

Latella, B.A., Triani, G., Zhang, Z., Short, K.T., Bartlett, J.R. & Ignat, M. (2007a). Enhanced adhesion of atomic layer deposited titania on polycarbonate substrates. *Thin Solid Films*, 515(5) pp. 3138-3145.

Latella, B.A., Ignat, M., Triani, G., Cassidy, D.J. & Bartlett, J.R. (2007b). Fracture and Adhesion of Thin Films on Ductile Substrates. In: *Adhesion Aspects of Thin Films, Volume 3*, Mittal, K.L., pp. 47-57, VSP, Leiden.

Latella, B.A., Gan, B.K., Barbé, C.J. & Cassidy, D.J. (2008a). Nanoindentation hardness, Young's modulus, and creep behavior of organic-inorganic silica-based sol-gel thin films on copper. *Journal of Materials Research*, 23(9) pp. 2357-2365.

Latella, B.A. (2008b). Indentation creep and adhesion of hybrid sol-gel coatings. *Advanced Materials Research*, 41-42 pp. 305-311.

Lawn, B.R. (1998). Indentation of ceramics with spheres: a century after Hertz. *Journal of the American Ceramic Society*, 81(8) pp. 1977-1994.

Lee, E.H. & Radok, J.R.M. (1960). The contact problem for viscoelastic bodies. *Journal of Applied Mechanics*, 27 pp. 438-444.

Letailleur, A., Ribot, F., Boissiere, C., Teisseire, J., Barthel, E., Desmazieres, B., Chemin, N. & Sanchez, C. (2011). Sol-gel derived hybrid thin films: the chemistry behind processing. *Chemistry of Materials*, 23(22) pp. 5082-5089.

Lucas, B. & Oliver, W. (1999). Indentation power-law creep of high-purity indium. *Metallurgical and Materials Transactions A-Physical Metallurgy and Materials Science* 30(3) pp. 601-610.

Mackenzie, J.D. (1994). Structures and properties of ormosils. *Journal of Sol-Gel Science and Technology*, 2(1-3) pp. 81-86.

Mackenzie, J.D. & Bescher, E. (1998). Structures, properties and potential applications of Ormosils. *Journal of Sol-Gel Science and Technology*, 13(1-3) pp. 371-377.

Mackenzie, J.D. & Bescher, E. (2000). Physical properties of sol-gel coatings. *Journal of Sol-Gel Science and Technology*, 19(1-3) pp. 23-29.

Mackenzie, J.D. & Bescher, E. (2003). Some factors governing the coating of organic polymers by sol-gel derived hybrid materials. *Journal of Sol-Gel Science and Technology*, 27(1) pp. 7-14.

Malzbender, J., den Toonder, J.M.J., Balkenende, A.R. & de With, G. (2002). Measuring mechanical properties of coatings: a methodology applied to nano-particle-filled sol-gel coatings on glass. *Materials Science & Engineering R-Reports*, 36(2-3) pp. 47-103.

Mammeri, F., Le Bourhis, E., Rozes, L. & Sanchez, C. (2005). Mechanical properties of hybrid organic-inorganic materials. *Journal of Materials Chemistry*, 15(35-36) pp. 3787-3811.

Metroke, T.L., Parkhill, R.L. & . Knobbe, E.T. (2001). Passivation of metal alloys using sol-gel derived materials – a review. *Progress in Organic Coatings,* 41 pp. 233-238.

Metroke, T.L. Kachurina, O. & Knobbe, E.T. (2002). Spectroscopic and corrosion resistance characterization of GLYMO-TEOS ormosil coatings for aluminum alloy corrosion inhibition. *Progress in Organic Coatings,* 44(4) pp. 295-305.

Oliver, W.C. & Pharr, G.M. (1992). An improved technique for determining hardness and elastic modulus using load and displacement sensing indentation experiments. *Journal of Materials Research,* 7(6) pp. 1564-1573.

Oyen, M.L. (2005). Spherical indentation following ramp loading. *Journal Of Materials Research,* 20(8) pp. 2094-2100.

Oyen, M.L. (2006). Analytical techniques for indentation of viscoelastic materials. *Philosophical Magazine,* 86(33-35) pp. 5625-5641.

Roest, R., Latella, B.A., Heness, G. & Ben-Nissan, B. (2011). Adhesion of sol-gel derived hydroxyapatite nanocoatings on anodised pure titanium and titanium (Ti6Al4V) alloy substrates. *Surface and Coatings Technology,* 205(11) pp. 3520-3529.

Samson, F. (1996). Ophthalmic lens coatings. *Surface and Coatings Technology,* 81(1) pp. 79-86.

Sanchez, C., Julian, B., Belleville, P. & Popall, M. (2005). Applications of hybrid organic-inorganic nanocomposites. *Journal of Materials Chemistry,* 15(35-36) pp. 3559-3592.

Scafidi, P. & Ignat, M. (1998). Cracking and loss of adhesion of Si_3N_4 and SiO_2:P films deposited on Al substrates. *Journal of Adhesion Science and Technology,* 12(11) pp. 1219-1242.

Schmidt, H. (1985). New type of non-crystalline solids between inorganic and organic materials. *Journal of Non-Crystalline Solids,* 73 pp. 681.

Schottner, G. (2001). Hybrid sol-gel-derived polymers: Applications of multifunctional materials. *Chemistry of Materials,* 13(10) pp. 3422-3435.

Volinsky, A., Vella, J.B. & Gerberich, W.W. (2003). Fracture toughness, adhesion and mechanical properties of low-k dielectric thin films measured by nanoindentation. *Thin Solid Films,* 429(1-2) pp. 201-210.

Wang, J.S., Sugimura, Y., Evans, A.G. & Tredway, W.K. (1998). The mechanical performance of DLC films on steel substrates. *Thin Solid Films,* 325(1-2) pp. 163-174.

Wen, J. & Wilkes, G. (1996). Organic/inorganic hybrid network materials by the sol-gel approach. *Chemistry of Materials,* 8(8) pp. 1667-1681.

Chapter 7

Nanoindentation as a Tool to Clarify the Mechanism Causing Variable Stiffness of a Silane Layer on Diamond

Ksenia Shcherbakova, Akiko Hatakeyama,
Yosuke Amemiya and Nobuo Shimamoto

Additional information is available at the end of the chapter

http://dx.doi.org/10.5772/50545

1. Introduction

For utilizing biological molecules as nanomachines, it is usually necessary to immobilize them on a solid surface of abiological materials. Covalent modification or physical adsorption is used for retaining the target molecules on the solid surface. The following four properties are required for successful immobilization: 1) the surface enables the immobilization with high enough surface density of the target molecules to exert their functions, unless an overly high density prevents interaction as in the case of hybridization; 2) such exerted functions are not significantly perturbed by non-target substances adsorbed on the same surface during use; 3) the surface is suitable for stable immobilization of many kinds of biological materials with high enough mechanical stiffness; 4) the perturbation caused by immobilization does not alter the functions of the immobilized molecules. Immobilization inevitably diminishes the accessibility to the immobilized molecules and changes the dynamic structure of water molecules surrounding the target biomolecules, affecting their structure and conformation as well as their interactions with other substances. Soluble proteins are generally much softer than structured nucleic acids, and thus some are easily flattened in the close vicinity of a hydrophobic surface due to the surface tension. In contrast to proteins, small molecules and structured nucleic acids tend to maintain their native structures, but the limitation of accessibility becomes significant in the close vicinity of the surface. The most problematic consequence of immobilization is the alteration of the function, while the loss of function, to a certain extent, is usually tolerable as long as a significant fraction of homogeneously active molecules is still present on the surface. Therefore, a direct fixing on a solid support often leads to denaturation of proteins due to the surface tension. To avoid these detrimental perturbations, linkers and porous intervening layers may be introduced onto the solid support to buffer the surface effects.

Here, we present an example of such an intervening layer, namely an aminosilane layer formed on a diamond surface. One of the useful characteristics of this layer is that its stiffness can be controlled by changing the solvent used for the deposition. To clarify the mechanism causing the varying stiffness, we used Atomic Force Microscopy (AFM) nanoindentation. Our method does not require sophisticated equipment and is suitable for a typical wet biochemical lab, which enables the immobilization to be easily performed in the same lab where biological material is purified and used.

2.1. Functionalization of diamond surface

The material for immobilization of biomolecules on its surface is selected according to its properties, such as mechanical, chemical, electrical, and optical properties, as well as availability. Diamond is at present drawing attention as a material for biological applications because of its hardness, absence of toxicity, and potential conductivity. It is now used for electrodes and biosensors, DNA and protein chips, and for coating of implants [1-3]. Though it is known to be chemically inert, there are many approaches, which allow modification of its surface [2,4,5]. The effects of immobilizing biomolecules on their activities have also been discussed [6,7].

Diamond has always been an expensive material, but the chemical vapor deposition (CVD) method, developed several decades ago for the synthesis of nano- and polycrystalline diamond films on silicon and other substrates on a large scale and at a reasonable cost, increased the commercial availability of diamond and facilitated its use in research and development [2]. Nonetheless, diamond surface modification often requires sophisticated equipment and facilities.

To introduce any chemical groups onto the diamond surface, at first, the entire surface must be either hydrogenated or oxidized, hereafter H- or O-terminated. The H_2 plasma treatment, used in the CVD method in a diamond film synthesis, automatically results in H-termination [5]. According to our observation, the water contact angle on an H-terminated surface gradually decreased at room temperature, probably because of the local oxidation to generate an inhomogeneous surface, or because of the contamination of the surface. The uniform oxidation of the diamond surface can be achieved by various processes, such as anodic polarization [8], treatment with oxygen plasma [9], treatment with thermally activated oxygen [10], UV irradiation in the presence of ozone [11], or treatment with acids [12-14].

An H-terminated surface can be directly functionalized, either photochemically by UV irradiation in halogen gases or alkenes [15-17], or chemically by thermal decomposition of benzoyl peroxide, activated by argon gas [18], or by electrochemical diazonium salts reduction in an inert atmosphere [19,20]. In contrast, an O-terminated surface can be chemically modified by silanization or by esterification [4], or, as recently reported, aminated by NH_3-plazma treatment [1].

Biological molecules can be covalently immobilized on a diamond surface either directly, or indirectly through an intervening layer. The former may be suitable for structured nucleic acids, if their lying down in parallel to the surface is prevented. Alternatively, their unstructured part can play as a linker, increasing the accessibility of the rest of the part of the molecule. In contrast, direct attachment is generally unadvisable for proteins, especially enzymes, which tend to lose their activities due to deformation caused by immobilization, though it has been actualized for particular proteins in several studies [1,21]. Diamond may be more biocompatible for direct protein immobilization than, for example, metal surfaces [22]. However, from our experience of single-molecule dynamics [23,24], as well as that of others' [10,25-30], we still recommend the use of an intervening layer for preserving enzymatic activities. Polymer layers have a distinct advantage compared with flat solid surfaces in terms of retaining protein activity. Their porous or mesh-like structure allows an easier access of water to the immobilized biological molecules, maintaining their conformations and accessibilities.

Diamond technology is still immature, and the biologically active interfaces on a diamond have not yet been extensively studied. Several pioneering works, reporting the properties of polymer layers, formed on a diamond surface, describe testing of the layers on H-terminated diamond by scratching with an AFM cantilever [28,29,31,32]. Since different proteins require different treatments for maintaining their activities, an intervening layer needs the adjustment of thickness, roughness, and stiffness. Therefore, control over the properties of the layer is important.

2.2. Deposition of aminosilane on the diamond surface

The most common functionalization used for immobilizing biological molecules is the introduction of amine residues on the surface or the use of an amine-rich intervening layer, which is fixed on the surface. The utility of amine residue arises from its activity as a nucleophile in coupling chemistries under moderate aqueous conditions [29,33]. Silanization is commonly used in a wide variety of both industrial and research-oriented applications as a coupling agent for creating intervening silane layers on different kinds of surfaces. The chemical formula of a typical silane molecule is X3Si-R, where X is the group leaving during its polymerization, and R is the hydrocarbon-containing functional group, which remains after the reaction. A wide range of various silanes, such as aminosilanes (-NH_2), mercaptosilanes (-SH), and glycidoxytrimethoxysilanes (epoxide), is commercially available, allowing different chemical functionalities to be incorporated on the surfaces [34]. Formation of aminosilane layers on a diamond surface has been reported in several studies [10,26,27]. In our work, we used 3-aminopropyltriethoxysilane (APTES), which is one of the commonly used aminosilanes.

Some silanes form uniform monolayers, while APTES tends to form multilayers on various substrates, and the layer's thickness increases during deposition [35,36]. In several studies, APTES monolayers have been reported, including those on GaN [37], silicon oxide [38], and porous silicon [39]. However, it is difficult to distinguish a monolayer from dispersed molecules of APTES deposited in a limited time [40,41]. Thus, in a long enough time period, a multilayer is expected to be formed. Such a multilayer is likely to have a disordered mesh-like structure as a consequence of the flexibility of APTES polymer molecules, which is suitable for protein attachment as it is.

It is generally recognized that aqueous conditions and a sufficient accessibility of water molecules to the protein are important for the functions of many soluble proteins which tend to be adsorbed on hydrophobic surfaces [42]. Therefore, the rough and disordered surface of an APTES multilayer is suitable for protein attachment as it creates an environment of high water accessibility for protein molecules. Using APTES for protein immobilization is also often considered to be a means of introducing linkers, intended to reduce steric hindrance, thus providing a greater freedom of movement to immobilized biological molecules [34]. Thus, the introduction of the APTES multilayer to the diamond surface contributes to keeping immobilized molecules active. The disordered mesh-like structure might not be the best choice in all cases. For example, in microfabrication of arrays for DNA attachment, a regular array could not be prepared with the mesh-like structured APTES, whereas another silane gave much better regularity and spatial resolution [35].

To model the reaction of protein immobilization, we attached biotin and then streptavidin onto the APTES layer (Fig. 1), which has formed on the O-terminated diamond surface by a wet method, which is described later. To enhance specificity of binding, we pretreated the surface with a mixture of polyethylene glycol (PEG) with different chain lengths according to a previously established method for preventing non-specific adsorption [43]. The best specificity of attachment was achieved, when APTES was consequently treated first with a long NHS-PEG-biotin and then with a short NHS-PEG.

We then immobilized fluorescently labeled streptavidin on an APTES layer and measured the density of its immobilization. By the measured density, we estimated the average distance between flanking streptavidin molecules. On the two layers with the highest densities this distance was estimated to be 5.6 and 5.3 nm. Since the size of the streptavidin tetramer, derived from the crystallographic data, was $5.4 \times 5.8 \times 4.8$ nm^3 [44], the streptavidin on these layers was immobilized in a state proximate to the hexagonal closest packing, if streptavidin molecules were assumed to distribute in a flat plane.

There is a strategic problem as to whether one should use the most refined methods or the just-satisfactory methods in an experiment. For the combination of two or more lineages of technologies, we chose the latter strategy, because preparing a new pure protein in the lab already requires a number of techniques. In this way, rather than absolute accuracy of the measurements by sophisticated instruments, we preferred familiar and common technologies giving a sufficient accuracy: we formed an APTES layer in a wet process and performed nanoindentation by AFM. Therefore, the developed method can be performed in a standard biological laboratory where target biomolecules are prepared, and does not require any commercially unavailable reagents or sophisticated equipment.

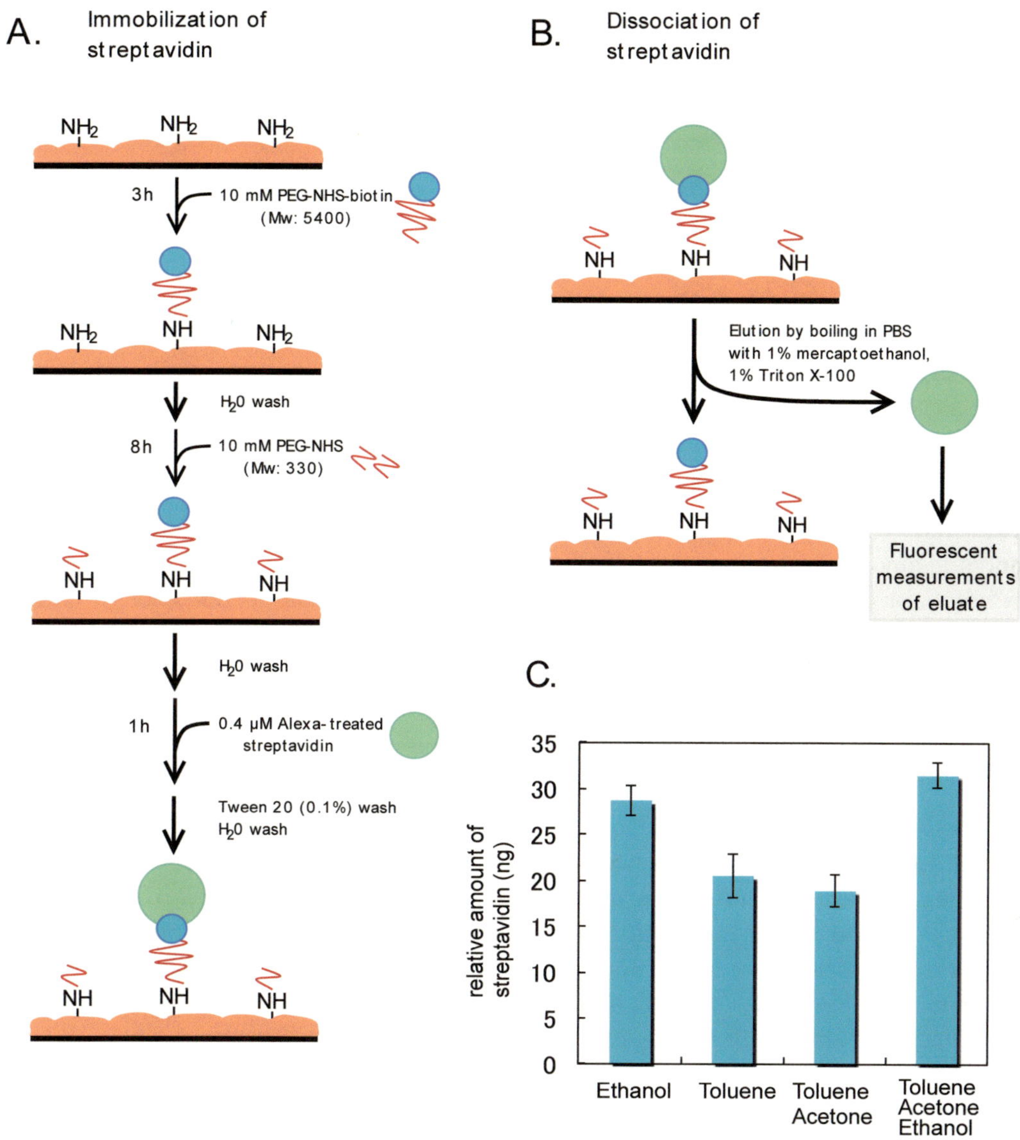

Figure 1. Immobilization of streptavidin on an APTES multilayer. A. The steps of the PEG treatment. B. Elution of streptavidin from the multilayer for measuring the immobilization density. C. The immobilization densities observed for the multilayer deposited in one or more steps in different solvents.

There is a consensus on the basic mechanisms of APTES polymerization. The reaction begins with the hydrolysis of silane molecules, resulting in siloxane bonds forming and attaching APTES molecules to the surface. Hydrolysis may occur both at the surface of the substrate, or in solution, depending on the water concentration. If water is predominantly present in solution, the polymerization will occur predominantly in the solution, resulting in formation of aggregates of APTES, which otherwise form covalent bonds with the substrate [40,41]. Therefore, a distribution of water can determine the size of aggregates and the degree of covalent linking to the surface. In our study, we used several solvents with different isoelectric constants for APTES deposition and found that the layers, formed in different solvents, showed various morphology and nanoscopic hardness, which is discussed below.

2.3. Nanoindentation

Nanoindentation is a powerful and useful method for assessing the mechanical properties of a material. In a typical indentation test, a load is applied to the sample examined with a hard indenter, and the analysis of the load-depth curve and the morphological changes in the indented material allow the measurement of such properties as hardness, Young's modulus, and stress relaxation data [45-47].

Over the recent few decades, AFM, initially invented as an imaging tool, has been extensively used for nanoindentation, especially on soft materials. The conventional nanoindention instruments, which utilize the Oliver and Pharr's procedure [48], are preferentially used for hard materials. However, they cannot offer a wide enough range of loads necessary for soft materials, which must be indented with less force [49]. AFM is a very useful tool to study the properties of biologically relevant materials [50,51]. By using AFM cantilevers as indenters, it is possible to measure nanomechanical properties with high force and depth resolutions. Imaging can be performed with the same tool right after the indentation, without resetting the sample. Furthermore, AFM can detect pile-up or sink-in effects, which conventional indenters cannot [52-54].

Several serious reservations have been asserted regarding the problem of whether or not AFM measurements are sufficiently quantitative in indentation [55]. Because the cantilever's apex is deformed in the indentation process, and may not be exactly vertical, the AFM cantilever requires more corrections of measurements compared with conventional indenters [54]. However, we still hold the view that AFM is usable for our purpose, according to our strategy, by taking into account the distortion of the apex.

2.4. Scratching manipulation and AFM nanoindentation

In our experiments, single-crystalline synthetic (100) type Ib diamond (Sumimoto Electric Industries) was etched with acids to prepare an O-diamond surface and then APTES was deposited. Then we performed the imaging with scratching manipulation [29] on ous APTES layers (Fig. 2), prepared under different deposition conditions. We used AFM (SPI3700, Seiko Instruments) for both topographic imaging and nanoindentation. We calibrated silicon AFM cantilevers (SI-DF-3) with the Cleveland method [56] and selected tips with a spring constant of 1.2-1.4 N/m.

We found that one can control the roughness of the layer by changing the polarity of the solvent. Deposition of APTES in the nonpolar solvents resulted in a rougher surface, and the use of polar solvents resulted in a smoother surface.

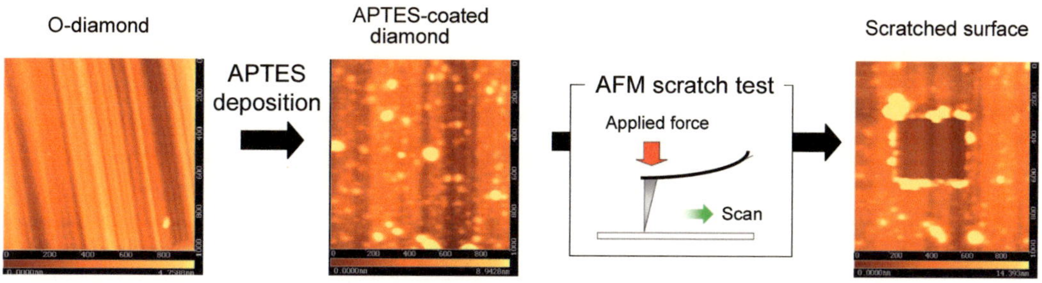

Figure 2. Scratching manipulation (Modified from [57]).

In the scratching manipulation, the stiffness was indexed by the minimum force essential to disrupt the layer and to remove it from the diamond support. The lower the dielectric constant of a solvent, the greater the force had to be applied to remove the layer: only 1 nN for the layers prepared in ethanol mixture or acetone, 100 nN for octanol, and even greater forces for more nonpolar solvents. In this way, we could draw arbitrary nanopatterns composed of lines and squares on softer layers, though we could not scratch the harder layers even at forces greater than 500 nN (Fig. 3). The composed patterns remained stably on the diamond surface. Therefore, both roughness and stiffness increased relative to the decreasing polarity.

It is important to clarify the relationship between scratching, which is terminologically a macroscopic examination, and microscopic nanoindentation. In the macroscopic world, scratching means scraping or removing a part of a substance by disrupting it with a sharp edge, which is a macroscopic process and is on a different level from molecular events. Therefore, the force, essential to break the APTES layer with the edge, is dependent on the sharpness and the shape of the edge, as well as its velocity and moving direction. However, the scratching manipulation in our experiment includes breaking molecular bonds between APTES and the diamond by vertical pressing with the cantilever, which we consider to be close to nanoindentation and a microscopic process. In the microscopic model, the break should be independent of the shape of the apex and dependent solely on the work of the pressing force. In other words, the macroscopic model predicts that the force breaking the layer depends on the sharpness of the edge and, thus, the horizontal movements in case of a pyramidal cantilever. Alternatively, the microscopic model asserts the independence of the cantilever used, its shape, and the horizontal direction of its movement (Fig. 4).

The experimental results demonstrated that an APTES layer could be scratched and removed from the diamond surface with a common pressuring force within one order of magnitude in different examinations and was independent of cantilevers used (Seiko SI-DF-3, SI-DF-20S, and Olympus OMCL-AC240TS-C3), the velocity of scratching, and the horizontal direction of scratching. These results indicate that the microscopic model is more accurate in describing the mechanism of scratching of the APTES layer.

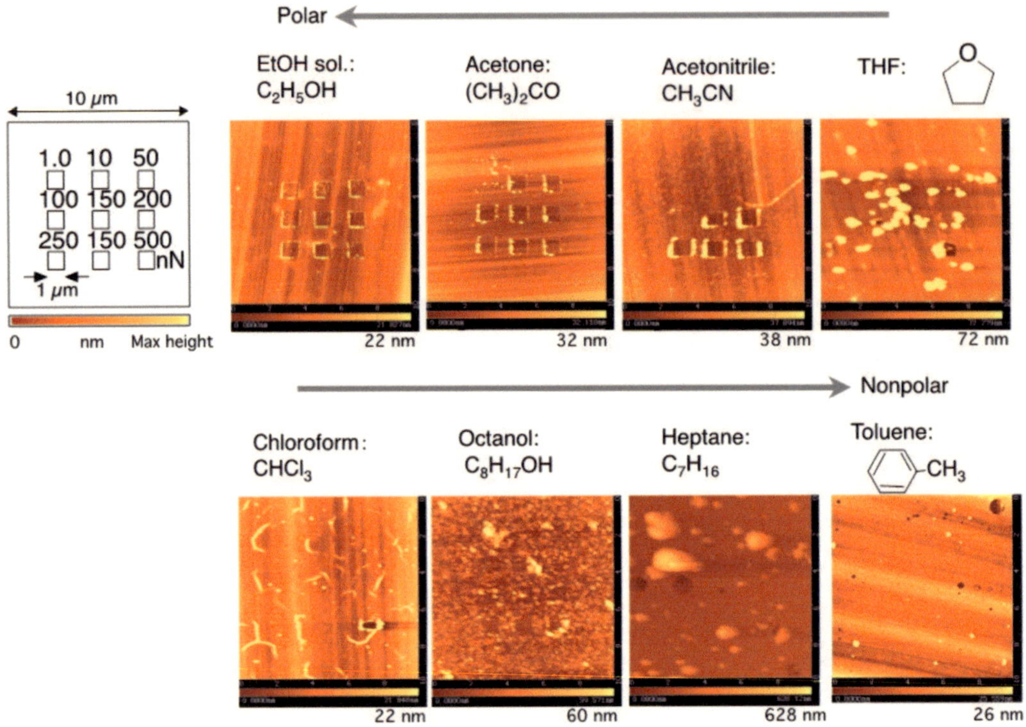

Figure 3. The results of the scratching manipulation of APTES multilayers formed in various solvents (Modified from [57]).

As a control for checking whether or not the result is independent of AFM instruments, we performed a scratching manipulation with our instrument on a layer of ω-unsaturated 10-trifluoroacetic amide-dec-1-ene (TFAAD), photo-chemically attached to the diamond surface, which had been inspected with the same manipulation [29]. The measured force was well aligned with the value (~100 nN) previously obtained. When toluene was used for the APTES deposition, the forces obtained in the scratching manipulation were greater than 500 nN, the upper limit of our instrument. The attachment of APTES to the diamond, therefore, is stronger than that of TFAAD, which is generally recognized to be covalently bound to diamond [29], suggesting that APTES is also covalently bound to the diamond. This circumstantial evidence for the microscopic model led us to compare the results of the scratching manipulation and the AFM indentation. The nanoindentation with no horizontal movements was performed by pressing the APTES layer in the contact mode. The image of the contact area was then obtained in the noncontact mode.

The results of nanoindentation showed an agreement of forces with the scratching manipulation within the same order of magnitude. The layer could not be scratched even at 500 nN, showing no trace after the indentation at 500 nN. Indentation of the softest and smoothest layer, which had been formed in the ethanol mixture, showed a force < 1 nN in the scratching manipulation. This semi-quantitative agreement again suggests that the microscopic model is more appropriate than the macroscopic one. Therefore, the force measured in the indentation test can be used for scratching and nanopatterning of the surface of a material. The indentation test sometimes required pressing for up to 10 min to yield reproducible results. This hysteresis indicates that the APTES layer is elastic in nanoscale when a force is rapidly removed.

2.5. Interpretation of the results

As discussed above, the result of the scratching manipulation suggests the existence of covalent bonds between APTES and diamond, at least for the multilayers deposited in nonpolar solvents. If APTES were physically adsorbed on the surface, the changes of the breaking force of more than 100-fold could not be explained, since more than a 100-fold change of the contacting area is unlikely for such a soft material as APTES polymer which is easily flattened. The likely explanation is the change in the surface density of the covalent bonds between APTES and diamond, although its evidence is required. Since the shape of the apex of the silicone cantilever used in the indentation should be deformed significantly at forces in the order of magnitude of 100 nN as shown in Fig. 4, a detailed analysis of the force curve of AFM indentation is not very informative.

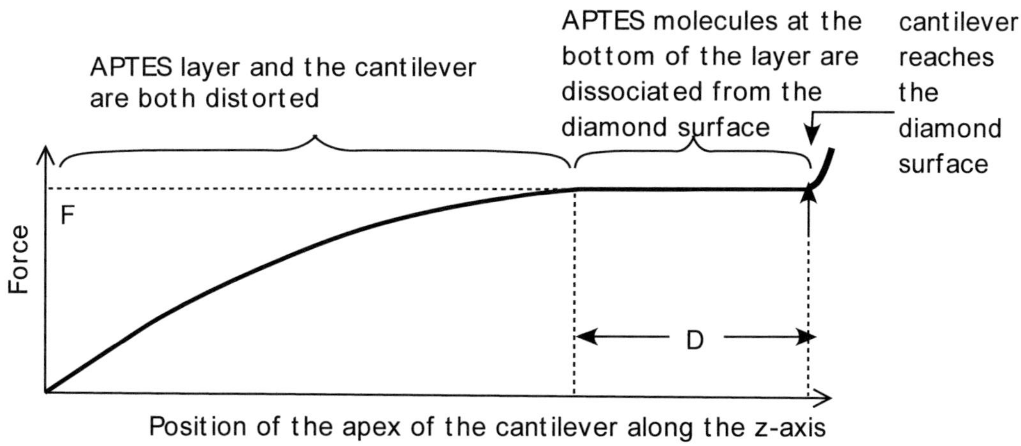

Figure 4. Schematic illustration of a force-position curve during AFM indentation.

However, the following consideration of energetics is still possible. The force giving the plateau (Fig. 4) is about 500 nN for chloroform and greater than 500 nN for heptane and toluene. The distance D is the thickness of the compressed multilayer and is close to 1 nm. This value was obtained from the AFM imaging of the surface with uncoated holes with exposed diamond. During this period when the force is constant, the energy is used for breaking the covalent bonds connecting between APTES and diamond for squeezing the layer out of its original position. Since a Si-O bond is stronger than a C-O bond, the energy is used for breaking the C-O bonds, which costs 6×10^{-19} J/bond [58]. A part of the energy may be used for frictional dissipation and eventual bond breakage within the APTES multilayer, too. Therefore, the work done by the force F is FD, and FD must satisfy Eq. 1, where S is the area of the place where APTES was removed.

$$FD \geq (density\ of\ the\ anchoring\, C-O\ bond) \times S \times (energy\ of\ a\ C-O\ bond) \tag{1}$$

Since F = (strain) x ES

$$(density\ of\ the\ anchoring\, C-O\ bond) \leq \frac{(strain) \times ED}{(energy\ of\ a\ C-O\ bond)} \tag{2}$$

where E is Young's modulus of silicone, 150 Gpa. It is noted that Eq. 2 is independent of F and *S*, which indirectly contribute through the strain.

As illustrated in Fig. 5, the strain is overestimated if the shape of the apex is substituted by an inscribing pyramid with the same base, while the strain is underestimated if it is substituted by another circumscribing pyramid with the same base. Therefore, the two pyramids give the maximum and the minimum values of the strain, respectively. If their values are within the same magnitude, the actual one should also be within it. Therefore, this approximation is sufficient in estimating the order of the strain and the density. Therefore, we consider the distortion of a pyramidal cantilever (Fig. 6).

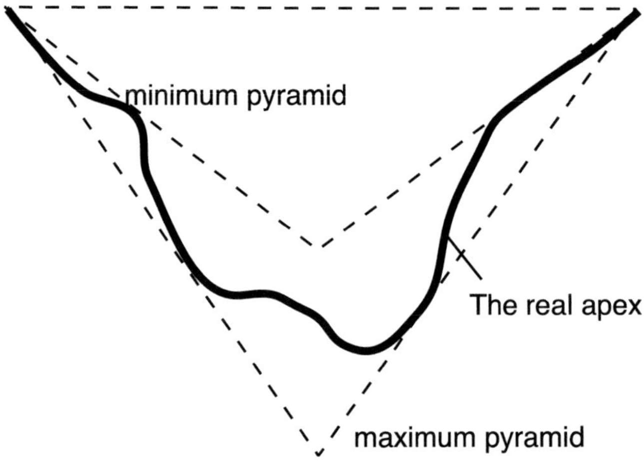

Figure 5. The relationship among the actual shape of the apex, the maximum pyramid, and the minimum pyramid. If the heights of two pyramids agree within the factor of 3.2, the actual value of the strain also agrees within the factor. Because a force is parallel to the square of the height, the factor 3.2 maintains the value of the force or the density within a factor of 10. This condition is likely to be established because of the use of different cantilevers and because of the loss of sharpness of the apexes used repeatedly.

If dF shown in Fig. 6 is integrated from 0 to a,

$$F = 8E\frac{L^2}{l^2}\int_o^a \frac{a-z}{l-z} zdz = 8EL^2\left(\frac{a}{l}\right)^2 = 8EL^2(strain)^2 \tag{3}$$

Because $a \ll l$, and $\int_o^a \frac{a-z}{l-z} zdz = \frac{a^2}{2} + (l-a-b)a + (l-a)(l-b)\ln\left(1-\frac{a}{l}\right)$

The strain obtained for the present parameters (L=4.5 μm, l=10 μm, and F=500 nN) is the order of magnitude of 10^{-4}, leading to the order-estimated maximum value of the density to be 3 C-O bonds in a square of 10 nm x 10 nm. Since there are about 10 carbon atoms per $(nm)^2$ of (100) plane, APTES is covalently linked to a maximum 0.3 % of the surface carbon. This value is for the APTES multilayer deposited in chloroform, while the one in toluene is more densely bound to the diamond surface.

These sparse covalent bonds on diamond can be explained by the sparseness of the reactive carbon of the diamond surface. The diamond surface used in this experiment is (100) crystal plane, and the carbon atoms on (100) plane have two dangling bonds. On the O-terminated surface, they form bridge-bonded ether bonds with their nearest surface carbon atoms, or become carbonyl carbon atoms [10]. The reactive carbon atom must have a single dangling bond, which is converted into the hydroxyl form during the O-termination [27,59]. Such carbon atoms will exist at defects of (100) plane, the edge of steps between two (100) planes, for example. The steps can be seen in Fig. 2 as stripes on the surface of uncoated O-diamond. Since APTES is known to react with a hydroxyl residue [40], a covalent bond between APTES and the O-diamond surface is limited to the carbon atoms with single dangling bonds.

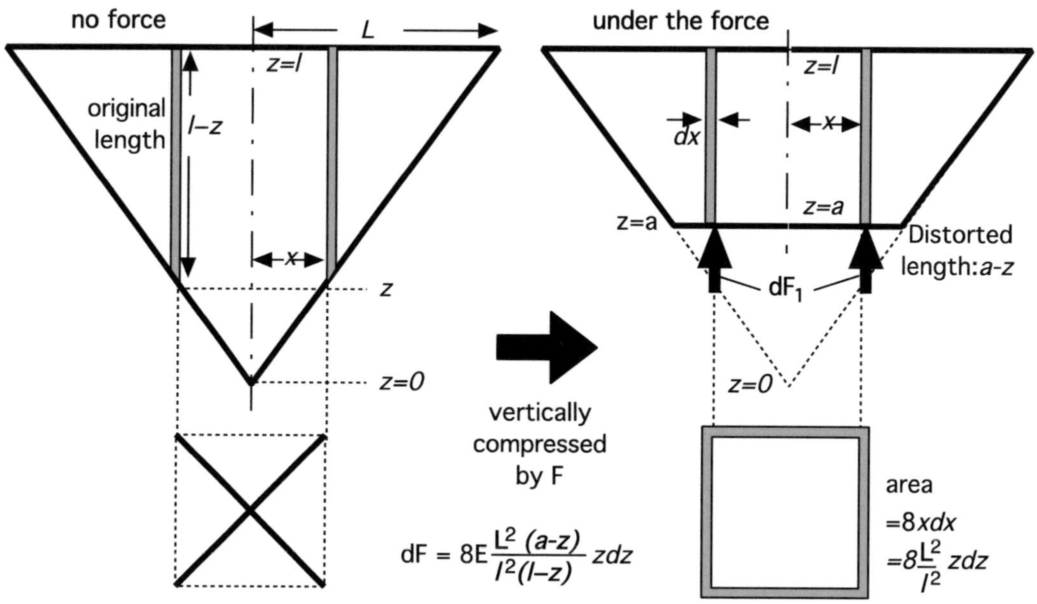

Figure 6. Pyramidal approximation of the apex of the cantilever.

In the above consideration, we assumed that the effect of different shapes of apexes is smaller than an order of magnitude. Such a difference was reported to be 2-3 fold in the scratching manipulation using dull apexes at 10-40 nN [62]. Since the effect becomes smaller at a larger force, the assumption is likely to be reasonable at a force larger than 100 nN. The estimation by an order of magnitude, therefore, is a productive criterion almost independent of the shape of the apex.

In our consideration, we took into account only C-O bond between diamond and APTES, but not the Si-O bonds crosslinking between APTES molecules. The contribution of the Si-O bonds could be significant in the scratching manipulation but not in the nanoindentation, because scratching may rip the APTES layer. However, the work consumed in ripping the layer must be done by a horizontal force, which is the driving force of the AFM stage, and would not be included in the vertical force required. Actually, the results of the scratching manipulation and the nanoindentation agreed with each other at the accuracy of an order of magnitude. Furthermore, we observed that a similar ripping happens also after the measurement of nanoindentation, namely when the force applied to the cantilever is removed. As Fig. 7 shows, a hole much larger than the apex of the cantilever was formed on the APTES layer deposited in ethanol. The APTES layer, which had been covering the hole, was ripped from the surface and remained around the hole. In conclusion, these results indicate that the observed variation of the stiffness is due to the various densities of the covalent bonds between APTES and diamond.

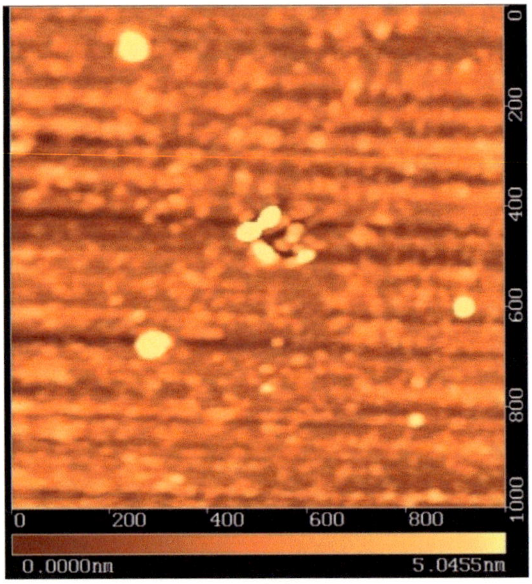

Figure 7. The trace after AFM nanoindentation of a multilayer deposited in ethanol at 500 nM. The trace is as great as 50 nm, which is much greater than the apex of the cantilever, ~ 10 nm.

2.6. Mechanism yielding the dependence of the stiffness on solvent polarity

The last problem to be addressed is why the density of the covalent bonds varies by 100-fold depending on the solvent used in the deposition. As already mentioned above, the polymerization of the APTES layer proceeds by reacting with water, indicating that there are two places for the polymerization as shown in Fig. 8: in the solvent and on the surface where water molecules are trapped. In a polar solvent like ethanol, water is homogeneously dispersed in the solvent but there is less water distributed on the surface. Therefore, homogeneous polymerization occurs in the bulk, while infrequent polymerization occurs on the surface. Thus, the size of aggregates is uniform, which makes a smooth surface of the multilayer, but it is sparsely bound to the diamond surface, making the stiffness less. In contrast, in a nonpolar solvent, water content is small and with water tending to exist as water clusters or small droplets. Since the C-OH is more hydrophilic than the solvent, it holds a water droplet. Therefore, APTES aggregates are much fewer, and their size is inhomogeneous, making a rough surface of the multilayer. Instead, more C-OH carbons are covalently bound to APTES, making the stiffness greater. In this case, a longer period would be required for deposition. In fact, the deposition in ethanol-water mixture lasts for 1 h, while the deposition in toluene requires longer than 10 h to cover the surface. This proposed mechanism [57] explains not only the variation of the stiffness but also the roughness and the required time period for deposition.

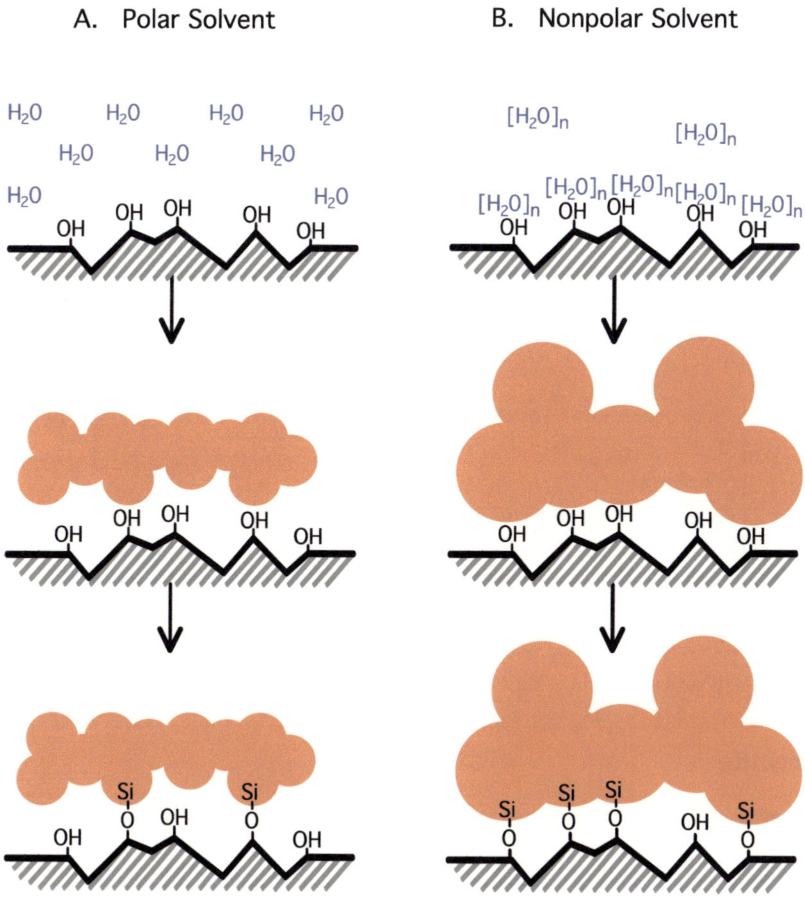

Figure 8. Mechanism yielding the dependence of stiffness on the solvent polarity.

Thus, when a stiff multilayer is required, we must deposit APTES in a nonpolar solvent, a time-consuming process. The proposed mechanism suggests that the time required will be shortened if we increase the number of water droplets in a nonpolar solvent. Therefore, we used water-saturated toluene, which may be a rare treatment of a nonpolar solvent in chemistry. In this way, we accelerated the deposition in toluene from overnight to an hour. This success supports our proposed model for APTES deposition. If the second deposition is made in ethanol-water, a smoother multilayer is obtained.

2.7. Efficient deposition of APTES multilayer with a great stiffness and on diamond

According to the described mechanism, a method for preparing a stiff APTES multilayer on a diamond surface has been developed. The 2-step deposition in toluene and ethanol has been repeated in quick steps, followed by baking to remove the water remaining inside the layer after each step. This procedure has yielded the best results so far of any tried in our lab in terms of stiffness, which is essential for long-term stability in practical use (Fig. 9).

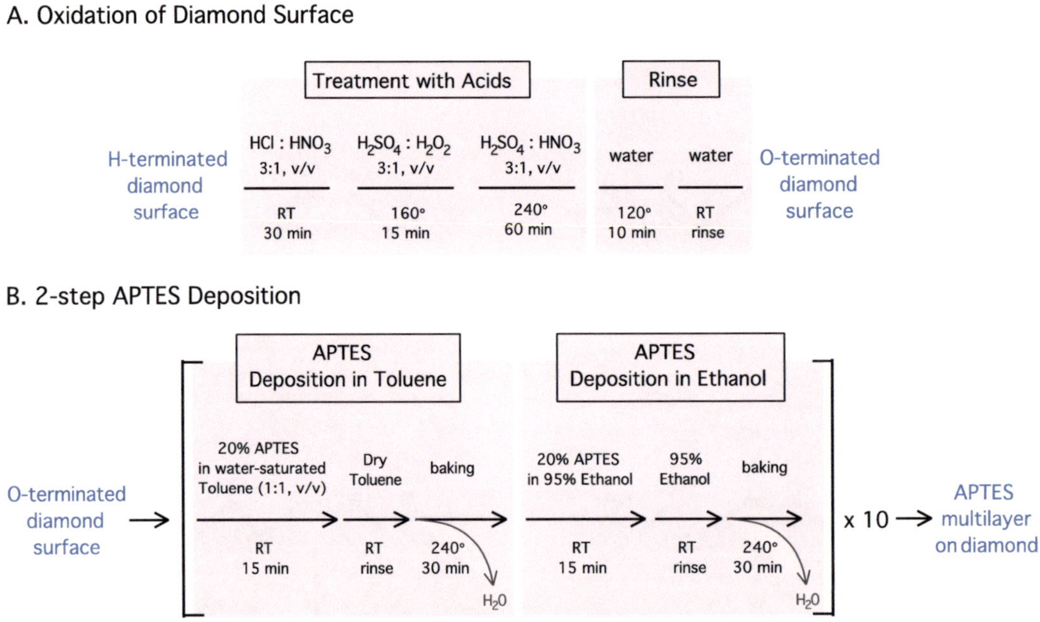

Figure 9. An efficient procedure forming a stiff APTES multilayer on diamond.

3. Conclusions

We have developed a new method for forming a bioactive APTES multilayer on an O-diamond surface with a controlled stiffness, roughness, and capacity for immobilizing streptavidin. This method does not require sophisticated machinery or expensive materials, and thus, it can be adopted for use in many biological laboratories where proteins and nucleic acids are prepared and immobilized.

Using AFM nanoindentation, the scratching manipulation with AFM, and AFM imaging, we examined the stiffness and morphology of the APTES multilayers formed on a CVD diamond. The results of nanoindentation and the scratching manipulation were interpreted semiquantitatively rather than fully quantitatively. This interpretation enables us to propose a model which explains the various levels of stiffness of APTES multilayers, its correlation with polarity of the solvent, and the time required for covering the surface with an APTES multilayer. The model also allows us to improve the method to make a stiff and smooth APTES surface.

Author details

Ksenia Shcherbakova
Faculty of Life Sciences, Kyoto Sangyo University, Kyoto, Japan

Akiko Hatakeyama and Nobuo Shimamoto
Faculty of Life Sciences, Kyoto Sangyo University, Kyoto, Japan
Structural Biology Center, National Institute of Genetics, Mishima, Japan

Yosuke Amemiya
Structural Biology Center, National Institute of Genetics, Mishima, Japan

Acknowledgement

The authors thank Dr. Christoph E. Nebel and Dr. Hiroshi Uetsuka, for the TFAAD-attached diamond, and Ms. Harriet Sallach and Ms. Sarah Stovalosky for reading the manuscript prior to publication.

4. References

[1] Wang Q, Kromka A, Houdkova J, Babchenko O, Rezek B, Li M, Boukherroub R, Szunerits S. Nanomolar Hydrogen Peroxide Detection Using Horseradish Peroxidase Covalently Linked to Undoped Nanocrystalline Diamond Surfaces. Langmuir 2012;28: 587–592.

[2] Nebel CE, Shin D, Rezek B, Tokuda N, Uetsuka H, Watanabe H. Diamond and biology. J. R. Soc. Interface. 2001;4: 439-461.

[3] Tryk DA, Tachibana H, Inoue H, Fujishima A. Boron-doped diamond electrodes: The role of surface termination in the oxidation of dopamine and ascorbic acid. Diamond Relat. Mater. 2007;16: 881–887.

[4] Szunerits S, Boukherroub R. Different strategies for functionalization of diamond surfaces. J. Solid State Electrochem. 2008;12: 1205–1218.

[5] Luong JHT, Maleb KB, Glennona JD. Boron-doped diamond electrode: synthesis, characterization, functionalization and analytical applications. Analyst 2009;134: 1965–1979.

[6] Rezek B, Michalıkova L, Ukraintsev E, Kromka A, Kalbacova M. Micro-Pattern Guided Ahesion of Osteoblasts on Diamond Surfaces. Sensors 2009;9: 3549-3562

[7] Hoffmann R, Kriele A, Obloh H, Tokuda N, Smirnov W, Yang N, Nebel CE. The creation of a biomimetic interface between boron-doped diamond and immobilized proteins. Biomaterials 2011;32: 7325-7332.

[8] Notsu H, Yagi I, Tatsuma T, Tryk DA, Fujishima A. Introduction of Oxygen-Containing Functional Groups onto Diamond Electrode Surfaces by Oxygen Plasma and Anodic Polarization. Electrochem. Solid-State Lett. 1999;2: 522–524.

[9] Yagi I, Notsu H, Kondo T, Tryk DA, Fujishima AJ. Electrochemical selectivity for redox systems at oxygen-terminated diamond electrodes. Electroanal. Chem. 1999;473: 173–178.
[10] Pehrsson PE, Mercer TW. Oxidation of the hydrogenated diamond (100) surface. Surface Science 2000;460: 49–66.
[11] Zhang G, Umezawa H, Hata H, Zako T, Funatsu T, Ohdomari I, Kawarada H. Micropatterning oligonucleotides on single-crystal diamond surface by photolithography. Jpn. J. Appl. Phys. 2005;44: L295-L298.
[12] Kawarada H. Hydrogen-terminated diamond surfaces and interfaces. Surf. Sci. Rep. 1996;26: 205-259.
[13] Ri SG, Nebel CE, Takeuchi D, Rezek B, Tokuda N, Yamasaki S, Okushi H. Surface conductive layers on (111) diamonds after oxygen treatments. Diamond. Relat. Mater. 2006;15: 692-697.
[14] Ushizawa K, Sato Y, Mitsumori T, Machinami T, Ueda T, Ando T. Covalent immobilization of DNA on diamond and its verification by diffuse reflectance infrared spectroscopy. Chem. Phys. Lett. 2002;351: 105-108.
[15] Miller JB. Amines and thiols on diamond surfaces. Surface Science 1999;439: 21–33.
[16] Ohtani B, Kim YH, Yano T, Hashimoto K, Fujishima A, Uosaki K. Surface Functionalization of Doped CVD Diamond via Covalent Bond. An XPS Study on the Formation of Surface-bound Quaternary Pyridinium Salt. Chemistry Letters 1998;27: 953-954.
[17] Strother T, Knickerbocker T, Russell JN, Butler JE, Smith LM, Hamers RJ. Photochemical Functionalization of Diamond Films. Langmuir 2002;18: 968–971.
[18] Tsubota T, Tanii S, Ida S, Nagata M, Matsumoto Y. Chemical modification of diamond surface with $CH_3(CH_2)_nCOOH$ using benzoyl peroxide. Physical Chemistry Chemical Physics 2003;5: 1474-1480.
[19] Kuo TC, McCreery RL, Swain GM. Electrochemical Modification of Boron-Doped Chemical Vapor Deposited Diamond Surfaces with Covalently Bonded Monolayers. Electrochem. Solid-State Lett. 1999;2: 288-290.
[20] Wang J, Firestone MA, Auciello O, Carlisle JA. Surface Functionalization of Ultrananocrystalline Diamond Films by Electrochemical Reduction of Aryldiazonium Salts. Langmuir 2004;20: 11450–11456.
[21] Härtl A, Schmich E, Garrido JA, Hernando J, Catharino SCR, Walter S, Feulner P, Kromka A, Steinmueller D, Stutzmann M. Protein-modified nanocrystalline diamond thin films for biosensor applications. M. Nat. Mater. 2004;3: 736-742.
[22]Iniesta J, Esclapez-Vicente MD, Heptinstall J, Walton DJ, Peterson IR, Mikhailov VA, Cooper HJ. Retention of enzyme activity with a boron-doped diamond electrode in the electro-oxidative nitration of lysozyme. Enzyme and Microbial Technology 2010;46: 472-478.
[23] Kabata H, Kurosawa O, Arai I, Washizu M, Margarson SA, Glass RE, Shimamoto N. Visualization of single molecules of RNA polymerase sliding along DNA. Science 1993;262: 1561-1563.
[24] Sakata-Sogawa K, Shimamoto N. RNA polymerase can track a DNA groove during promoter search. PNAS 2004;101: 14731–14735.
[25] Matrab T, Chehimi MM, Boudou JP, Benedic F, Wang J, Naguib NN, Carlisle JA. Surface functionalization of ultrananocrystalline diamond using atom transfer radical

polymerization (ATRP) initiated by electro-grafted aryldiazonium salts. Diamond. Related Materials 2006;15: 639–644.

[26] Hernando J, Pourrostami T, Garrido LA, Williams OA, Gruen DM, Kromka A, Steinmüller D, Stutzmann M. Immobilization of horseradish peroxidase via an amino silane on oxidized ultrananocrystalline diamond. Diamond. Relat. Mater. 2007;16: 138–143.

[27] Notsu H, Fukazawa T, Tatsuma T, Tryk DA, Fujishima A. Hydroxyl Groups on Boron-Doped Diamond Electrodes and Their Modification with a Silane Coupling Agent. Electrochem. Solid-State Lett. 2001;4: H1-H3.

[28] Uetsuka H, Shin D, Tokuda N, Saeki K, Nebel CE. Electrochemical grafting of boron-doped single-crystalline chemical vapor deposition diamond with nitrophenyl molecules. Langmuir 2007;23: 3466–3472.

[29] Yang N, Uetsuka H, Watanabe H, Nakamura T, Nebel CE. Photochemical amine layer formation on H-terminated single-crystalline, CVD diamond. Chem. Mater. 2007;19: 2852–2859.

[30] Lasseter TL, Clare BH, Abbott NL, Hamers RJ (2004) Covalently Modified Silicon and Diamond Surfaces: Resistance to Nonspecific Protein Adsorption and Optimization for Biosensing. J. Am. Chem. Soc. 2004;126: 10220-10221.

[31] Rezek B, Shin D, Nakamura T, Nebel CE. Geometric Properties of Covalently Bonded DNA on Single-Crystalline Diamond. J. Am. Chem. Soc. 2006;128: 3884-3885.

[32] Rezek B, Shin D, Uetsuka H, Nebel CE. Microscopic diagnostics of DNA molecules on mono-crystalline diamond. Phys. Stat. Sol. (a) 2007;204(9): 2888 –2897.

[33] Wong SS. Chemistry of Protein Conjugation and Cross-Linking (1st Ed.). Boca Raton, FL: CRC Press; 1993

[34] Davis DH, Giannoulisa CS, Johnson RW, Desai TA. Immobilization of RGD to <111> silicon surfaces for enhanced cell adhesion and proliferation. Biomaterials 2002;23: 4019–4027.

[35] Moon JH, Shin JW, Kim SY, Park JW. Formation of Uniform Aminosilane Thin Layers: An Imine Formation To Measure Relative Surface Density of the Amine Group. Langmuir 1996;12: 4621-4624.

[36]Zhang G, Tanii T, Miyake T, Funatsu T, Ohdomari I. Attachment of DNA to microfabricated arrays with self-assembled monolayer. Thin Solid Films 2004;464-465: 452-455.

[37] Baur B, Steinhoff G, Hernando J, Purrucker O, Tanaka M. Chemical functionalization of GaN and AlN surfaces. Appl. Phys. Lett. 2005;87: 263901-1-3.

[38] Zheng J, Zhu Z, Chen H, Liu Z. Nanopatterned Assembling of Colloidal Gold Nanoparticles on Silicon. Langmuir 2000;16: 4409–4412.

[39] Ouyang H, Striemer CC, Fauchet FM. Quantitative analysis of the sensitivity of porous silicon optical biosensors. Appl. Phys. Lett. 2006;88: 163108-1-3.

[40] Vandenberg ET, Bertilsson L, Liedberg B, Uvdal K, Erlandsson R, Elwing H, Lundström IJ. Structure of 3-aminopropyl triethoxy silane on silicon oxide. Colloid Interface Sci. 1993;147: 103–118.

[41] Howarter JA, Youngblood JP. Optimization of Silica Silanization by 3-Aminopropyltriethoxysilane. Langmuir 2006;22: 11142-11147.

[42] Sigal GB, Mrksich M, Whitesides GM. Effect of Surface Wettability on the Adsorption of Proteins and Detergents. J. Am. Chem. Soc. 1998;120: 3464-3473.

[43] Uchida K, Otsuka H, Kaneko M, Kataoka K, Nagasaki Y. A reactive poly(ethylene glycol) layer to achieve specific surface plasmon resonance sensing with a high S/N ratio: the substantial role of a short underbrushed PEG layer in minimizing nonspecific adsorption. Anal. Chem. 2005;77: 1075–1080.

[44] Hendrickson WA, Pahler A, Smith JL, Satow Y, Merritt EA, Phizackerley RP. Crystal structure of core streptavidin determined from multiwavelength anomalous diffraction of synchrotron radiation. PNAS 1989;86: 2190-2194.

[45]Oliver WC, Pharr GM. Measurement of hardness and elastic modulus by instrumented indentation: advances in understanding and refinements to methodology. J. Mat. Res. 2004;19: 3–20.

[46] Ahearne M, Yang Y, Liu KK. Mechanical characterization of hydrogels for tissue engineering applications. In: Topics in Tissue Engineering. Available from http://www.oulu.fi/spareparts/ebook_topics_in_t_e_vol4/abstracts/ahearne.pdf

[47] Kurland NE, Drira Z, Yadavalli VK. Measurement of nanomechanical properties of biomolecules using atomic force microscopy. Micron 2012;43: 116-128.

[48] Oliver WC, Pharr GM. An improved technique for determining hardness and elastic modulus using load and displacement sensing indentation experiments. J. Mater. Res. 1992;7: 1564–83.

[49] Jee AY, Lee M. Comparative analysis on the nanoindentation of polymers using atomic force microscopy. Polymer Testing 2010;29: 95–99.

[50] Bowen WR, Lovitt RW, Wright CJ. Application of atomic force microscopy to the study of micromechanical properties of biological materials. Biotechnology Letters 2000;22: 893–903.

[51] Ebenstein DM, Pruitt LA. Nanoindentation of biological materials. Nano Today 2006; 1: 26–33.

[52] Chowdhury S, Laugier MT. Non-contact AFM with a nanoindentation technique for measuring the mechanical properties of thin films. Nanotechnology 2004;15: 1017–1022.

[53] Beegan D, Chowdhury S, Laugier MT. A nanoindentation study of copper films on oxidised silicon substrates. Surface and Coatings Technology 2003;176: 124-130.

[54] Clifford CA, Seah MP. Quantification issues in the identification of nanoscale regions of homopolymers using modulus measurement via AFM nanoindentation. Applied Surface Science 2005;252: 1915–1933.

[55] Tranchida D, Piccarolo S, Loos J, Alexeev A. Accurately evaluating Young's modulus of polymers through nanoindentations: A phenomenological correction factor to the Oliver and Pharr procedure. Appl. Phys. Lett. 2005;89: 171905-171905-3.

[56] Cleveland JP, Manne S, Bocek D, Hansma PK. A nondestructive method for determining the spring constant of cantilevers for scanning force microscopy. ReV. Sci. Instrum. 1993;64: 403–405.

[57] Amemiya Y, Hatakeyama A, Shimamoto N. Aminosilane multilayer formed on a single-crystalline diamond surface with controlled nanoscopic hardness and bioactivity by a wet process. Langmuir 2008;25: 203-209.

[58]Atkins P, de Paula J. Atkin's physiscal chemistry (7th ed). NY: Oxford University Press; 2002.

[59] Notsu H, Yagi I, Tatsuma T, Tryk DA, Fujishima AJ. Surface carbonyl groups on oxidized diamond electrodes. Electroanal. Chem. 2000;492: 31–37.

Chapter 8

Mechanical and Tribological Properties of Plasma Deposited a-C:H:Si:O Films

Bruno B. Lopes, Rita C.C. Rangel, César A. Antonio,
Steven F. Durrant, Nilson C. Cruz, Elidiane C. Rangel

Additional information is available at the end of the chapter

http://dx.doi.org/10.5772/50278

1. Introduction

Plasma deposited organosilicon films (a-C:H:Si:O) have attracted increasingly attention due to superior properties such as good adhesion (Morent *et al.*, 2009), optical transparency (Zajíčková et al., 2001), corrosion resistance (Rangel et al., 2012) and selective permeability (Czeremuszkin et al., 2001). Moreover, highly hydrophilic to extremely hydrophobic films can be prepared by properly adjusting the plasma excitation conditions (Schwarz et al., 1998). Such selectable wettability combined with the optical transparency and chemical inertness make these films potential candidates for a series of applications including haemocompatible (Ong et al., 2007), biocompatible (Ong et al., 2008), and corrosion protective (Fracassi et al., 2003) coatings, fire retardants (Quédé et al., 2002), diffusion barriers (Görbig et al., 1998), adherent (Pihan et al., 2009) and non-adherent surfaces (Navabpour et al., 2010), and dielectric layers (Borvon *et al.*, 2002). Although the literature is abundant in reports of improvements upon the application of a-C:H:Si:O films, only a few studies dealt with the mechanical and tribological properties of such coatings, characteristics of fundamental relevance to the durability of the coating. In this context, this article describes an investigation of the mechanical and tribological properties of silicon containing organic coatings deposited by Plasma Enhanced Chemical Vapor Deposition from hexamethyldisiloxane, HMDSO, and oxygen mixtures. The effect of the dilution of HMDSO in oxygen on the hardness and elastic modulus was accessed by nanoindentation. Friction coefficient, scratching and wear resistances of the films were also evaluated via scratching tests performed with the same instrument used for nanoindentation. Topography and morphology of the films as well as their average roughnesses were determined from scanning probe microscopy images.

2. Experimental details

Figure 1 illustrates the experimental setup employed in the treatments. It consists of a cylindrical glass chamber, of 190 mm internal diameter and 250 mm high, sealed by aluminum plates and fitted with two parallel electrodes. Prior to the depositions glass substrates were sonicated in detergent solution and acetone baths and then dried using a hot air gun. Subsequently, they were sputter-cleaned for 600 s in plasmas established in atmospheres of 50% Ar /50% H_2 at a total pressure of 1.33 Pa. The sputtering plasmas were produced by the application of radiofrequency power, RF, (13.56 MHz, 150 W) to the lower electrode which also serves as substrate holder while the steel mesh used as the upper electrode was grounded. The depositions were performed for 3.600 s in atmospheres of pure HMDSO or 30% of HMDSO and 70% of O_2 by applying RF (13.56 MHz, 200 W) to the substrate holder. In all the experiments, the total gas pressure was kept at 20 Pa.

Film thickness was determined from film step heights, which were delineated on glass surfaces using a mask of Kapton tape during the depositions. The height of the steps was measured in three different regions using a Veeco Dektak 150 profilometer. Chemical composition and structure of films deposited onto aluminum-coated glass slides were investigated by infrared reflectance-absorbance spectroscopy using a Jasco FTIR-410 spectrometer. Each spectrum is the result of the addition of 124 spectra acquired with a resolution of 4 cm^{-1}. The coating mass was measured by weighing the substrates before and after the deposition, while the coating volume was evaluated from the substrate area and film thickness, thus allowing the determination of its density. The effect of the percentage of O_2 on the hardness and the elastic modulus of the coatings was assessed by the Oliver and Pharr method (Oliver et al., 1992) using load displacement curves obtained from nanoindentation experiments undertaken using a Hysitron TriboIndenter. At least 9 measurements were conducted on each sample using a multiple load function applied to a diamond Berkovich tip with maximum loads from 1000 to 10000 μN. The application rate at the maximum load varied from 36 (1000 μN) to 360 μN/s (10000 μN). A dwell time of 1 s was used throughout. Scratch experiments were conducted by associating the normal and lateral force capabilities of the nanoindenter. The friction coefficient was determined from 10 μm-long scratches produced by the application of a normal load of 300 μN on the indenter tip. The tip speed and the load application rate were 1 μm/s and 150 μN/s, respectively. A second batch of 10 μm-long scratches was obtained to investigate the scratching resistance of the films. Prior to the test, to evaluate the initial surface condition, the 10 μm region was scanned at a normal load of 200 μN applied at 100 μN/s. Subsequently, 5.000 μN was applied to the stationary nanoindenter tip and, as the maximum load was reached, scratching was produced over the 10 μm region for 10 s. The normal load was then reduced to 200 μN and a new scan was performed to probe the tested surface. The wear experiments were conducted using a 19-cycle tooth saw-like function with maximum loads of 1000, 2000, 3000 and 4000 μN. For all the scratching tests mentioned here, at least three experiments were conducted at different positions of each sample. Scanning probe microscopy (SPM) images were acquired in the Hysitron Triboindenter system before and after the nanoindentation, scratching and wear experiments, undertaken by applying 4.0 μN of normal load to the Berkovich tip. The topography and morphology of the films as well as their average roughnesses were also determined from SPM images of the non-scratched/non-indented areas. Mechanical, tribological and topographical characterizations were all performed in samples prepared on glass plates.

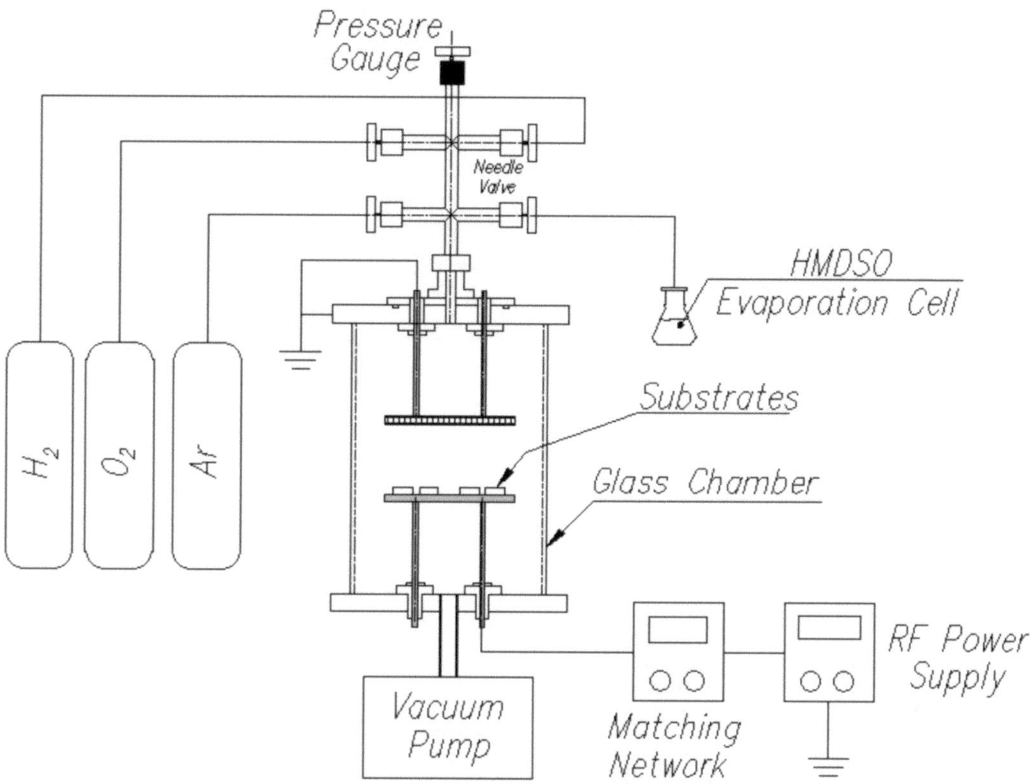

Figure 1. Experimental setup used for substrate cleaning and film deposition.

3. Results and discussions

Figure 2 shows the infrared spectra of films deposited with 0 and 70% of oxygen in the gas feed. Considering the spectrum of the film prepared from a pure HMDSO (0% O_2) plasma, the most prominent absorptions are related to symmetrical C-H stretching vibrations in CH_3 (2959 and 2899 cm^{-1}), to rocking of $(CH)_x$ in $Si(CH_3)_x$ (1262 cm^{-1}), to symmetrical and asymmetrical stretching of Si-O in Si-O-Si (1100 and 1020 cm^{-1}, respectively) and to rocking of CH_3 in $Si(CH_3)_3$ (845 and 750 cm^{-1}) and $Si(CH_3)_2$ (808 cm^{-1}) groups. Lower intensity bands are detected in 2100-2250 (ν Si-H), 1408 (δ C-H in $Si(CH_3)_x$) and 687 (ν $Si\text{-}H_n$) cm^{-1}. All the above mentioned absorptions were also observed in similar films studied in previous works (Gengenbach & Griesser, 1999, Rangel et al., 2012, Ul et al., 2002).

A general widening of the bands (3000, 1408, 1262 and 800 cm^{-1}) is observed as oxygen is introduced to the plasma, indicating a higher degree of fragmentation of the organic compound in the presence of higher proportions of active oxygen. Furthermore, the detection of the stretching mode of the silanol group (3644 cm^{-1}), which is not an original bond of the organometallic molecule, indicates multiple step reactions in the plasma phase. In addition, oxygen incorporation by trapped free-radical grows, as suggested by the emergence of the contributions lying at 3281 (hydrogen bonded O-H) (Ricci et. al., 2011) and 1725 (ν C=O) cm^{-1}. All these changes indicate that oxygen catalyses film deposition, which is readily confirmed by considering the results presented in Fig. 3, that shows an increase in thickness (from 2.2 to 8.3 μm) upon oxygen incorporation. Despite the enlargement of the layer, the intensity of the band related to methylsilyl groups (1262 cm^{-1}) decreases for the film deposited with 70% of oxygen. The loss of Si-connected CH_3 groups enables new points of crosslinking to be established between the Si backbones, increasing the structure density (Pfuch et. al., 2006, Ul et al., 2000). The downshift of the Si-O asymmetric component from 1020 to 993 cm^{-1} is further evidence of film densification (Choudhury et.al., 2010) upon oxygen incorporation. Therefore, all the above discussion suggests that films prepared from pure HMDSO possess a silicone-like structure with Si-H terminations while that deposited from the HMDSO/O_2 mixture is more inorganic, denser and therefore more susceptible to atmospheric oxygen uptake than the former.

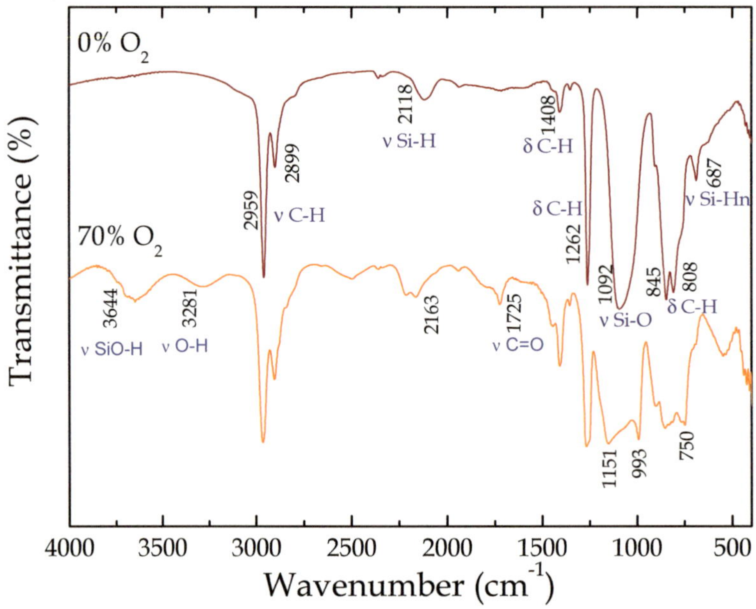

Figure 2. Infrared spectra of films deposited in plasmas with 0 and 70% oxygen.

Besides thickness, Fig. 3 also shows the density of the films. The simultaneous rise in thickness and density suggests that oxygen has a strong influence on the plasma kinetics. Density increases by around 7% (1.83 - 1.96 g/cm^3) as the proportion of O_2 enhances from 0 to 70%. This result is consistent with the loss of the light methyl groups, inducing both crosslinking and enrichment with the inorganic Si-O group. Films derived from HMDSO/O_2 plasma mixtures with densities ranging from 1.4 (polymer-like) to 2.2 g/cm^3 (oxide-like) have already been obtained in a previous study (Ul et al., 2000) by varying the proportion of oxygen from 0 to 80%. The lower density of the film prepared with the highest oxygen proportion in the present work is attributed to the acceleration in the deposition rate as oxygen is incorporated in the process.

Figure 4 shows 2D topographic images of the samples prepared in this work. Below each image there is a cross-section profile representing the line in the correspondent image. The left-hand images were taken from the sample prepared in pure HMDSO plasmas while those on the right, are from films obtained from HMDSO-O_2 plasmas. Comparing the topmost images (50 X 50 μm²), the presence of particulates with different diameters spread over a more uniform region is readily observed. The proportion and diameter of the particles substantially rise upon oxygen incorporation, which is reinforced by the 20 X 20 μm² profiles (middle position images). As magnification is further increased (lowermost 10 X 10 μm² images) it is possible to verify the non-existence of the uniform matrix connecting the particles. In both samples, the structure is granular and the dimension of the particulates increases with increasing oxygen incorporation (line profiles).

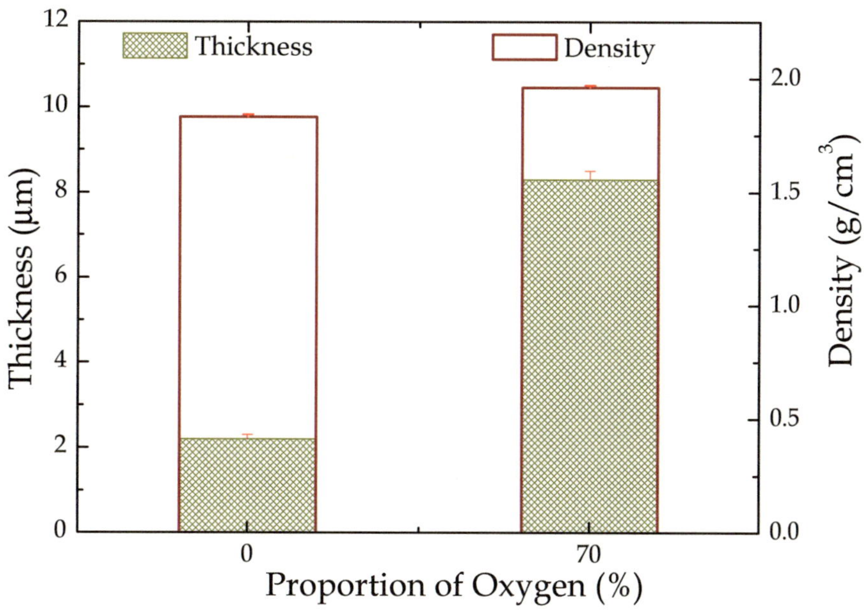

Figure 3. Thickness and density of the films as a function of the proportion of oxygen in the plasma feed.

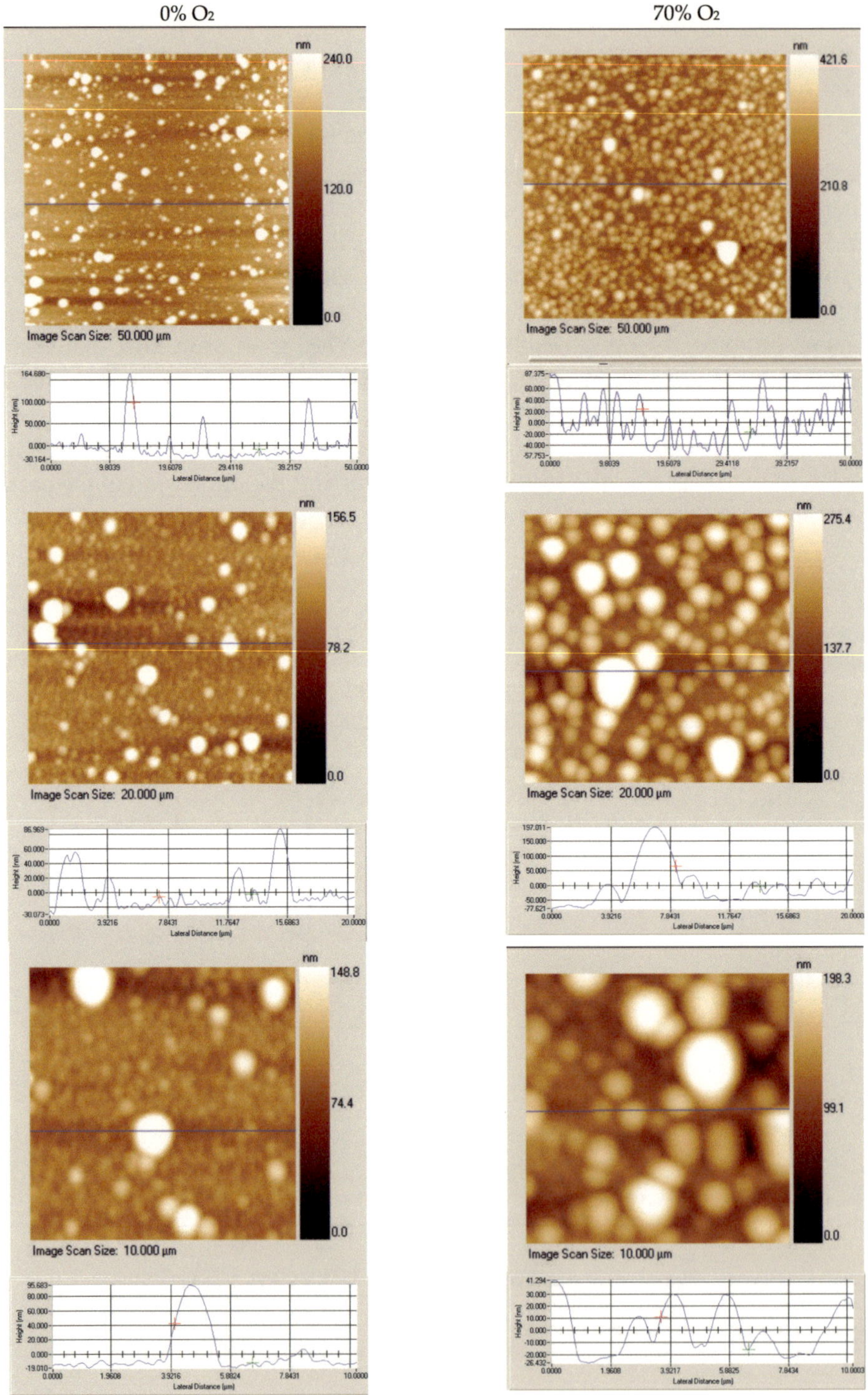

Figure 4. Surface topographical images of the sample deposited in pure HMDSO (left) and in $HMDSO/O_2$ (right) plasmas acquired at different magnifications.

The analysis of the images suggests that particulates are formed in the highly reactive plasma atmosphere by recombination of the HMDSO fragments. It has already been demonstrated that, under the same plasma condition, the size of the particles evolves with increasing deposition time (Ricci et al., 2011), giving rise to a structure composed of small diameter particles underneath the large diameter ones (lowermost image of Fig.4). The deposition of the macromolecules originates the film in which connections are established by radicals trapped in neighboring structures. As the reactivity of the plasma increases upon oxygen incorporation, the coalescence process accelerates, explaining the deposition of still higher dimensional structures, in equivalent time intervals. Ricci has also demonstrated that oxygen stimulates powder formation in HMDSO/O_2 plasmas (Ricci et al., 2011).

Figure 5 depicts the 3D topographical images of the surfaces derived from the images presented in Fig.4. Even though the height scales are not the same in the graphs, the differences on the surface topography caused by oxygen incorporation are clearly detected as one compares the left-hand to the right-hand images in this picture.

The root mean square roughness of the films, RMS, was evaluated from the 50 X 50 μm^2 total image areas. Figure 6 shows the results as a function of the proportion of oxygen in the plasma feed. The elevation of RMS from 25 to 35 nm is consistent with the growth in the particulate diameters induced by the increase in the deposition rate.

Several works in the literature (Boscher et. al., 2010, Rangel et al., 2012, Rouessac et al., 2011, Ul et al., 2002) report the production of a ball-like structure in films deposited from HMDSO under specific conditions. In none of those works, however, is a detailed characterization of the mechanical properties of this kind of structure provided. Thus, associating the scanning probe microscopy and nanoindentation capabilities of the equipment employed here, the mechanical properties of the films were derived using the Oliver and Pharr method (Oliver et al., 1991).

The images showing the indentation positions before and after the tests and the corresponding multiple load functions generated are depicted in Fig. 7 (0% O_2) and 8 (70% O_2). Although a high load was applied to the indenter (5000 μN), there is no sign of plastic deformation remaining in the images of the surfaces after the tests. The low scattering of the curves in Fig. 7 is surprising as one considers the non uniform nature of the structures studied here. From such curves it is observed that material deforms upon the application of the force but it recovers rapidly as the indenter is withdrawn from the film. From the unloading portion of the curves it can be seen that the material stops pushing the indenter tip, that is, the tip senses a null force, at around 350 nm (Fig.7), which then represents the permanent deformation caused by the indentation. It is also interesting to observe that the residual strain shown in the curves of Fig. 8 is around half (< 150 nm) of that produced in coatings prepared in pure HMDSO plasmas.

Despite its higher roughness, the film deposited in oxygen containing plasmas presented, in general, lower dispersion of the curves (Fig. 8) even if all the 14 load-displacement curves generated for this sample are placed together (Fig. 9a). The good repeatability of the curves then justifies the number of indentation experiments (at least 9) employed. In addition, comparing the areas inside the indentation curves presented in Fig. 9b, reveals a higher recovery rate for the film deposited using 70% of O_2, suggesting a more elastic structure.

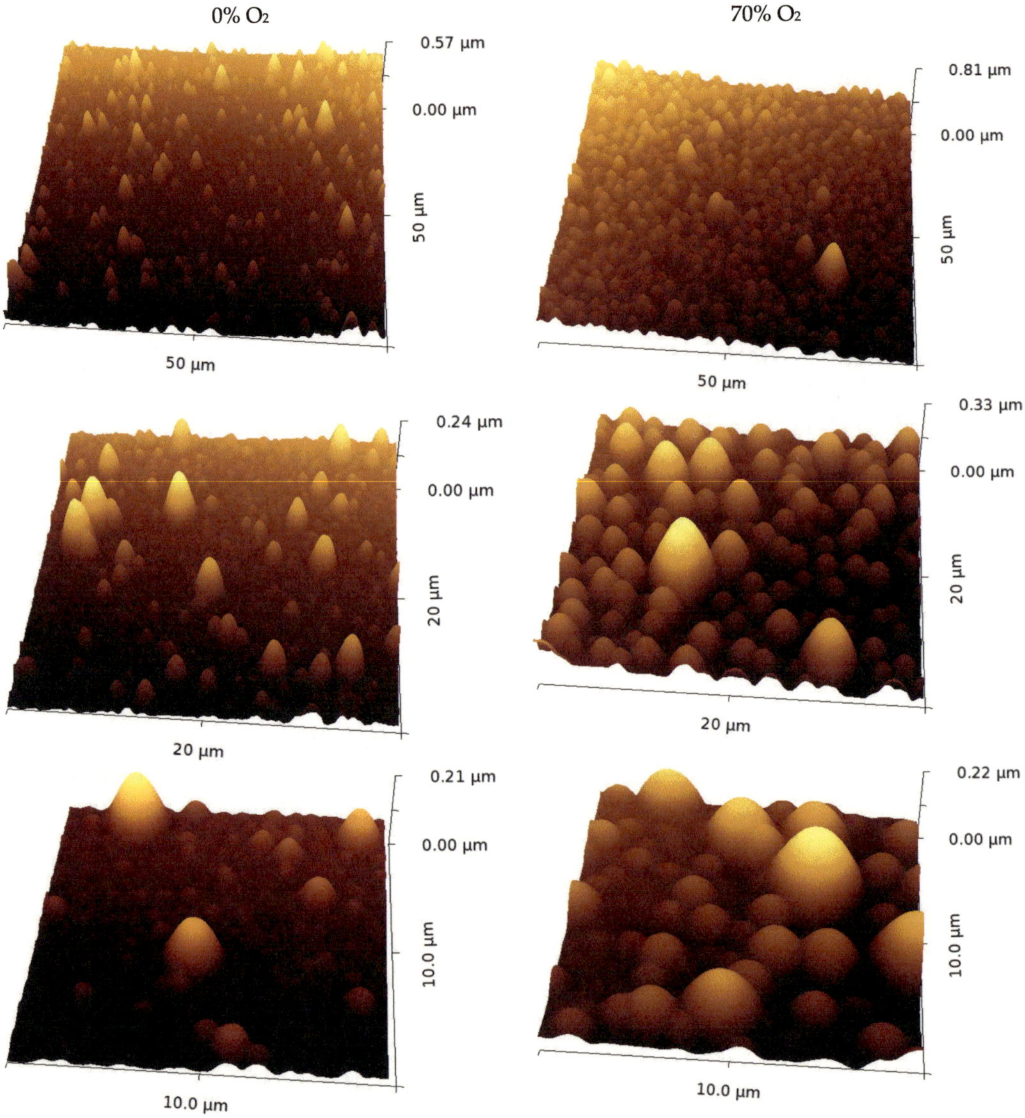

Figure 5. 3D topographical images of the sample deposited in pure HMDSO (left) and in HMDSO/O_2 plasmas (right) taken with different magnifications.

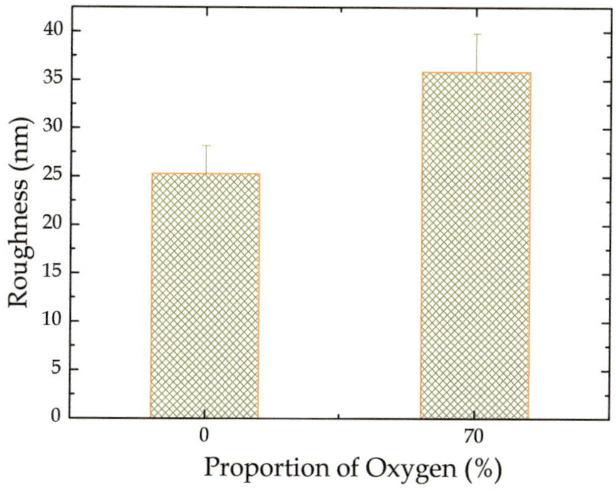

Figure 6. Root mean square roughness the films as a function of the oxygen proportion in the plasma phase.

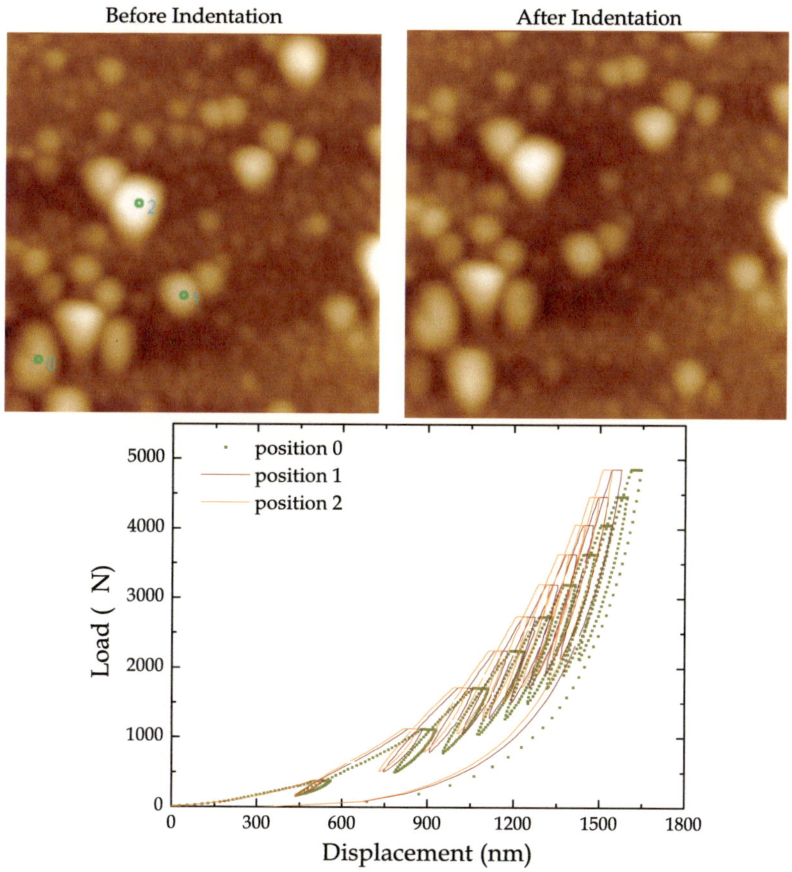

Figure 7. 10 X 10 μm² atomic force microscopy images of the film deposited in plasmas of pure HMDSO: (left) before and (right) after indentation. The points in the left image correspond to the location where the indentations were conducted using 5000 μN maximum load in a multiple load function. The graph at the bottom presents the loading-displacement functions generated at each corresponding point.

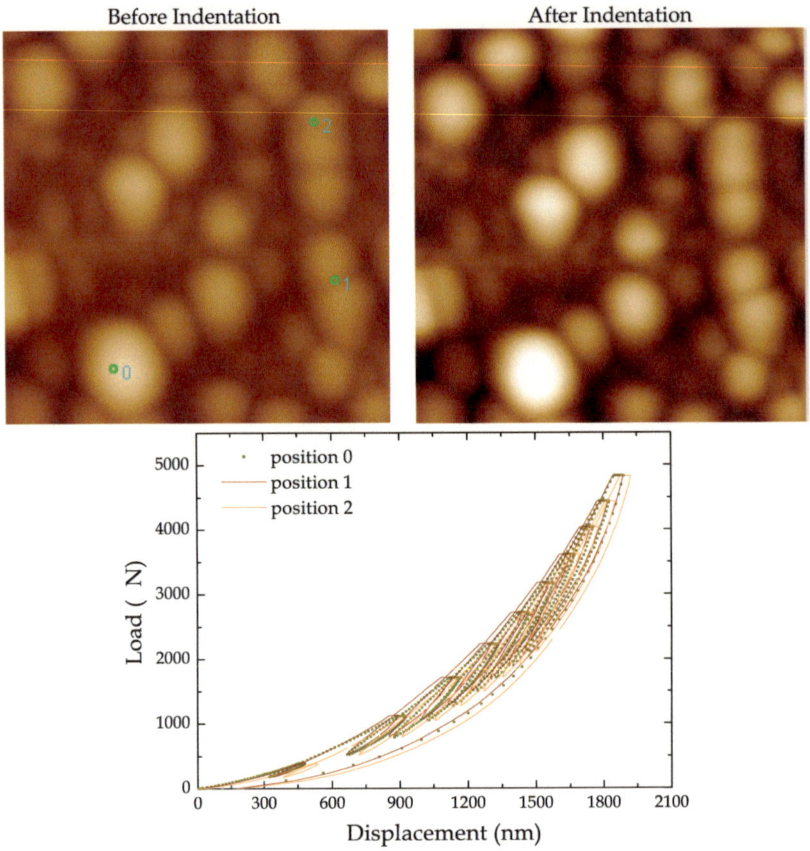

Figure 8. 10 X 10 μm² atomic force microscopy images of the film deposited in plasmas of HMDSO/O_2 mixtures: (left) prior and (right) after indentation. The points in the left image correspond to the location where the indentations were conducted using 5000 μN of maximum load in a multiple load function. The lower graph presents the loading-displacement functions generated in each corresponding point.

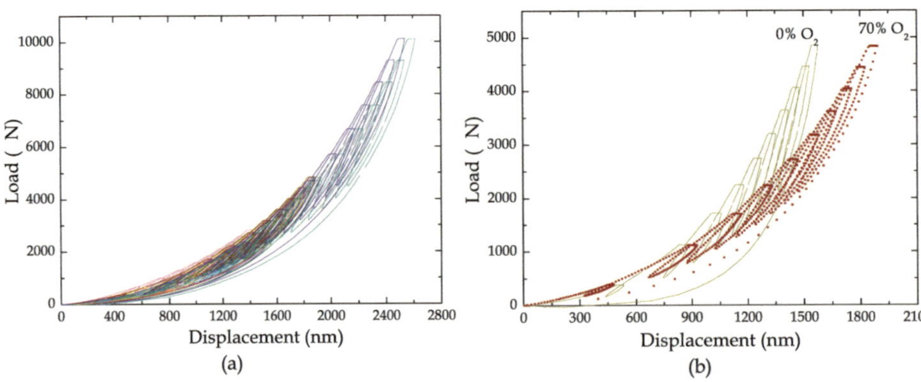

Figure 9. (a) Multiple load functions obtained from the indentations of the sample prepared in oxygen containing plasmas. The maximum loads employed in the experiments were 1000, 3000, 5000 and 10000 μN. (b) Load displacement curves of the samples prepared with 0 and 70% of oxygen in the plasma acquired with 5000 μN of maximum load.

The hardness of the structures was evaluated and is depicted in Fig. 10 as a function of depth for films prepared with 0 and 70% of oxygen in the plasma. Each curve represents data acquired using different loads. For the sample prepared in pure HMDSO plasma, the hardness tends to increase with increasing penetration depth, except for the curve taken using the lowest load (2000 μN). This behavior is ascribed to the growing interference of the mechanical properties of the glass substrate with the results, since the film thickness was 2.2 μm in this case. As the sample deposited with 70% of oxygen is thicker (8.3 μm), no substrate interference is evident but the hardness tends to a fixed value (~ 0.10 GPa) for penetration depths greater than 600 nm. In both graphs, the error bars are observed to decrease with increasing depth due to the greater uncertainties in depth determination as shallower penetrations are employed. Comparing the results obtained at the lowest depth in each sample, presented in the lowermost graph of Fig. 10, one finds that hardness is, on average, greater for the films deposited with oxygen in the plasma phase, which is consistent with the increase in the proportion of oxide groups in the structure.

The hardness values encountered here are around one order of magnitude smaller than those reported for silica-like films (Bewilogua et al., 2011, Choudhury et al., 2011, Jin et al., 2011), but are very similar to those reported for HMDSO plasma deposited films with silicone-like structures (Choudhurya, et al., 2011). Considering such findings, it is possible to infer that the ball-structures are essentially polymeric. This interpretation is corroborated by the results obtained on the glass substrate, presented in Fig.11, which is essentially composed of silica and presents hardness values at least 40 times greater than those encountered for the materials studied here.

Figure 12 shows the reduced elastic modulus as a function of depth for both samples investigated here. Data acquired using different maximum loads are represented by the distinct curves in the graphs. Since the Young's modulus of the film is very low compared to that of the indenter tip, the effect of the tip material on the results will be neglected (Schilde et al., 2012). Therefore, from now on, the reduced elastic modulus will be used as the elastic modulus of the structure. For the thinnest film (0% O_2), the elastic modulus tends to increase with increasing depth in all the curves while the same behavior is detected in the curves of the thicker layer (70% O_2) only at the highest indentation load (10000 μN). Considering the smallest penetration depth in each sample, it is observed that elastic modulus decreases upon oxygen incorporation (lowermost graph); that is, the film gains elasticity as the proportion of oxide groups in the structure increases. The elastic modulus measured for the glass plate, depicted in Fig. 13, presents values more than 40 times greater than those obtained for the films, confirming the polymeric nature of the Si-O containing particles.

Considering that the particulates are finite structures with diameters of some micrometers and that the penetration depth reaches this order of magnitude in some cases, the values found in this work for the mechanical properties do not correspond to those of the particulates but, instead, to the overall film structure. This means that the mechanical properties of the neighboring particulates affect the indentation results obtained in a particular structure and no attempt was made to correct this interference. For depths greater than 15% of the total film thickness, there is also interference owing to the mechanical properties of the glass substrate.

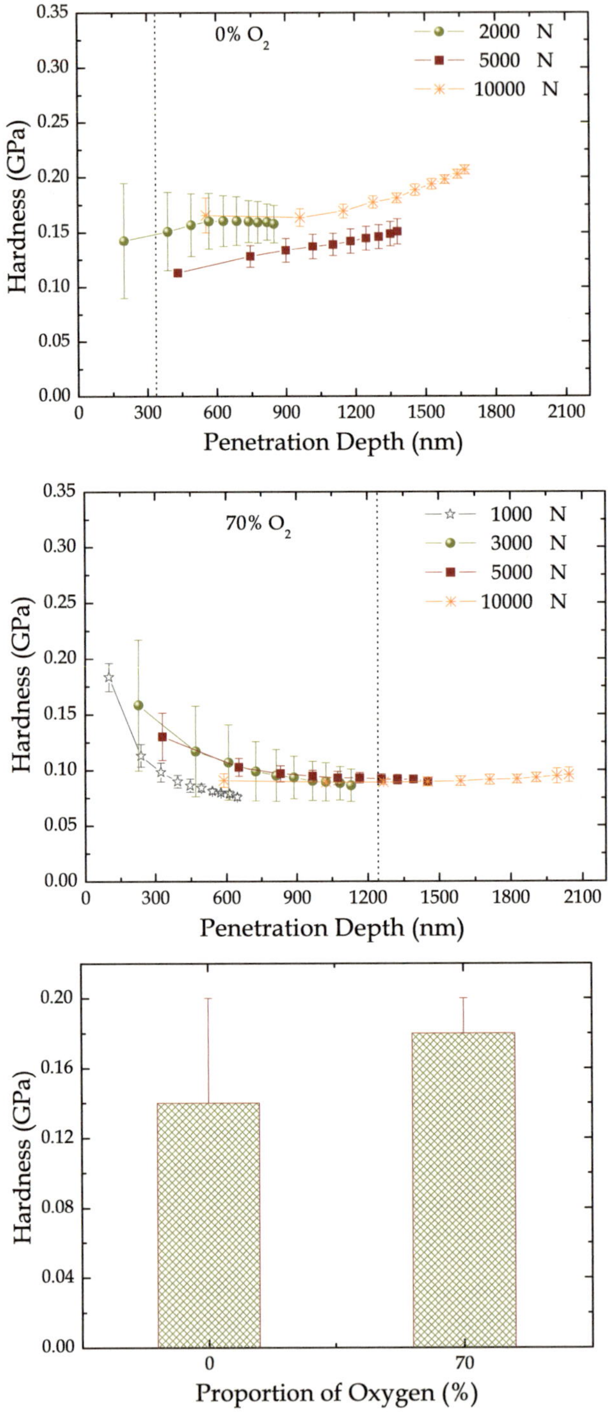

Figure 10. Hardness as a function of depth for films deposited using 0 and 70% of oxygen in the plasma. The vertical dotted lines in the graphs represent 15% of the total film thickness. The lowermost graph presents the hardness of the shallowest region probed in the different samples.

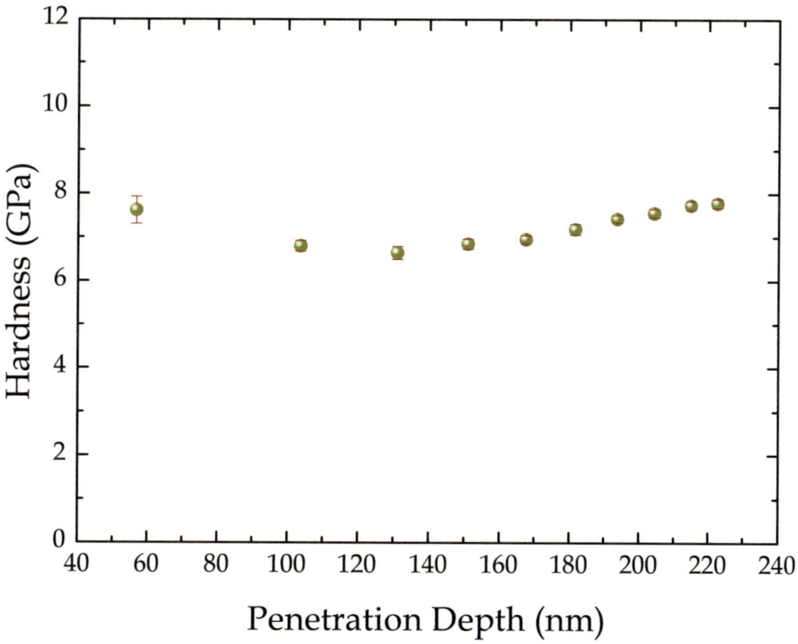

Figure 11. Hardness as a function of the penetration depth for the glass substrate.

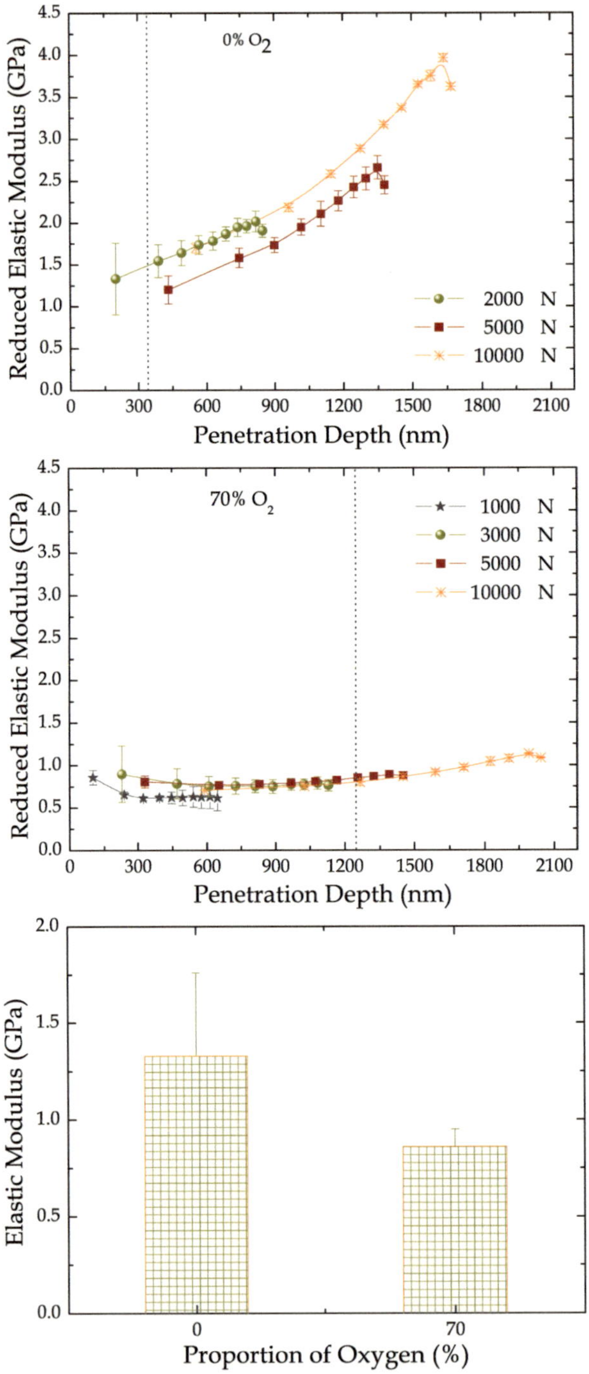

Figure 12. Elastic modulus as a function of penetration depth for films deposited using 0 and 70% of oxygen in the plasma. The vertical dotted lines in the graphs represent 15% of the total film thickness. In the lowermost graph, the elastic modulus measured at the shallowest probed depth is provided for both samples.

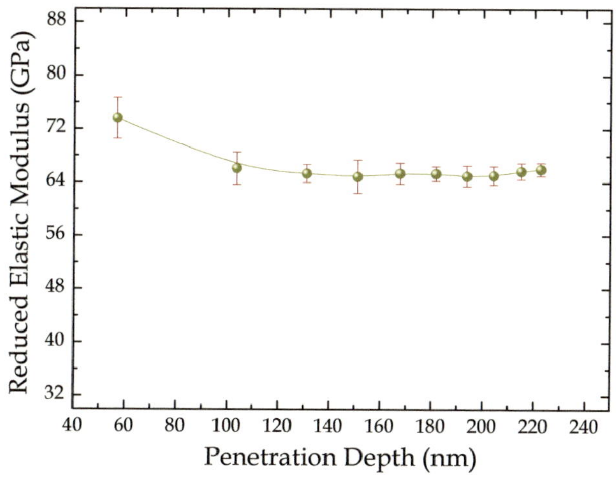

Figure 13. Reduced elastic modulus as a function of penetration depth for the glass plate used as the substrate.

It is generally accepted that wear resistance of a solid can be adjusted by tailoring its elasto-plastic properties, either by increasing H or decreasing E (Leyland & Matthews, 2000). To predict the wear resistance of the films, the elasticity index (H/E) was evaluated from data obtained at the shallowest indentation depth and is depicted in Fig. 14 as a function of the oxygen proportion. From the previous H and E results (Figs. 10 and 12) and from the H/E behavior, shown in Fig. 14, it is observed that oxygen incorporation during deposition tends to increase the wear resistance of the films. Figure 14 also depicts the resistance to plastic deformation (H^3/E^2), evaluated at the shallowest indentation depth for both samples. H^3/E^2 increases from 1.6 to 7.9 MPa as oxygen is incorporated into the plasma. Even though the elasticity index agrees well with values reported in the literature for a-Si:C:H films deposited by PECVD (Guruvenket et al., 2010), the plastic deformation resistance found here is around 50 times lower.

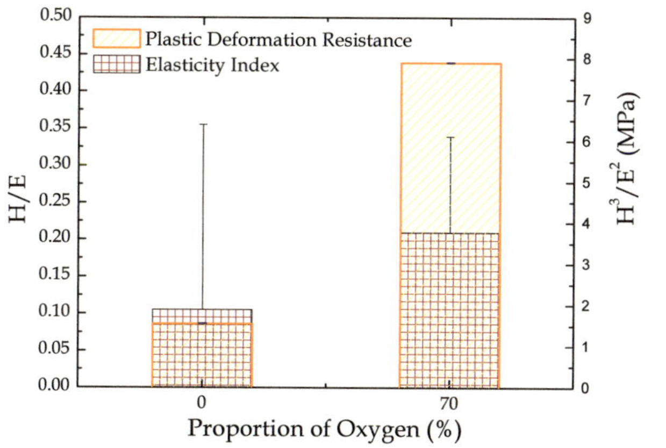

Figure 14. Elasticity index (H/E) and plastic deformation resistance (H^3/E^2) as a function of the proportion of oxygen in the depositing plasma.

According to the Johnson analysis (Tsui et al., 1995), H^3/E^2, is proportional to the load at which the material starts deforming plastically under the action of a rigid sphere of radius r. Therefore, to obtain a material highly resistant to plastic deformation H should be high and E low, consistently with the H/E parameter. This association allows the applied load to be dissipated over a larger area. On the other hand, if high plastic deformation is required, as in materials employed in nanoindentation lithography (Sirghi et al., 2009), low H^3/E^2 values are desirable. Even though the films prepared in this work resulted in low H^3/E^2 values, no residual deformation was detected after the tests, confirming that, at the loads employed here, deformation was mainly elastic due to the high H/E ratios of the films.

Therefore, from the nanoindentation results discussed here it is possible to infer that the structure gained in hardness and elasticity upon oxygen incorporation. This association improved the elasticity index and the resistance to plastic deformation, suggesting a better tribological performance of the coatings. The results indicate that even with 70% of oxygen in the mixture, carbon was effectively incorporated into the structure, resulting in a granular polymeric-like material. Oxygen has catalyzed the HMDSO deposition in the plasma environment. Still higher proportions of oxygen should be considered for the formation of a uniform oxide-like structure.

Figure 15 and 16 show SPM images of the surfaces before (left) and after (right) the wear tests. Cross-section profiles taken from specific points of the images are also presented. For the sample prepared in the pure HMDSO plasma (Fig. 15) there was no signal of wear as normal loads of up to 1000 μN were used. Slight tracks start appearing at loads greater than 1000 μN. Consistently, depressions are detected in the line profile around 19, 29 and 39 μm of the scan length, with depths increasing with the normal load. It is interesting to note, however, the preservation of the particulates over the track (3D images), suggesting a compaction of the overall tested region without material removal. For the sample deposited in the HMDSO-oxygen mixture (Fig. 16), no sign of track is detected in the images or in the cross-section profile, independently of the load used during the test. These results thus confirm the prediction of the elasticity index (Fig. 14) of a better wear resistance for the film deposited in the presence of oxygen.

Figure 17 shows the friction coefficient between the diamond nanoindenter tip and the film prepared in plasmas of HMDSO (top) and HMDSO/O_2 mixture (down). For the film deposited with no oxygen in the plasma, the friction coefficient changed from 0.1 to 0.3 along the scan length. The range of oscillation was slightly higher for the samples deposited with oxygen in the plasma, since it presented roughness values around 40% higher than the former (Fig. 6). For both samples, the major friction mechanism is attributed to the adhesion since in fully elastic deformations, this is the main interaction mode (Wang & Kato, 1998).

Figures 18 and 19 show representative scratch profiles taken from the films deposited with different proportions of oxygen in the plasma. The "Pre-scan" and "Post-scan" profiles in the graphs represent the surface condition just before and after the scratching test. Their comparison allows inference to be made concerning the plastic deformation remaining on the scratched surface. The SPM image presented below each graph was taken immediately before (left) and after (right) the scratching experiment.

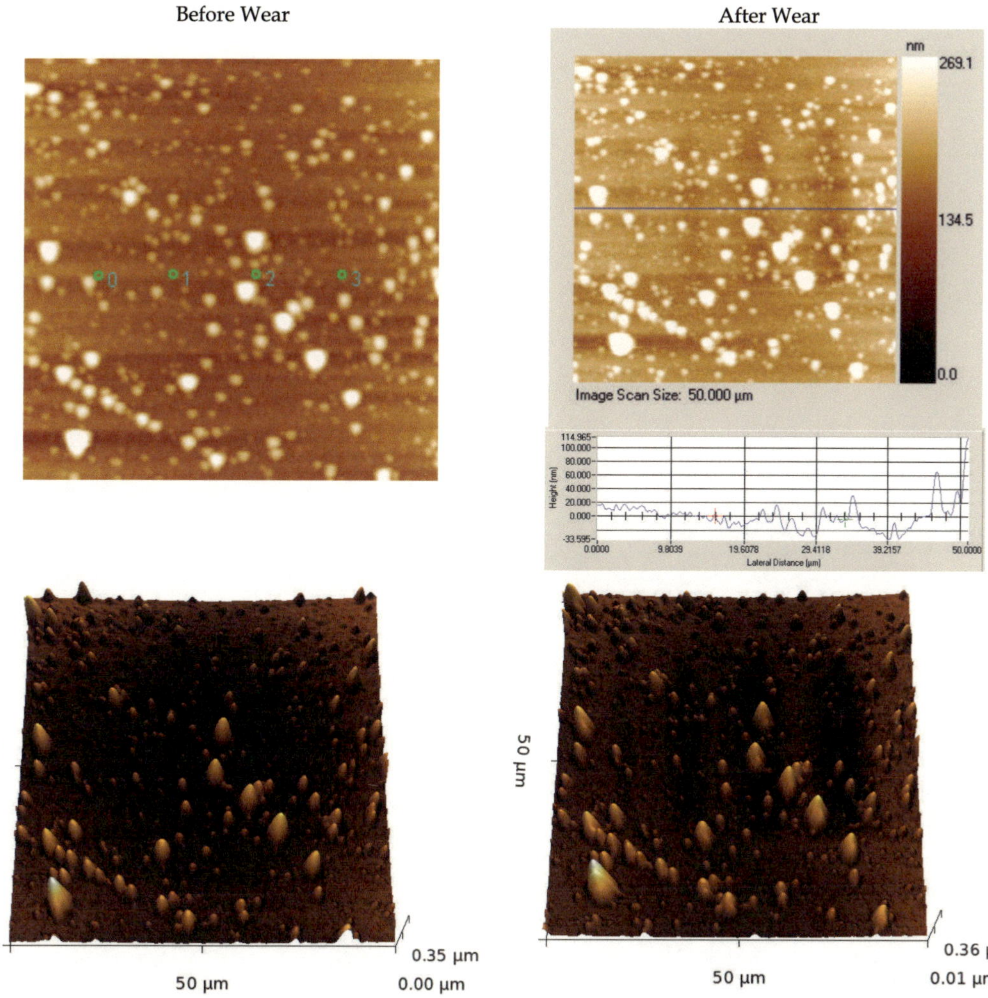

Figure 15. Bi- (top) and Tri-dimensional (bottom) scanning probe microscopy images of the film deposited in plasmas of pure HMDSO: (left) before and (right) after the wear cycle. The points in the left top image correspond to the location where the tests were conducted using 19 cycles saw-tooth function of 1000 (0), 2000 (1), 3000 (2) and 4000 µN (3) of maximum load.

All the profiles of Fig. 18 present a small slope which may be a consequence of the lack of planicity of the system. During the scratching cycle, the material undergoes deep penetration (1300 - 1100 µm) but a fast recovery is suggested by comparison of the "Scratching" and the "Post-scan" profiles, confirming the high elasticity of the structure. Interestingly, the time elapsed from the scratching test up to the end of the post-scan acquisition was only 27 s. An almost complete recovery of the deep strain produced during the scratching test is evidenced as one compares the pre- and post- scan curves. Probably owing to the large tip diameter (~ 200 nm), no plastic deformation was observed in any region of the SPM image taken after the test

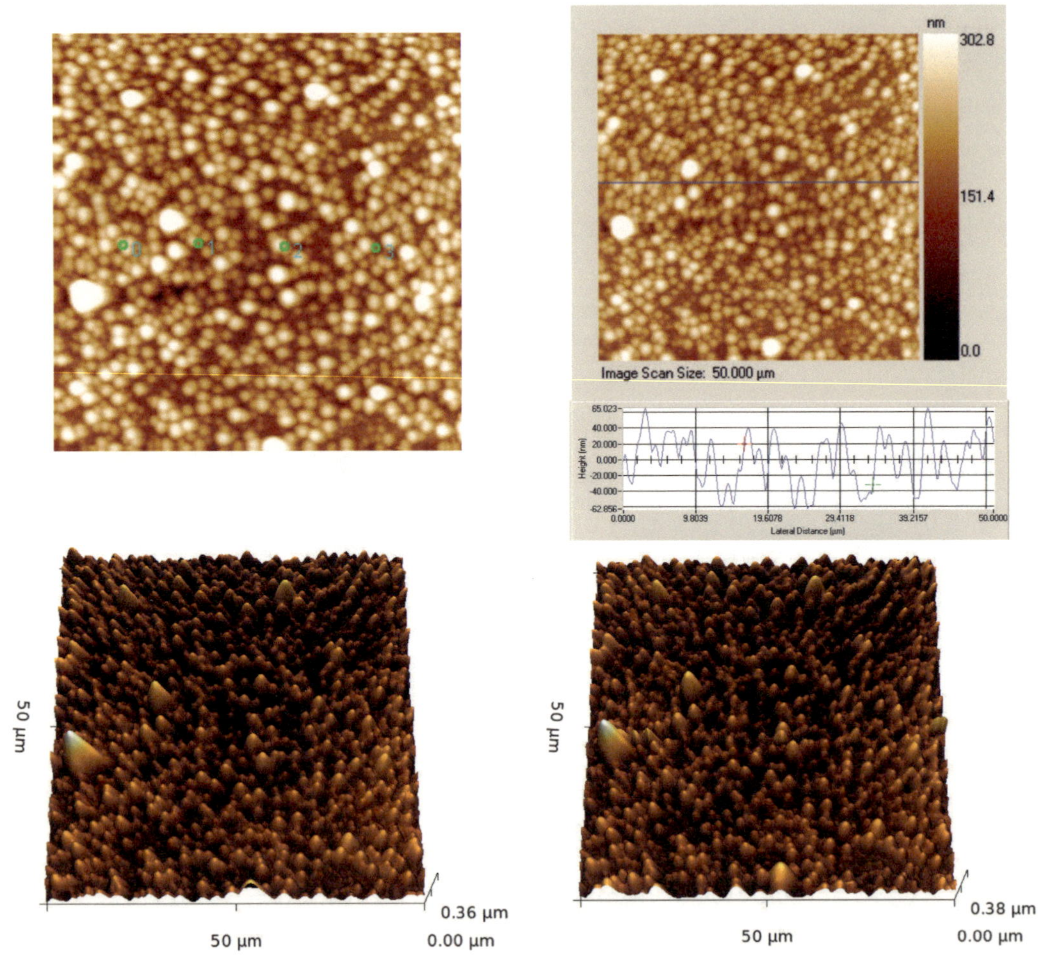

Figure 16. Bi- (top) and Tri-dimensional (bottom) scanning probe microscopy images of the film deposited in plasmas of the HMDSO/O_2 mixture: (left) before and (right) after the wear cycle. The points in the left top image correspond to the location where the tests were conducted using 19 cycles tooth saw-like function of 1000 (0), 2000 (1), 3000 (2) and 4000 µN (3) of maximum load.

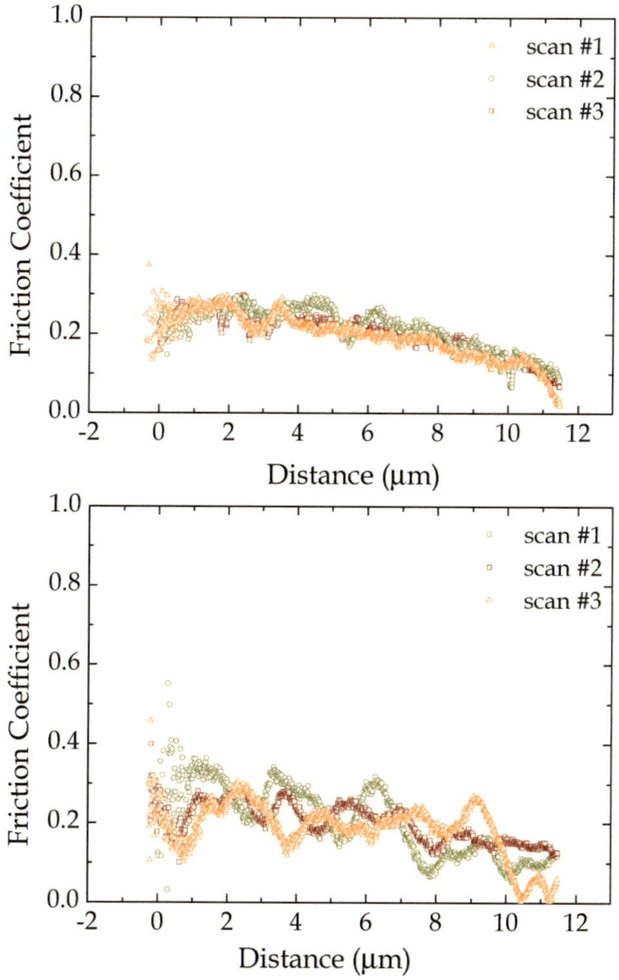

Figure 17. Friction coefficient as a function of the position for the films prepared in pure HMDSO (left) and HMDSO/O_2 (right) plasma. The scans were acquired at 300 μN of normal load and at a velocity of 1.0 μm/s.

A very similar behavior is revealed when the curves and images of Fig. 19 are considered, except that now a complete elastic recovery is seen, which is consistent with the higher elasticity index presented by this sample (Fig. 14). Although the film deposited in the presence of oxygen had a higher friction coefficient it also presented a higher scratching resistance.

Therefore, it should be pointed out that, even though a soft material was produced in any of the conditions employed here, they are wear and scratch resistant due to the high elasticity. In a general way, oxygen introduction improved the tribological properties of the films, despite the increase in the friction coefficient. The high capability of recovery of the scratched region and the low wear induced in the samples suggest the application of the films as anti-scratching coatings for a series of practical applications.

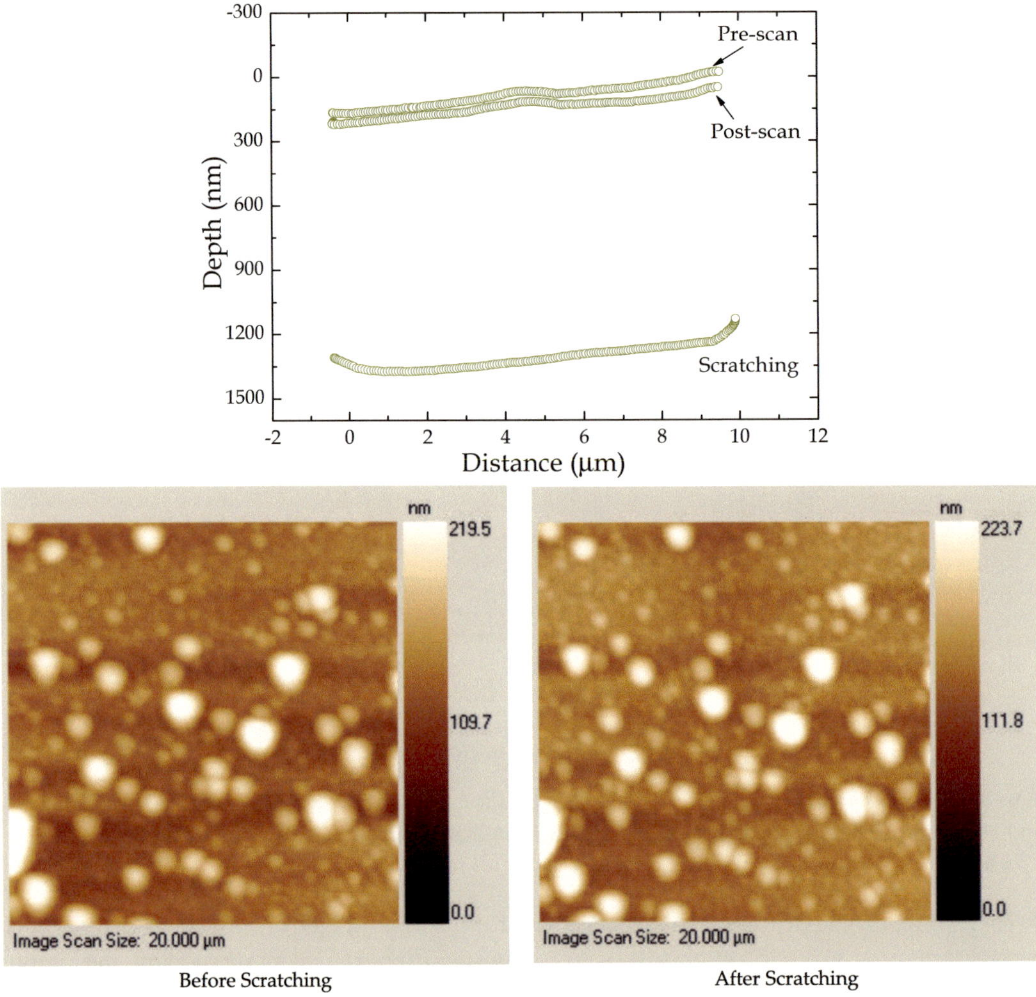

Figure 18. Profiles acquired before, during and after the scratching tests from the film deposited in a pure HMDSO plasma. The scans acquired before and after the tests were conducted with 200 μN of normal load. The lowermost pictures represent 20 X 20 mm² scanning probe microscopy images of the film prior (left) and after (right) the scratching tests conducted with a constant load of 5000 μN.

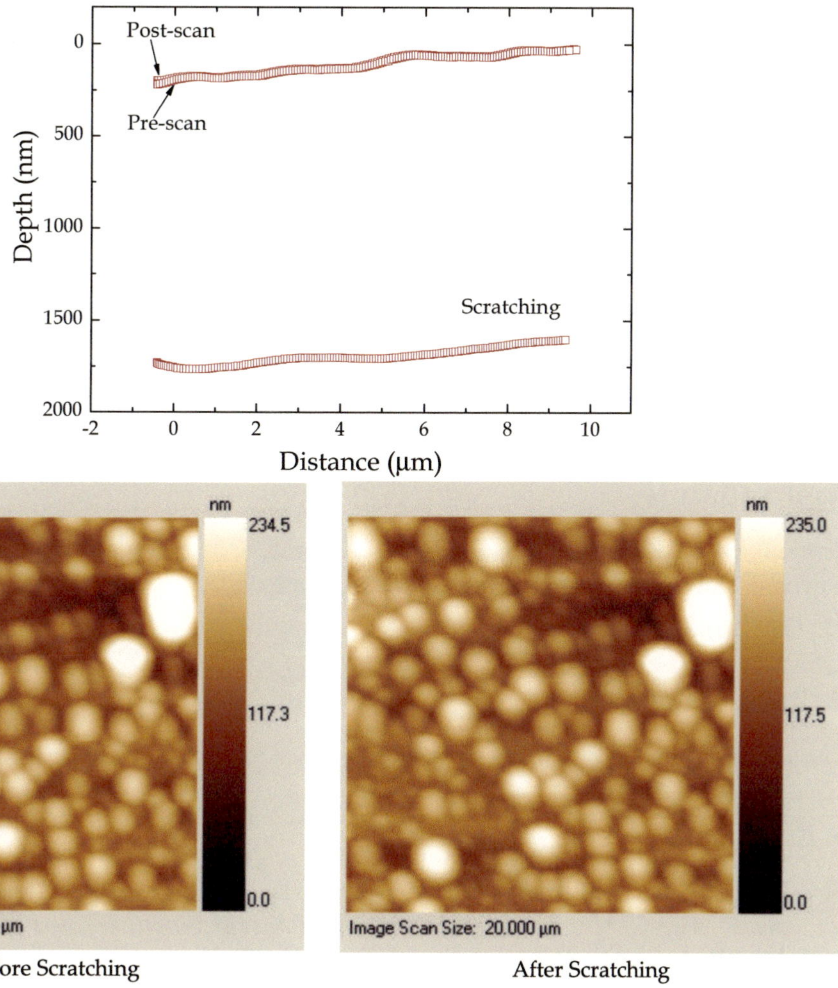

Figure 19. Profiles acquired before, during and after the scratching tests from the film prepared in the HMDSO/O_2 plasma mixture. The scans acquired before and after the test were conducted with 200 µN of normal load. The lowermost pictures represent 20 X 20 mm^2 scanning probe microscopy images of the film before and after the scratching tests conducted with a constant load of 5.000 µN.

4. Conclusion

The physical, chemical, mechanical and tribological properties of films deposited from the HMDSO precursor are dependent on the oxygen proportion in the plasma phase. The proportion of oxide inclusions increased upon oxygen incorporation, enhancing hardness and decreasing elastic modulus. Friction slightly increased due to better adhesion of the surfaces promoted by chemical and topographical alterations. Independently of the deposition condition, films are scratch and wear resistant due to the high elasticity of the structure. The coating with best tribological performance was produced as oxygen was incorporated during the deposition process. Associating the results obtained for the chemical structure with the mechanical and tribological properties it was possible to infer the polymeric nature of the coatings. Still higher oxygen proportions should be necessary to effectively scavenge carbon from the structure during depositions. In the proportion used here, however, oxygen played a crucial role in the plasma kinetics since it increased the deposition rate and the coalescence of particulates in the plasma, affecting the overall structure of the film. Furthermore, the indentation, scratching and images capabilities of the nanoindenter employed as well as of the nanoindentation theory were shown to provide important results or predictions of the material properties on the nanometric scale. The determination of the mechanical and tribological properties of the films revealed their potential, despite their initially being investigated specifically owing to their excellent corrosion barrier properties, as anti-scratch transparent coatings for optical devices.

Author details

Bruno B. Lopes, Rita C.C. Rangel, César A. Antonio,
Steven F. Durrant, Nilson C. Cruz, Elidiane C. Rangel
São Paulo State University, Brazil

Acknowledgement

The authors would like to thank Brazilian agencies FAPESP and CNPq for financial support.

5. References

Bewilogua, K.; Bialuch, I.; Ruske, H. & Weigel, K. (2011). Preparation of a-C:H/a-C:H:Si:O and a-C:H/a-C:H:Si multilayer coatings by PACVD. *Surface & Coatings Technology*. 2011, Vol. 206, 4, pp. 623–629

Borvon, G.; Goullet, A.; Mellhaoui, X.; Charrouf, N. & Granier, A. (2002). Electrical properties of low-dielectric-constant films prepared by PECVD in O_2/CH_4/HMDSO. *Materials Science in Semiconductor Processing*. Apr-Jun 2002, Vol. 5, 2-3, pp. 279-284

Boscher, N.D.; Choquet, P.; Duday, D. & Verdier, S. (2010). Chemical compositions of organosilicon thin films deposited on aluminium foil by atmospheric pressure dielectric barrier discharge and their electrochemical behaviour. *Surface & Coatings Technology*. Dec 2010, Vol. 205, 7, pp. 2438–2448

Choudhury, A.J.; Chutia, J.; Kakati, H.; Barve, S.A.; Pal, A.R.; Sarma, N.S.; Chowdhury, D. & Patil, D.S. (2010). Studies of radiofrequency plasma deposition of hexamethyldisiloxane films and their thermal stability and corrosion resistance behavior. *Vacuum*. Jun 2010, Vol. 84, 11, pp. 1327-1333

Choudhury, A.J.; Barve, S.A.; Chutia, J.; Pal, A.R.; Kishore, R.; Jagannath; Pande, M. & Patil, D.S. (2011). RF-PACVD of water repellent and protective HMDSO coatings on bell metal surfaces: Correlation between discharge parameters and film properties. *Appl. Surf. Sci.* Aug 2011, Vol. 257, 1, pp. 8469-8477

Czeremuszkin, G.; Latreche, M.; Wertheimer, M.R. & Sobrinho da Silva, A.S. (2001). Ultrathin Silicon-Compound Barrier Coatings for Polymeric Packaging Materials: An Industrial Perspective. *Plasmas Polym.* Jun 2001, Vol. 6, 1-2, pp. 107-120

Fracassi, F.; d'Agostino, R.; Palumbo, F.; Angelini, E.; Grassini, S. & Rosalbino, F. (2003). Application of plasma deposited organosilicon thin films for the corrosion protection of metals. *Surface and Coatings Technology*, Sep-Oct 2003, Vol. 174–175, pp. 107-111

Gengenbach, T.R. & Griesser, H.J. (1999). Post-deposition ageing reactions differ markedly between plasma polymers deposited from siloxane and silazane monomers. *Polymer.* Aug 1999, Vol. 40, 18, pp. 5079-5094

Görbig, O.; Nehlsen, S. & Mtiller, J. (1998). Hydrophobic properties of plasma polymerized thin film gas selective membranes. *Journal of Membrane Science.* Jan 1998, Vol. 138, 1, pp. 115-121

Guruvenket, S.; Azzi, M.; Li, D.; Szpunar, J.A.; Martinu, L. & Klemberg-Sapieha, J.E. (2010). Structural, mechanical, tribological, and corrosion properties of a-SiC:H coatings prepared by PECVD. *Surface & Coatings Technology.* Aug 2010, Vol. 204, 21-22, pp. 3358–3365

Jin, S.B.; Lee, J.S.; Choi, Y.S.; Choi, I.S. & Han, J.G. (2011). High-rate deposition and mechanical properties of SiO(x) film at low temperature by plasma enhanced chemical vapor deposition with the dual frequencies ultra high frequency and high frequency. *Thin Solid Films.* Jul 2011, Vol. 519, 19, pp. 6334–6338

Leyland, A. & Matthews, A. (2000). On the significance of the H/E ratio in wear control: a nanocomposite coating approach to optimised tribological behaviour. *Wear.* Nov 2000, Vol. 246, 1-2, pp. 1–11

Morent, R.; Geyter, N.D.; Vlierberghe, S.V.; Dubruel, P.; Leys, C. & Schacht, E. (2009). Organic-inorganic behaviour of HMDSO films plasma-polymerized at atmospheric pressure. *Surface and Coatings Technology.* Feb 2009, Vol. 203, 10-11, pp. 1366-1372

Navabpour, P.; Teer, D.; Su, X.; Liu, C.; Wang, S.; Zhao, Q.; Donik, C.; Kocijan, A. & Jenko, M. (2010). Optimisation of the properties of siloxane coatings as anti-biofouling coatings: Comparison of PACVD and hybrid PACVD-PVD coatings. *Surface and Coatings Technology.* Jul 2010, Vol. 204, 20, pp. 3188-3195

Oliver, W.C.; Pharr, G.M. (1992). An improved technique for determining hardness and elastic modulus using load and displacement sensing indentation experiments. *Journal of Material Research.* Jan 1992, Vol. 7, 6, pp. 1564-1583.

Ong, S.E.; Zhang, S.; Du, H.; Too, H.C. & Aung, K.N. (2007). Influence of silicon concentration on the haemocompatibility of amorphous carbon. *Biomaterials.* Oct 2007, Vol. 28, 28, pp. 4033-4038

Ong, S.E.; Zhang, S.; Du, H.; Wang, Y. & Ma, L.L. (2008). In-vitro cellular behavior on amorphous carbon containing silicon. *Thin Solid Films.* Jun 2008, Vol. 516, 16, pp. 5152-5156

Pfuch, A.; Heft, A.; Weidl, R. & Lang, K. (2006). Characterization of SiO2 thin films prepared by plasma-activated chemical vapour deposition. *Surface & Coating Technology.* Sep 2006, Vol. 201, 1-2, pp.189-196

Pihan, S.A.; Tsukruk, T. & Förch, R. (2009). Plasma polymerized hexamethyl disiloxane in adhesion applications. *Surface and Coatings Technology*. Mar 2009, Vol. 203, 13, pp. 1856-1862

Quédé, A.; Jama, C.; Supiot, P.; Le Bras, M.; Delobel, R.; Dessaux, O. & Goudmand, P. (2002). Elaboration of fire retardant coatings on polyamide-6 using a cold plasma polymerization process. *Surface and Coatings Technology*. Mar 2002, Vol. 151–152, pp. 424-428

Rangel, R.C.C.; Pompeu, T.C.; Barros Jr., J.L.S.; Antonio, C.A.; Santos, N.M.; Pelici, B.O.; Freire, C.M.A.; Cruz, N.C. & Rangel, E.C. (2012). Improvement of the Corrosion Resistance of Carbon Steel by Plasma Deposited Thin Films, In: *Recent Researches in Corrosion Evaluation and Protection*, Reza Shoja Razavi, pp. 91-116, Intech Open Access Publisher, Retrieved from http://www.intechopen.com/articles/show/title/improvement-of-the-corrosion-resistance-of-carbon-steel-by-plasma-deposited-thin-films

Ricci, M.; Dorier, J.; Hollestein, C. & Fayet, P. (2011). Influence of Argon and Nitrogen Admixture in HMDSO/O2 Plasmas onto Powder Formation. *Plasma Process. Polym.* Feb 2011, Vol. 8, 2, pp. 108-117

Rouessac, V.; Ungureanu, A.; Bangarda, S.; Deratani, A.; Lo, C.; Wei, T.; Lee, K. & Lai, J. (2011). Fluorine-Free Superhydrophobic Microstructured Films Grown by PECVD. *Chemical Vapor Deposition*. Sep 2011, Vol. 17, 7-9, pp. 198–203

Schilde, C.; Westphal, B.; Kwade, A. (2012). Effect of the primary particle morphology on the micromechanical properties of nanostructured alumina agglomerates. *J. Nanopart. Res.* Feb 2012, Vol 14, 3, pp.1-11.

Schwarz, J.; Schmidt, M. & Ohl, A. (1998). Synthesis of plasma-polymerized hexamethyldisiloxane (HMDSO) films by microwave discharge. *Surface and Coatings Technology*. Jan 1998, Vol. 98, 1-3, pp. 859-864

Sirghi, L.; Ruiz, A. ; Colpo, P. & Rossi, F. (2009). Atomic force microscopy indentation of fluorocarbon thin films fabricated by plasma enhanced chemical deposition at low radio frequency power. *Thin Solid Films.* Apr 2009, Vol. 517, 11, pp. 3310–3314

Tsui, T.Y. ; Pharr, G.M.; Oliver, W.C.; Bhatia, C.S.; White, R.L.; Anders, S.; Anders, A. & Brown, I.G. (1995). Nanoindentation and nanoscratching of hard carbon coatings for magnetic disks. *Material Research Society Symposium Procedings*. 1995, Vol. 383, pp. 447-452

Ul, C.V.; Laporte, C.B.; Benissad, N.; Chausse, A.; Leprince, P. & Messina, R. (2000). Plasma-polymerized coatings using HMDSO precursor for iron protection. *Progress in Organic Coatings.* Feb 2000, Vol. 38, 1, pp. 9-15

Ul, C.V.; Roux, F.; Laporte, C.B.; Pastol, J.L. & Chausse, A. (2002). Hexamethyldisiloxane (HMDSO)-plasma polymerised coatings as primer for iron corrosion protection: influence of RF bias. *J. Mater. Chem.* Jun 2002, Vol. 12, 8, pp. 2318-2324

Wang, D. F. & Kato, K. (1998). Friction studies of ion beam assisted carbon nitride coating sliding against diamond pin in water vapor. *Wear.* May 1998, Vol. 217, 2, pp. 307-311

Zajíčková, L.; Buršíková, V.; Peřina, V.; Macková, A.; Subedi, D.; Janča, J. & Smirnov, S. (2001). Plasma modification of polycarbonates. *Surface and Coatings Technology*. Jul 2001, Vol. 142–144, pp. 449-454

Section 4

Characterization of Microdevices and Integrated Circuits

Chapter 9

Characterization of Microdevices by Nanoindentation

Mamadou Diobet DIOP

Additional information is available at the end of the chapter

http://dx.doi.org/10.5772/50997

1. Introduction

In the last two decades, the study of microdevices such as micro-electro-mechanical systems (MEMS) has rapidly grown into a critical area of technology with several applications that impact many industrial sectors including automotive, consumer electronics, telecommunication, aerospace, and medical.

Microdevices can undergo early failure during the operation due to mechanical stresses, which can be induced by a single high stress or cyclic low stresses compared with the strength of the component. An accurate knowledge of the mechanical properties of micro and nanomaterials, especially thin films, which form mechanical structures of MEMS systems, is necessary to minimize or virtually eliminate failures. Mechanical properties of bulk materials were widely investigated as compared to thin films. In many cases, thin films properties differ from that of the bulk materials owing to the presence of residual stresses, preferred orientations of crystallographic planes, and the morphology of the microstructure [1, 2]. The studies of thin films properties were mostly done for the semiconductor industry, but they were mainly on the electrical properties [2]. Usually, when the mechanical properties of micro and nanomaterials are studied, thin continuous films deposited on a silicon wafer are used. This differs to patterned thin film deposited on structures of microdevices where film exists as individual part with different shapes and various cross-section geometries. Contrary to the continuous films use case, the deformation field in the line structure is dominated by the edge effect [3]. So, extracting the realistic behavior of materials for preventing failures dictates a direct mechanical characterization of the thin film structures in a configuration similar to the actual microdevices final design.

This chapter is centered around presenting the mechanical characterization of such microdevices using nanoindentation technique. These microdevices studied here are not limited to MEMS structures but also include micropillars and cylindrical bumps used in flip chip technology. Nanoindentation has been used so far to locally extract the Young's modulus and the hardness of materials. Here, we present the extension of this technique to characterize the microdevices structures. Beyond the high resolutions in load and displacement, nanoindentation offers the possibility to vertically load the structures similarly to their actual functioning in the components. The main experimental difficulty in characterizing the microdevices by nanoindentation is to ensure that the measured load-deflection data reflect the true deformation of the sample under test. We have intentionally based this study mainly on the experimental challenges; theoretical analysis is given for understanding of the evaluation of the mechanical properties.

For MEMS structures study, the conventional nanoindentation is used without any modification, as reported in the first section, with the intention of determining the elastic and plastic constants as well as the fatigue strength of cantilever and bridge microbeams. The second section is devoted to the uniaxial compression of micropillars in order to investigate the size effects on their plastic flow. The last section presents a study of the insertion of cylindrical bumps for flip-chip interconnection applications. To achieve this goal, major modifications related to sample holders both on the nanoindenter column extremity and the stage, are performed. The reliability of MEMS structures and the cylindrical bumps require both mechanical and electrical studies. So, electrical measurements can be synchronized with the nanoindenter in order to couple the mechanical and electrical mechanisms.

2. MEMS characterization by nanoindentation

The International Technology Roadmap for Semiconductors (ITRS), 2011 edition described MEMS as devices that are composed of micrometer-sized mechanical structures (suspended bridges, cantilevers, membranes, fluid channels, etc.) and often integrated with analog and digital circuitry. MEMS can act as sensors, receiving information from their environment, or as actuators, responding to a decision from the control system to change the environment. The majority of today's MEMS products include accelerometers, pressure and chemical flow sensors, micromirrors, gyroscopes, fluid pumps, and inkjet print heads. All MEMS sectors such automotive, optical MEMS, magnetometers for consumers and industrial high-end applications exceeded growth expectations in 2010. To sustain an annual market growth, the development of new innovative MEMS must be pursued and great efforts must be dedicated to reliability issues. In this section we focus on the mechanical reliability of micromaterials for MEMS structures as it is the principle part of the MEMS reliability [2].

Several reliability test methods have been employed in MEMS and they all differ depending on the user interest. Mechanical testing of MEMS using nanoindentation is related to the bending tests of samples such cantilever and bridge beams. The purpose being to extract the elastic and plastic characteristics as well as the failure mechanism of the mobile parts of the MEMS.

2.1. Specimen manufacturing

The common MEMS processes named surface manufacturing and bulk manufacturing are mostly employed to build the mobile parts of MEMS structures (cantilever and bridges microbeams). These two manufacturing techniques use processing steps that are compatible with the silicon Integrated Circuits (ICs) technology. The surface micromachining uses two thin layers—a structural material (polysilicon, Al, SiO_2, ...) and a sacrificial material (SiO_2,polymers, Al, ...) deposited on the surface of a silicon substrate to fabricate the mobile parts of the MEMS. The bulk micromachining technique employs the whole thickness of a silicon wafer to build the micro-mechanical structures [4]. In bulk micromachining, the air gap underneath the mobile parts are limited by the silicon thickness (up to 530μm for a standard wafer) while the surface micromachining leads to small air gap defined by the sacrificial layer which is less than 1.5μm thick [4]. Deep gap air allows large deformations during mechanical bending; this explains the wide use of nanoindentation for MEMS structures made by bulk micromachining compared to those obtained by surface micromachining.

2.2. Bending test by nanoindentation

Two experimental configurations of the bending test are presented here: bending of cantilever and bridge microbeams. As in the conventional nanoindenter, the recorded load-displacement data are used to determine the mechanical properties of the cantilever and the bridge microbeams through theoretical models. However the model employed in the bending tests is different to the standard Oliver and Pharr model [5].

2.2.1. Bending test of single-layer cantilever

The principle of the mechanical bending of cantilever microbeams by nanoindentation was firstly introduced in [6]. The test was done by applying a load with the nanoindenter tip at the edge of the free extremity of the cantilever while the load-deflection was continuously recorded. The principle of the bending test is shown in Figure 1; where e, l and L are the thickness, the width and the length of the cantilever, respectively. The dimensions of the structures in MEMS devices have wide ranges, from submicrometers to millimeters. Thus, evaluations of the mechanical properties of thin films cover a very wide range of measurement scale.

-Young modulus

According to the beam bending theory, the relationship between the Young's modulus, E, the cantilever geometry, the deflection, δ, at an arbitrary point "B" and the contacting load, P, is described as follow [7]:

$$E = \frac{P}{\delta} \frac{4(1-\upsilon^2)L^3}{le^3} \quad (1)$$

where υ is the Poisson's ratio of the cantilever material. E is obtained by extracting the initial linear slope, P/δ, from the load-deflection curve recorded by the nanoindentater [6, 8, 9]. An example of a load-deflection curve is seen in Figure 2. The above relation is applicable only for small beam deflections ($\delta \ll L$).

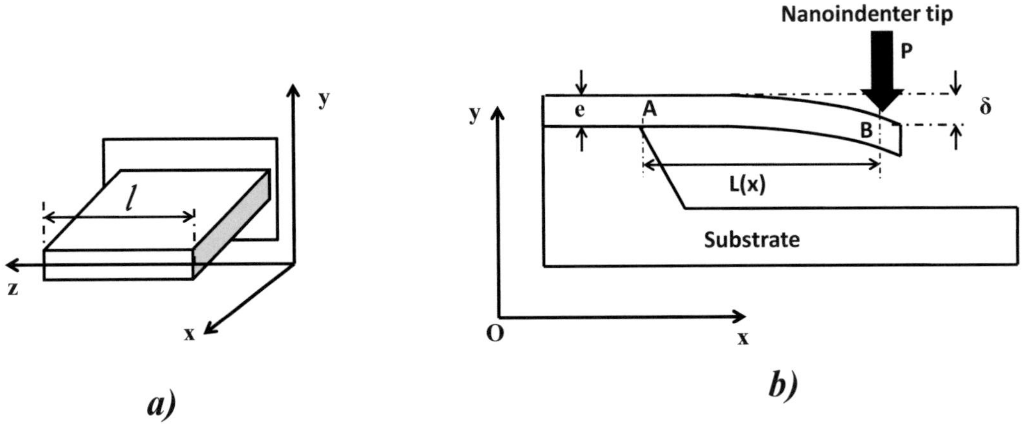

Figure 1. a) Cantilever geometry and b) principle of the bending test of cantilever beam by nanoindentation.

-Yield strength

Another interesting parameter to be extracted by bending test is the yielding strength of the cantilever material. This is obtained using the stress formula in the cantilever which reaches its maximum value at the fixed end of the beam where the applied moment is greatest (point "A" in Figure 1b). When this maximum stress reaches the yield strength of the material, the cantilever begins to deform plastically. This permanent deformation can be recognized in the plot of load-deflection curve by a deviation from linearity (Figure 2). The load at this deviation is noted the yield load, P_y, and the yield strength is given by [8]:

$$\sigma_y = \frac{6P_yL}{le^2} \tag{2}$$

The Table 1 summarizes the Young's modulus and the yield strength obtained after bending tests of Al and Au materials [8]. In parallel, standard nanoindentation was carried out on the same samples in order to verify the bending tests results. The comparison between Young's modulus obtained with the two methods show close agreements with errors within 5%. For the yield strength, the values obtained by the authors are much higher than the bulk values. This fact is not surprising because of the dependency of the plastic behavior of thin films with microstructure morphology and residual stresses. These studies have proven that the cantilever bending by nanoindentation can be used to accurately measure the mechanical properties of thin films.

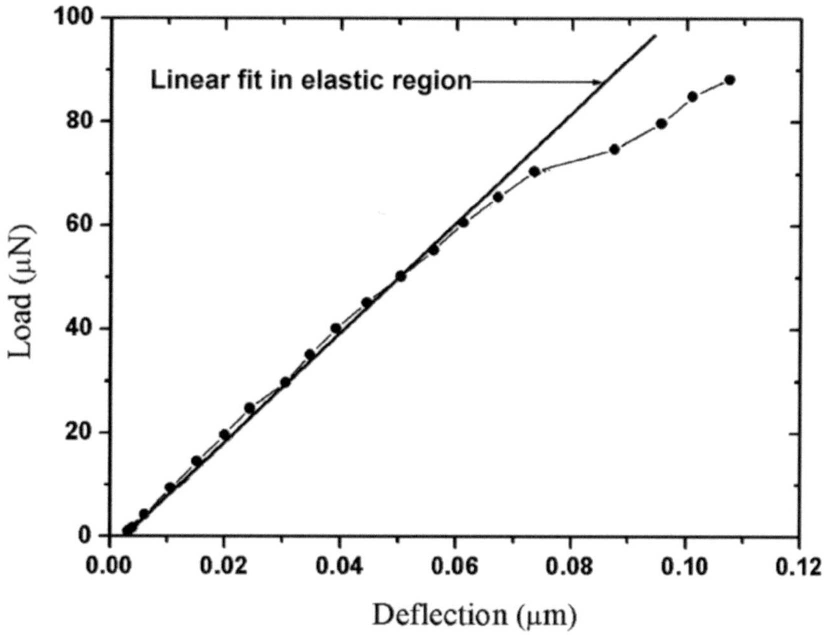

Figure 2. Load versus deflection of an Al cantilever in elastic-plastic region by nanoindentation technique [8].

Film	Thickness (μm)	Young's modulus (GPa)		Yield strength (MPa)
		Cantilever beam	Nanoindentation	Cantilever beam
Al	0.51	80.9	79.3	200.1
	1.03	79.6	73.7	135.7
	1.5	77.6	74.3	114.3
Au	0.56	130.3	122.7	391.4
	0.99	108.7	112.5	309.9
	1.26	107.2	110.4	114.7

Table 1. Experimentally Young's modulus and yield strength of Al and Au films [8].

2.2.2. Bending test of single-layer bridge microbeam

Young's modulus and bending strength can be measured by nanoindentation for a fixed bridge microbeam loaded at its center (Figure 3a). For bridge beams that follow linear elastic theory of an isotropic material, the Young's modulus under given load can be expressed as [10]

$$E = \frac{l^3}{192I}\frac{P}{\delta} \tag{3}$$

where l is the beam length, I is the area moment of inertia for the beam cross-section and P/δ is the slope of the load–deflection curve during bending. The area moment of inertia is calculated from the following equation [10]:

$$I = \frac{{w_1}^2 + 4w_1w_2 + {w_2}^2}{36(w_1 + w_2)}t^3 \tag{4}$$

where w_1 and w_2 are the upper and lower widths, respectively, and t is the thickness of the beam.

The bending strength, σ_b, corresponds to the maximum tensile stress which is produced on the top surface at the two ends and it is given by [10]:

$$\sigma_b = \frac{P_{max} l e_1}{8I} \tag{5}$$

where P_{max} is the applied load at the failure and e_1 is the distance of the top surface from the neutral plane of the beam cross-section and is given by [10]:

$$e_1 = \frac{t(w_1 + 2w_2)}{3(w_1 + w_2)} \tag{6}$$

An example of a linear loading followed by an abrupt failure of a Si nanobeam bent for failure analysis is presented in Figure 3b [11].

2.3. Fatigue behavior

Fatigue tests of MEMS are also of interest especially for the MEMS involving vibrating structures such as oscillators, vibratory gyroscopes [12], comb drive actuators [2] and hinges in digital micromirror devices (DMD) [13]. For example of DMD, the mirror in normal operating mode switches once every 200 microseconds for 5 years at 1000 operating hours per year. This means each mirror element needed to rotate, or switch, more than $90\text{x}10^9$ times to ensure a reliable MEMS product [13]. Fatigue tests have been presented in [11] where nanobeams were deformed with cyclic stresses using a nanoindenter through the "continuous stiffness measurement" (CSM) method; in which stiffness is measured continuously during the loading. The CSM technique has been described in detail in reference [5]. In reference [11], the fatigue behavior of Si nanobeams was characterized by monitoring the change in contact stiffness versus the number of cycles as shown in Figure 3c. Fatigue damage was determined by the sudden decrease in contact stiffness which occurred for the Si nanobeams at $0.6\text{x}10^4$ cycles.

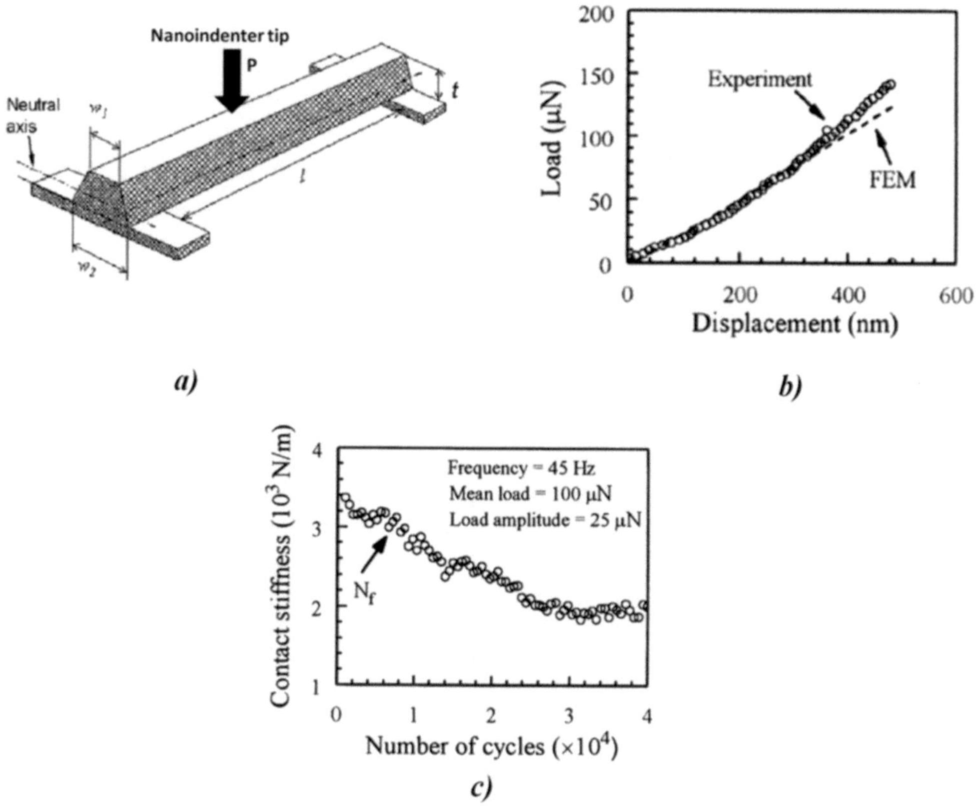

Figure 3. a) A schematic of a typical nanobeam under central loading, b) Load–displacement curve of a Si nanobeam obtained from a nanoindentation experiment, and c) Contact stiffness as a function of the number of cycles for a Si nanobeam cyclically deformed by an oscillation load amplitude of 25 mN with a mean load of 100 mN at a frequency of 45 Hz [11].

2.4. MEMS switches microcontacts characterization by nanoindentation

MEMS switches are switching device that are fabricated using the micromachining technology, where the switching between the on- and off-states is achieved via a pair of microcontacts. One microcontact is typically rigid in space and the other microcontact is movable since it is a part of the deformable elastic structure of the switch. The displacement is induced by various actuation mechanisms such electrostatic, electrothermal, magnetostatic, and electromagnetic. So, the reliability of MEMS switches depends on both the mechanical and the electrical aspects and the main failure of switches occurs in the microcontacts. Therefore, a realistic characterization of MEMS switches requires a simultaneous study of the mechanical and electrical aspects. To achieve this goal, a nanoindentater is combined and synchronized with an electrical resistance measurement technique (see Figure 4). The nanoindenter controls the movement of the deformable structure and also records the load-displacement or the contact stiffness data while the electrical apparatus measures the variations of the electrical resistance of the microcontacts.

Typical contact loads used in MEMS switches range from a few μN to several hundred μN [14].

An example of the evolution of the contact stiffness and the contact resistance versus applied load for a 175μm long Au cantilever is presented in Figure 5 [15]. The main result that can be deduced from the Figure 5 is that the electrical contact was not simultaneously established with the mechanical actuation of the switch. A gap of 20μN is noticed and can be explained by the analysis of the contact interface between the two contacting materials. Contact between two bodies occurs at discrete spots produced by the mechanical contact asperities of the two surfaces.

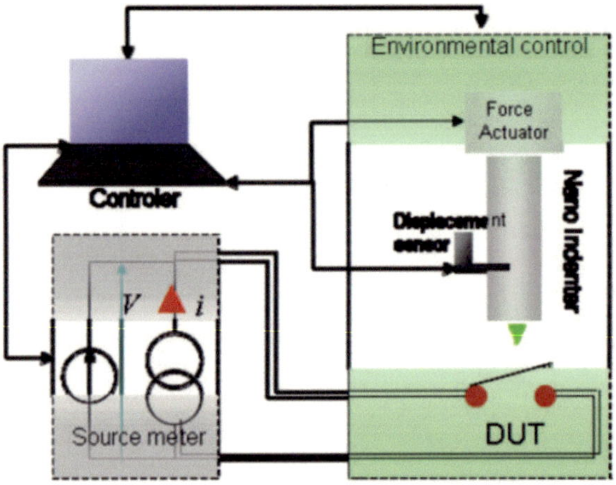

Figure 4. Principle of mechanical and electrical tests of MEMS switches by nanoindentation [9].

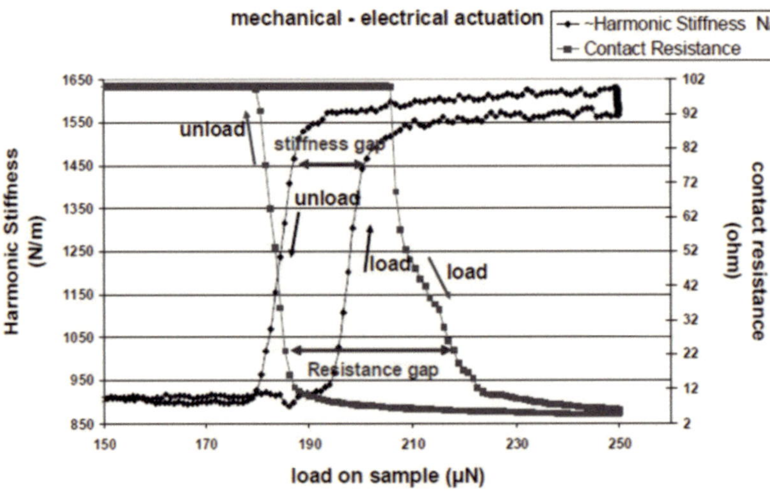

Figure 5. Stiffness and contact resistance versus applied load [15].

This true contact area is smaller than the nominal contact area, leading to high resistance values at the first stage of the actuation.

The calculation of the contact resistance between two bodies is usually done by using the Holm's electric contact theory [16]. In this theory, the contact resistance, R_c, between two contacting bodies of resistivity ρ is equal to:

$$R_c = \frac{\rho}{2a} \tag{7}$$

where a is the radius of the contact area which is assumed to be a unique circular spot. However, this model is not realistic because it does not take into account the multiple contact spots within the area of contact due to both the large-scale waviness and the surface roughness of the contacting materials. Other assumptions of the Holm's theory are related on the fact that the contacting bodies are considered as semi-infinite compared to the area of contact and the surface contamination on the contacts such as oxide film is ignored. The radius of the contact area is linked to the mechanical parameters such as the applied load, P, and the hardness, H, of the softer material and it is expressed as follow [16]:

$$a = \sqrt{\frac{P}{\pi H}} \tag{8}$$

Failure analysis such adhesion of the microcontacts can also be investigated by measuring the contact resistance variation over a mechanical cycling of the switch [17]. For a rounded microcontact bump coated with 300 nm of gold studied in [17], the adhesion failure was noticed after 60,000 contact cycles. This is illustrated by significant changes in pull-off force and contact stiffness (Figure 6a). Also, the adhesion failure is illustrated by scanning electron microscope (SEM) images in Figure 6. Figure 6b and 6c show the images of the bump microcontact prior and after 60,000 cycles, respectively. The bottom contact pad after 60,000 cycles is seen in Figure 6d.

Other studies on thermomechanical failures such changes of the mechanical properties of contact material, the modifications of the contact topology and the diminution of the time dependence creep effect have been investigated for Au-to-Au microcontacts by [18].

2.5. Experimental limitations

The MEMS micro-fabrication techniques like the surface manufacturing and the bulk manufacturing do not guarantee uniform specimen dimensions. Cantilevers made by bulk manufacturing can result to dimensional deviations of 1 to 2μm [19]. Thickness is particularly the major sources of error in the determination of the Young's modulus and yield strength. The dependency of the cantilever Young's modulus and the yield strength to the thickness third power and second power, respectively, can impair the measurements of the MEMS bending tests (equations 1 and 2). Furthermore, the measurement technique of the cantilever dimensions may also be an additional source of error. The dimensional variations may cause 13% deviation in the Young's modulus and 9% deviation in the yield strength [6].

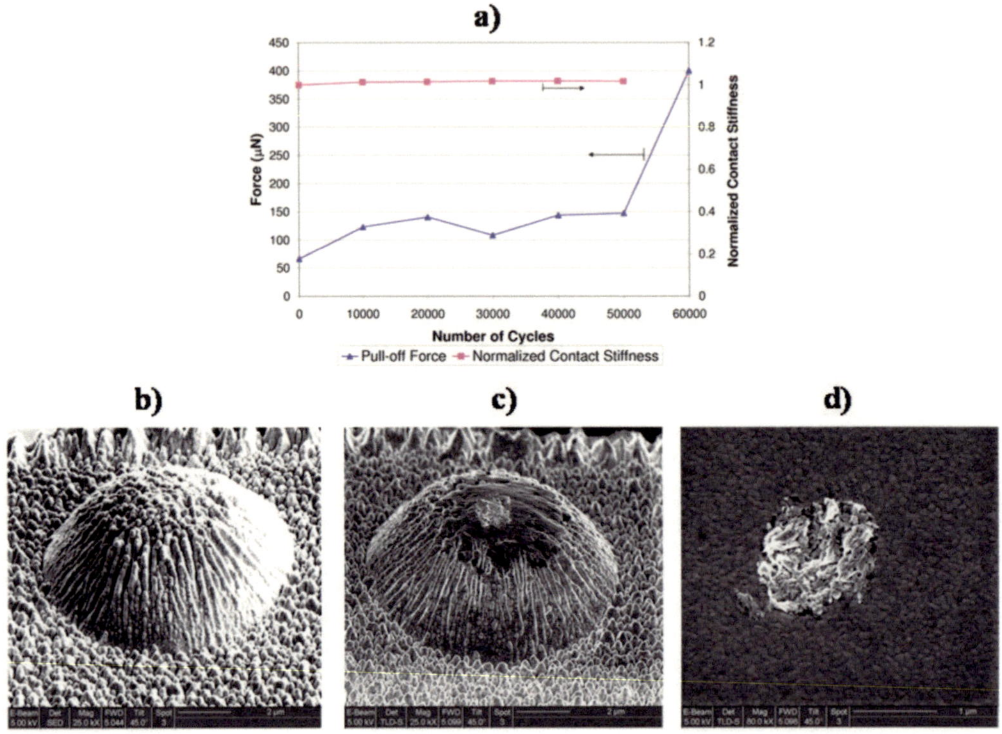

Figure 6. Images of microcontacts during catilever tests a) gold contact bump prior cycling b) gold contact bump after 60,000 cycles and adhesive failure and c) bottom contact after 60,000 cycles adhesive failure [17].

To reduce the impact of cantilever dimensions on the mechanical properties, the use of focused ion beam (FIB) was suggested as a manufacturing technique since 0.03μm thickness variation was achieved for thin Al film ranged from 0.8 to 1.2μm [20].

Other aspects such as tip geometry and cantilever with strong texture must be considered during the MEMS bending test by nanoindentation. The standard Berkovich pyramidal tip can penetrate the cantilever material leading to an overestimation of the recorded deflection and therefore inaccurate mechanical properties based on equations 1 to 6. Removing the tip penetration from the recorded displacement or using blunt tip must be considered. However, in some cases, the tip penetration depth can be neglected when it is so much smaller than bending displacement. For highly anisotropic thin film such electroplated Cu which has strong texture, the corresponding Young's modulus is accordingly anisotropic. In this case, the measured Young's modulus of the cantilever by nanoindentation bending test represents the elastic modulus in parallel direction to the surface of Cu thin film which differs to the Young's modulus in the perpendicular direction [9].

3. Size effects study by uniaxial compression of micropillars

The nanoindentation technique can also be used to perform the uniaxial compression of micropillars for the purpose of investigating the influence of the size scale effects on the plastic response of the material constituting the micropillars. The first work of the uniaxial compression of micropillars by nanoindentation was done by Uchic et al. [21]. Micropillars have diameter varying from hundreds of nanometers to tens of micrometers with an aspect ratio ranging from 2:1 to 4:1. Most of micropillars are fabricated using focused ion beam milling to create samples of uniform cross-section that remain attached to the bulk. Compression testing is carried out using a nanoindenter ended with a flat tip. The extracted load-displacement data are converted into stress–strain curves with the intention of investigating the plastic response of micropillar samples of different materials.

3.1. Size effects on plastic deformation

A size-scale effect can be defined as a change in material properties due to a variation in either the dimensions of an internal feature or structure or in the overall physical dimensions of a sample. When a crystal deforms plastically, phenomena such as dislocation storage, multiplication, motion, pinning, and nucleationt are generally active over different length scales. So the understanding of the material deformation in micro or submicron scale is a main issue that allows designing reliable small devices [21].

An example of stress-strain curves of [111] Ni pillars with diameters ranging from 165 nm to 2 µm is presented in Figure 7 [22]. Two interesting observations can be found from the analysis of these curves. The first one is related to the evolution of the yield strength during compression as function of the micropillar diameters. At micron and submicron sizes, the yield strength, σ_y, was found to rise significantly when the micropillar diameter, *d*, decreases for various Ni and Ni alloy [21-23], Au micro/nano pillars [24,25], and Al [26] micropillars.

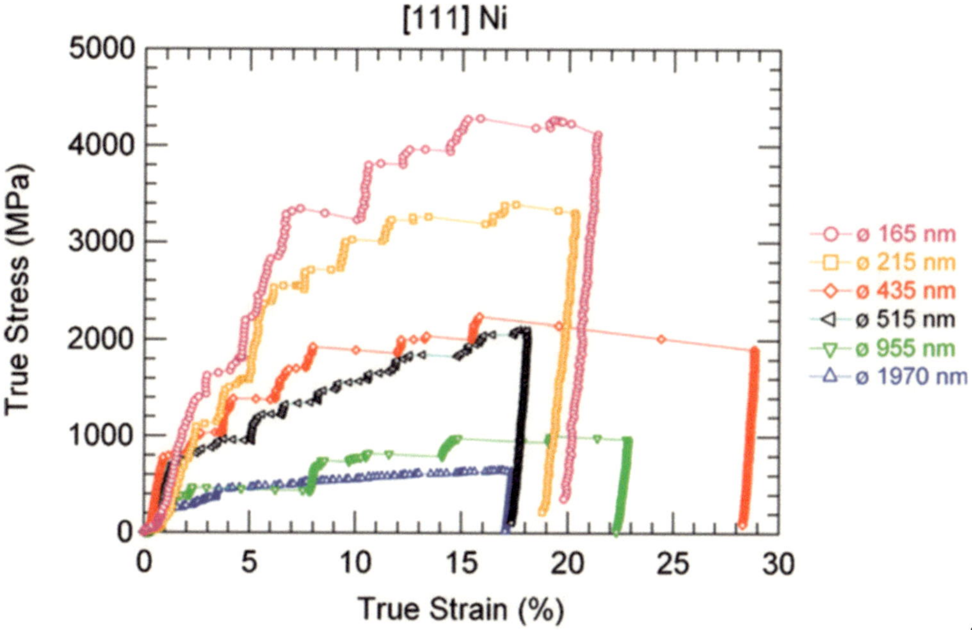

Figure 7. Stress–strain curves for [1 1 1] Ni pillars of various diameters [22].

The relationship between the two parameters can be expressed as follow:

$$\sigma_y \approx d^{-n} \tag{9}$$

Although the parameter n depends on the micropillar size and material as well as the compression strain of the micropillar under test, its value is comprised between 0.6 and 1 for all the studies reported in this work. As a result, the micron and submicron pillars yielded at stresses much higher than the typical yield strength of bulk materials where the yield strength, given by Hall-Petch relation, is proportional to the inverse square root of the average grain diameter. Furthermore, the size effect on the plastic yielding is insensitive to the micropillar fabrication method [24, 25].

The second observation seen from the compression of micropillars (Figure 7) is that all curves display during compression fast transients of strain, i.e., rapid bursts of strain separated by elastic loading segments. These bursts have been widely reported in several works [21-26].

The interpretation of the high flow stresses as well as the strain bursts of the micropillars under uniaxial compression remains an open question. One popular theory suggested a lack in dislocation multiplication events during deformation due to the small distances travelled by the dislocations before reaching the free surface of the micropillar. This leads to a dislocation free test structure and correspondingly a high flow stresses necessary to nucleate new dislocations and continue deformation. This theory is termed dislocation starvation theory [24, 25]. The dislocation starvation model was corroborated by the work in reference [26], where the authors investigated transmission electron microscopy (TEM) images of the dislocation structures which showed that the dislocation density of the micropillars did not grow significantly after severe deformations.

3.2. Experimental limitations

During nanoindentation tests, great efforts must be made to ensure that the measured deformation is due to the actual compression of the micropillars. To achieve this, firstly, the flat tip of the nanoindenter must be as parallel as possible to the top surface of the micropillar (perfect alignment); otherwise, the contact stiffness of the micropillar under test will be underestimating and as a result, the corresponding measured Young's modulus would be smaller than it should has been. In addition, severe misalignment will result in plastic buckling of the micropillars especially those with high aspect ratio. This will consequently affect the experimental stress-strain curves [27]. The misalignment between the flat tip and the micropillar top surface can be minimized by using a tiltable stage to hold the micropillar samples. Secondly, friction between the flat tip and the micropillar is also another issue and it may increase the yield strength, as well as prevent crystallographic rotation of the pillar during deformation [28]. However, some friction has been shown to be necessary to suppress plastic buckling of the sample [27]. Finally, another main concern that may have significant impact on the accuracy of the stress-strain curves is the actual measurement of the deformation of the micropillars. Indeed, the nanoindenter is sensitive to any deformation of all the structures under test including the elastic deformation of the substrate where the micropillar is attached. Accurate determination of the true compression of the micropillar requires accounting for the elastic displacement, h_s, of the substrate under the micropillar which is calculated by using the Sneddon's solution of a cylindrical punch with diameter, d, pressed into an elastic half-space [29]:

$$h_s = \frac{1}{E_{eff}}\left(\frac{P}{d}\right) \tag{10}$$

P and E_{eff} are the applied load and the effective Young's modulus of the substrate material, respectively. The actual micropillar displacement, h, is deduced by subtracting the substrate displacement from the measured displacement, h_{meas} [22, 28]:

$$h = h_{meas} - \frac{1}{E_{eff}}\left(\frac{P}{d_b}\right) \tag{11}$$

where d_b represents the diameter of the micropillar base.

4. Analysis of cylindrical bump insertion using the nanoindentation technique for fine pitch interconnection applications

The insertion technology of cylindrical bumps was proposed as an alternative solution for ultra fine pitch flip-chip 3D interconnection technology in order to overcome process limitations caused by planarity defects, high process loads and temperature, or exhibit low hybridization speeds [30-33]. The insertion concept consists of the penetration of hard cylindrical bumps or microinserts into a softer thin film material in order to achieve mechanically and electrically reliable contacts. The insertion process allows finer pitch possibilities than the conventional flip chip because 2μm diameter microinserts have been achieved [31]. The standard flip chip and the insertion processes are illustrated in Figure 8a and 8b, respectively. Nanoindentation is used to clarify the contact formation mechanisms, the performance and the reliability during insertion. Unlike the previous sections, here the standard nanoindenter tip is replaced by a holder which receives the cylindrical bump sample. The study is focused on the insertion test of a single microinsert.

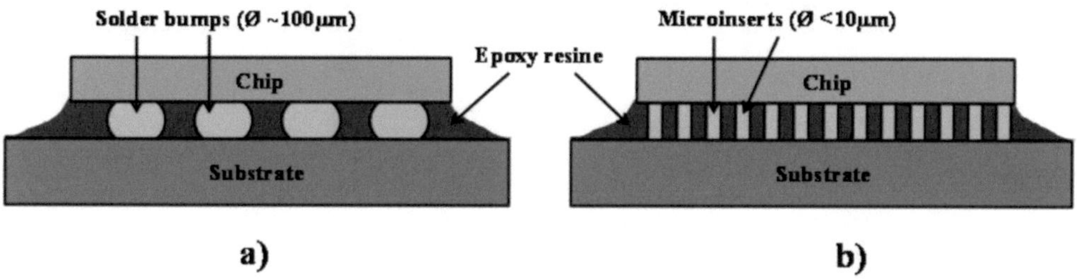

Figure 8. Schematic of flip chip interconnection technique a) Standard microbump and b) Insertion process.

4.1. Experimental details

4.1.1. Sample description

The microinsert used in this study is based on Ni material due its high hardness value and the thin film is made of Al which is a common layer of the substrate bond pad in flip chip technology. Therefore, two kinds of samples are fabricated. In both cases, Si wafer with 0.5 µm thick thermal oxide is used as starting substrate. The Ni microinsert is electroplated after photolithography steps on an Al sputtered thin film (1000 nm) covered by a Cu seed layer (500 nm). Experiments are performed with three nominal Ni microinsert diameters of 6 µm, 8.5 µm and 12.5 µm having the same height of 5 µm. The second test sample is a 450 nm Al thin film deposited on a 10 nm adhesion layer of W and a 250 nm diffusion barrier of WN. The two substrates are diced into 1.4 x1.4 mm² and 1x1 cm² chips for Ni microinsert and Al film samples, respectively.

4.1.2. Modified nanoindenter

For the experiments presented here, the standard nanoindenter tip is replaced by a holder which receives the Ni microinsert test sample. The second modification is related to the use of a ± 2° tiltable stage to support the Al sample instead of the conventional substrate holder. To perform the electrical measurement during the insertion, Ni microinsert and Al samples are directly wire-bonded to contact pads integrated into the sample holders which are connected to a voltage/current sourcing and measurement unit. Figure 9 shows the modifications made in the nanoindenter for insertion tests. Four-point probing technique is used in order to eliminate resistance contribution of the wiring and the connections.

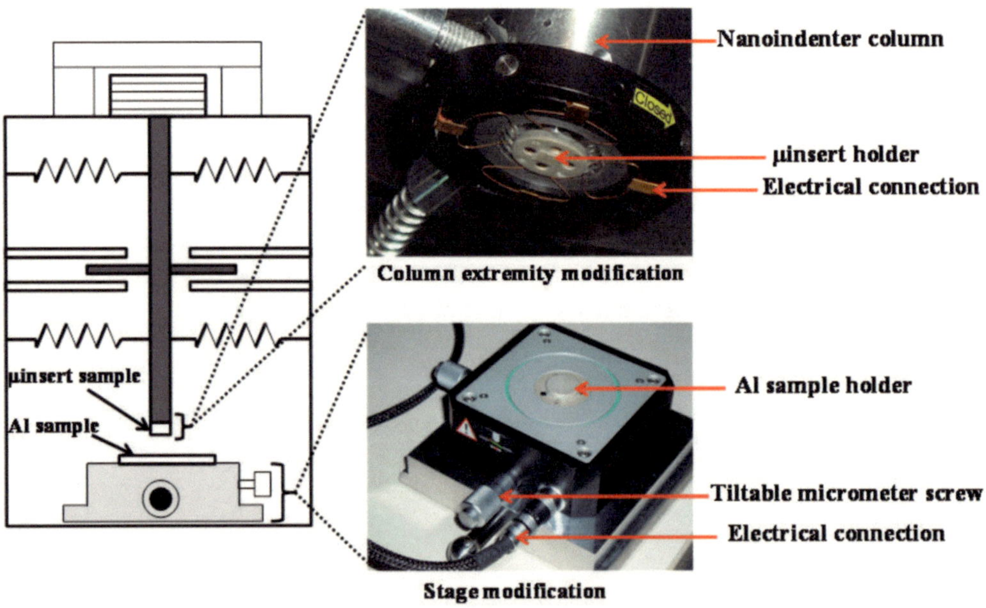

Figure 9. Principle of insertion test by modified nanoindentation.

4.1.3. *Preliminary operations*

The microinsert sample dimensions and the Ni and Al surfaces misalignment can be an obstacle to successfully performing insertion experiments. Hence, prior to experiments, we first carried out a mechanical polishing of Ni microinsert sample edges to avoid their contact with the target Al thin film. Then the sample misalignment with respect to the indentation axis is estimated by standard nanoindentation and correction is made. The misalignment of the Al surface can be reduced below 0.04° with the tiltable stage and the Ni microinsert sample misalignment is found to be less than 1°. Another operation which must be done before performing the experiments is the estimation of the additional frame compliance due to the Ni microinsert holder and the Al holder contributions. The new frame compliance is estimated using the Oliver and Pharr method [5] and its value is equal to 10^6 N/m which corresponds to a system less stiff than the classical nanoindentation frame (~8 $\times 10^6$ N/m). This new compliance allows subtracting the displacement due to the deformation of the sample holders from the raw displacement data recorded during insertion.

These preliminary operations ensure that the displacement measurement is only related to the insertion of the Ni microinsert into the Al film.

4.1.4. *Test procedure*

The experiments are performed at room temperature with a single Ni microinsert. The insertion of Ni microinsert into Al film is made with a single loading/unloading cycle separated by a hold period of 600 secs at maximum load. This hold period is called first creep and it allows studying the creep phenomenon during insertion process. The unloading is stopped just before its end at low load for a second creep stage in order to follow an eventual deformation recovery. A thermal drift correction not described here was performed prior to the loading step. For each Ni microinsert diameter (6 μm, 8.5 μm and 12.5 μm), three maximum loads are applied corresponding to the equivalent uniaxial stresses of 0.8 GPa, 1.6 GPa and 3.2 GPa. Table 2 summarizes the experimental parameters.

Microinsert diameters	6μm			8.5μm			12.5μm		
Maximum load (mN)	23	46	92	46	92	184	100	200	400
Uniaxial stress (GPa)	0.8	1.6	3.2	0.8	1.6	3.2	0.8	1.6	3.2
Loading time=Unloading time (sec)	0.5	1	2	0.5	1	2	0.5	1	2
1st holding time = 2nd holding time (sec)	600								

Table 2. Experimental parameters used for insertion tests.

4.2. Results and discussion

4.2.1. Mechanical behavior during insertion

The insertion curves (the plots of the load versus displacement) for each microinsert diameter at different maximum loads are presented in Figure 10, 11 and 12. It is important to point out that these curves represent the behavior of the global system in contact during the insertion test taking into account Al target film, Ni microinsert, the microinsert landing pad and the elastic deformation of both Si substrates. The goal of the authors was to use these curves to understanding the failure of the materials during the insertion for process validation; so less focus was dedicated to the mechanical parameters extraction. Only creep study was done and will be presented further.

According to diameters, the curves show similar evolution. So, for sake of simplicity, the results and analysis will be done only for 6 μm diameter microinsert. Sample characterizations with AFM, optical profiler and SEM are made on each Ni microinsert prior and after tests as well as the residual impression on Al thin film. Figure 13 shows the images of the 6 μm diameter microinserts with their corresponding residual impressions on Al films at the three maximum stresses.

Results from the three microinsert diameters show that the microinserts deform elastically for the applied maximum stresses of 0.8 GPa and 1.6 GPa, while at 3.2 GPa, a large plastic compression is noticed. A penetration of the microinserts into their landing pad which is constituted of Cu(500nm)/Al(1000 nm) layers is observed at 1.6 GPa and 3.2 GPa (Figure 13a(2) and 13a(3)). This is confirmed by the presence of pile-up around the microinsert base. The combination of the large plastic compression of the microinserts and the penetration into their landing pad, produce a large deformation that corresponds to an abrupt slope change of the insertion curves at applied maximum stress of 3.2 GPa (see Figure 10, 11 and 12). For Al thin film, the deformation is elastic-plastic and the insertion height increases with the maximum applied stresses for all microinsert diameters. At 0.8 GPa, the residual impression is superficial and is due to the roughness of the microinsert surface. At 1.6 GPa and 3.2 GPa, the average insertion heights are 60 nm and 280 nm, respectively. These values are lower than the penetration depths of the microinserts into their landing pad which are 260 nm and 710 nm at 1.6 GPa and 3.2 GPa, respectively. So, the microinserts penetrate mostly into their landing pad rather than into the facing thin Al film.

4.2.2. Al creep parameters

The creep parameters of the thin Al film were extracted from the insertion curves in Figure 10, 11 and 12. It was necessary to deduce from these curves the only displacement which occurs within the Al film since displacement measured during the first creep stage takes into account the displacement of the whole system. To achieve this goal, insertion tests with the experimental parameters presented in table 2 were reproduced by replacing the Al film by a Si substrate. As the deformation within Si substrate is perfectly elastic during the contact with Ni microinsert, Al creep curves are obtained by subtracting the creep curves of Ni/Si contact from the creep curves of Ni/Al contact. This approach was found to be correct only with 12.5 μm diameter Ni inserts. For smaller diameter, Ni deformation at Si contact was not similar to Ni deformation at Al contact and results could not be used confidently.

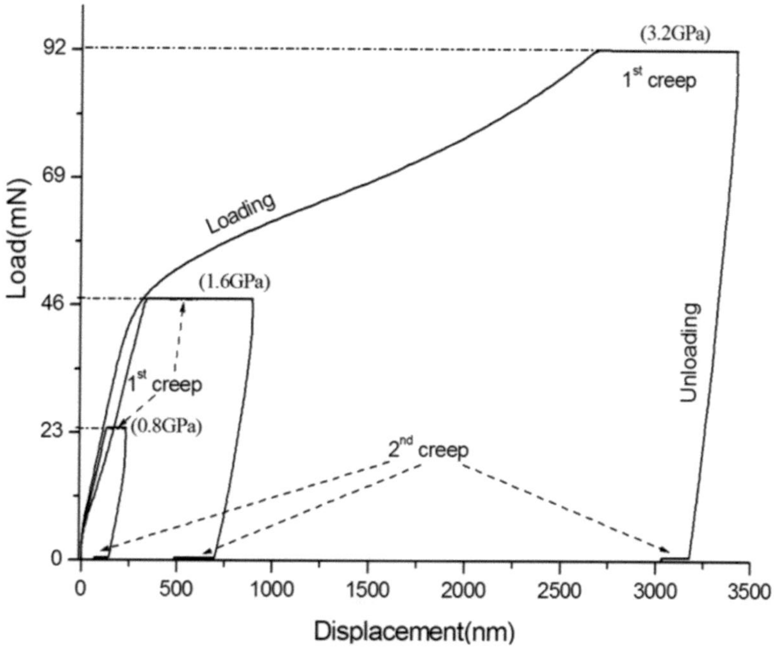

Figure 10. Insertion curve of 6 μm Ni diameter at different maximum loads.

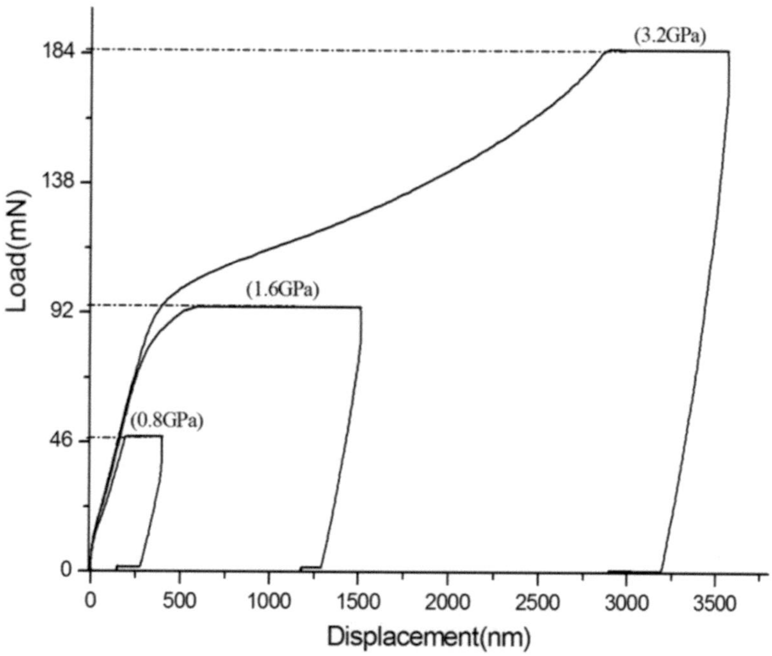

Figure 11. Insertion curve of 8.5 μm Ni diameter at different maximum loads

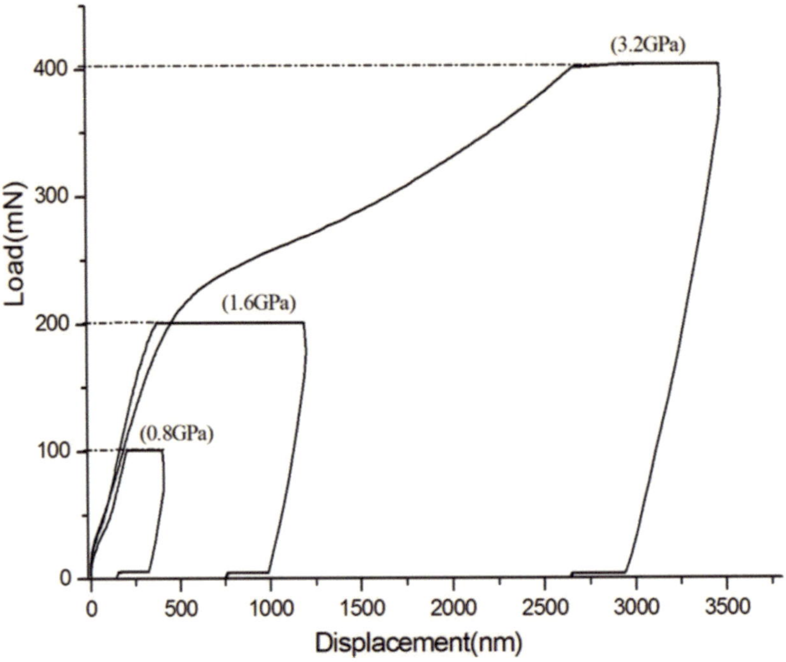

Figure 12. Insertion curve of 12.5 μm Ni diameter at different maximum loads.

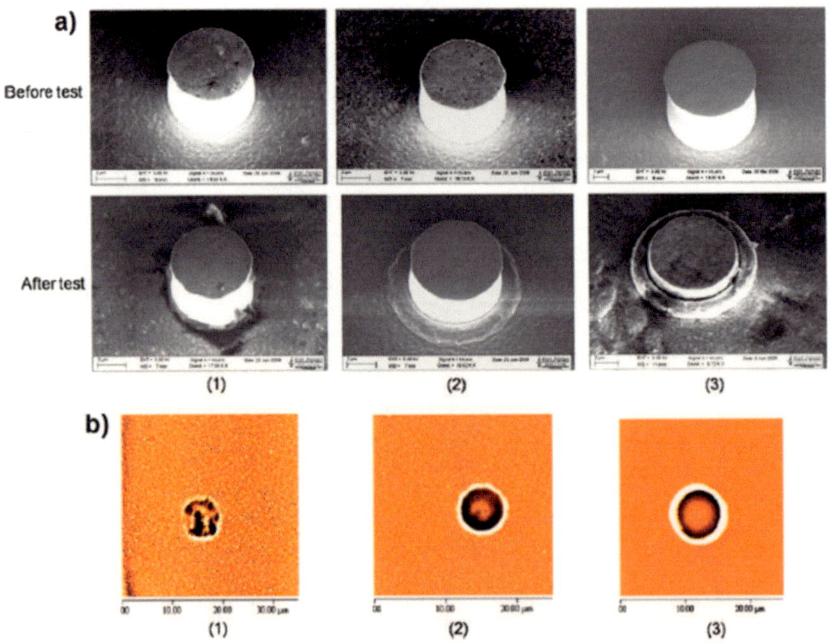

Figure 13. Samples characterization of insertion tests with 6 μm diameter Ni microinserts (a) microinserts before and after tests and (b) residual impression on Al film after test. (1–3) referred to maximum load of 23 mN (0.8 GPa), 46 mN (1.6 GPa) and 92 mN (3.2 GPa), respectively.

Figure 14 presents Al creep curves obtained after insertion with 12.5 μm diameter Ni microinsert for the first creep stages. Primary or delayed creep followed by secondary or steady creep is observed. One can notice a larger displacement variation of the curve obtained with a load of 200 mN compared to the one obtained with 400 mN. An explanation could be related to the fact that at 400 mN the microinsert showed a large plastic deformation prior the creep stage. This result to a larger contact area and a significant influence of the presence of the Si substrate underneath the Al film which both limit the creep deformation.

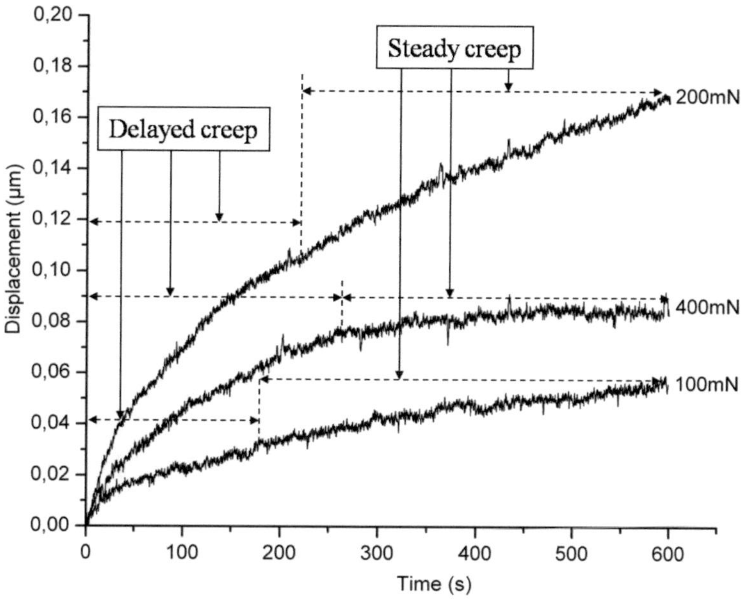

Figure 14. Al creep curves with 12.5μm Ni microinsert at maximum loads of 100mN, 200mN and 400mN.

To extract Al creep parameters, a Maxwell/Kelvin-Voigt phenomenological model was used. In the case of a Ni microinsert with a radius *r* inserted into Al, displacement change is given as function of time under a constant load *P* by:

$$h(t) = \frac{P}{r}\left[\frac{1}{E_1} + \frac{1}{E_2}\left(1 - e^{\frac{-tE_2}{\eta_2}}\right) + \frac{1}{\eta_1}t\right] \quad (12)$$

where E and η are phenomenological parameters related to the spring and the dashpot viscosity coefficient of the model, respectively. The indexes (1) and (2) refer to Maxwell element and Voigt-Kelvin element, respectively. The equation (12) is fitted to the experimental Al creep curves for parameter extraction. The $1/E_1$ term corresponds to Al young modulus and it is not included in the creep curves; so the extracted parameters here are E_2, η_1 and η_2 (Table 3).

Creep load (mN)	100	200	400
E_2 (GPa)	637	412	614
η_1(GPa s)	3.10^5	2.10^5	6.10^{28}
η_2(GPa s)	5.10^4	3.10^4	8.10^4

Table 3. Al creep parameters after insertion with 12.5μm Ni microinsert at different maximum loads

Creep at maximum loads reveals a weak variation of parameters as function of load except the value of η_1 for a 400mN load which tends towards infinity. This could be explained by the large contact area and the influence of the Si substrate resulting from the large plastic deformation prior the creep stage.

The creep parameters extracted here can be helpful for insertion process analysis and prediction.

4.2.3. Materials optimization and electrical resistance results

The results presented in section 3.1.1 reveal the mechanical failures of the material layers underneath the microinserts at certain stresses as well as the insufficient penetration of the microinsert into the facing Al film. Design optimization of the microinserts as well as material and thickness selections of the microinsert landing pad are required to achieve reliable insertion interconnection technology. Thus, a microinsert based on microtube geometry with 10μm pitch has been proposed by [30, 33] in order to reduce the cross sectional area of the microinsert, hence the assembly insertion load. These works have demonstrated higher insertion penetration of microtubes of different materials and dimensions into a thicker film compared to the 450 nm Al film used in section 3.2.1. In addition, since the lateral contact surface increases during the microtube insertion, the corresponded resistance is lower than the measured resistance of the insertion of Ni cylindrical bumps. Figure 15 shows the evolution of the insertion and resistance curves versus time of the two microinsert geometries. The Figure 15a and 15b do not have similar test parameters. On the one hand Figure 15a was extracted by applying a load (92 mN, 300 secs)/unload (300 secs) cycle without any holding stage; on the other hand, Figure 15b was obtained with a faster loading rate (300 μN/sec) at a maximum load below 9 mN. Another noticed difference is that the displacement presented in Figure 15a also takes into account the plastic deformation of the microinsert and its landing pad while the Figure 15b was based exclusively on the insertion of the microtube. Despite this difference in testing conditions, one can find that the interconnection resistance was less than 4 Ω in less than 10 seconds and a maximum insertion higher than 1.6 μm was achieved with the microtube insertion (see Figure 15b). Unlike the insertion of the standard microinsert (see Figure 15a) where a stable resistance of 20 Ω was measured and the maximum insertion was 224 nm. Besides the lateral contact surface, a 240 nm gold layer deposited onto the microtubes greatly influences the lower resistance during microtube insertion.

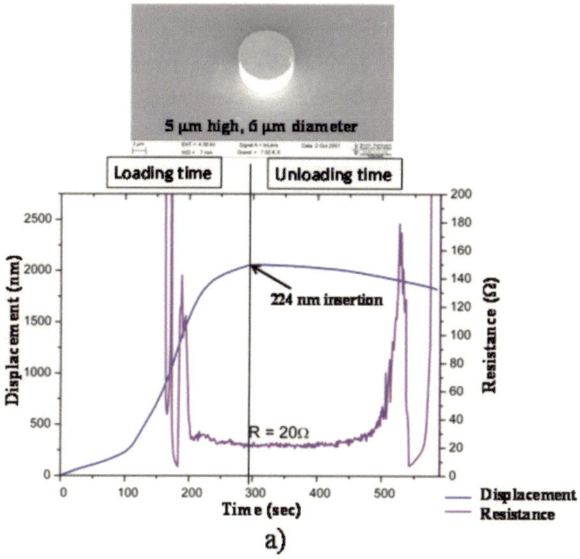

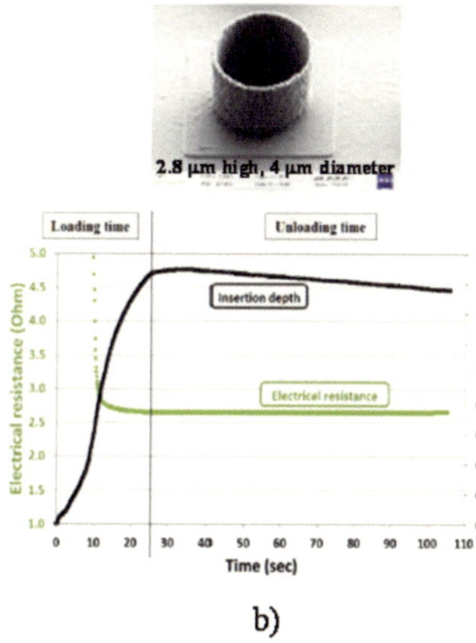

Figure 15. Insertion and resistance measurements versus time a) 6μm diameter microinsert and b) 4 μm diameter microtube [32, 33].

The results extracted from the single microtube were used for process parameters to carry out reliable Die To Wafer (D2W) integrations taking into account industrial requirements. The Figure 16 illustrates 183 D2W integrations carried out on a 200 mm wafer; each chip that has about 95,000 interconnections was assembled using the results of single microtube insertions [33].

Figure 16. Populated 200mm wafer with different insertion conditions for process validation. [33].

5. Conclusion

The use of the nanoindentation technique beyond continuous thin films characterization offers many possibilities for evaluating materials properties such as elastic-plastic constants and fatigue strength at micrometer and submicrometer scales of patterned structures. The experimental data are essential not only in materials selection and designing structures, but also for ensuring the reliability during operation of the microdevices. The research potential of the nanoindentation should be considered beyond the case studies presented here. As an example, carbon nanotubes (CNTs) mechanical characterization by nanoindentation is another attractive subject since few experimental results exist despite its fundamental importance and applications. Whatever the samples under test is and whatever the modifications on nanoindentation tool are, a particular attention should be paid on the extracting data by ensuring that they reflect the deformation of the structures under test. However, nanoindentation tests suffer from low accuracy in displacement at high temperature loading despite recent development made in this area. Another limitations also encountered in continuous thin films tests is the difficulty to dissociate thin film characteristics from the underlying layers or substrate contributions. Overcoming these limitations can be made through theoretical models and finite element simulations. As the need for accurate reliability data on materials used in microdevices expands, so will characterization techniques such as nanoindentation.

Author details

Mamadou Diobet DIOP
Faculty of Engineering, Department of Electrical and Computer Engineering, University of Sherbrooke, Sherbrooke, Quebec, Canada

6. References

[1] Fischer-Cripps AC. Introduction to Contact Mechanics. Mechanical Engineering Series, Springer-Verlag New York, Inc.; 2000.

[2] Tabata O., Tsuchiya T. Reliability of MEMS: Testing of materials and devices. Advanced Micro & Nanosystems.Volume 6. WILEY-VCH Verlag. Weinheim, Germany; 2008.

[3] Shen YL. Constrained Deformation of Materials Devices. Heterogeneous Structures and Thermo-Mechanical Modeling. Springer Sciences, New York; 2010.

[4] Tummala RR. Fundamentals of Microsystems packaging. McGraw-Hill; 2001.

[5] Oliver WC., Pharr GM. Measurement of hardness and elastic modulus by instrumented indentation: Advances in understanding and refinements to methodology. Journal of Materials Research 2004; 19(1) 3-20.

[6] Bravman JC., Weihs TP., Hong S. Nix WD. Mechanical deflection of cantilever microbeams: A new technique for testing the mechanical properties of thin films. Journal of Materials Research 1988; 3(5) 931-942.

[7] Timoshenko SP., Gere JM. Mechanics of Materials. Van Nostrand, New York; 1972.

[8] Jeong JH., Son D, Kwon D. Film-thickness considerations in microcantilever-beam test in measuring mechanical properties of metal thin film. Thin solid Films 2003; 437(1) 182-187.

[9] Hong SH., Kim KS., Kim YM, Hahn JH., Lee CS., Park JH. Characterization of elastic moduli of cu thin films using nanoindentation technique. Composites Science and Technology 2005; 65(9) 1401-1408.

[10] Bhushan B. Nanotribology and Nanomechanics: An Introduction. 2nd edition, Springer-Verlag Heidelberg, Berlin; 2008.

[11] Li X., Bhushan B. Fatigue studies of nanoscale structures for MEMS/NEMS applications using nanoindentation techniques. Surface and Coatings Technology 2003; 163 521–526.
[12] Acar C., Shkel A. MEMS Vibratory Gyroscopes Structural Approaches to Improve Robustness. Springer publisher, New York; 2009.
[13] Douglass MR. Lifetime Estimates and Unique Failure Mechanisms of the Digital Micromirror Device (DMD): proceedings of 36th IEEE International Symposium on Reliability Physics, March 31-April 2, 1998, Reno, Nevada, USA.
[14] Majumder S., McGruer NE., Adams GG. Contact Resistance and Adhesion in a MEMS Microswitch: proceedings of the ASME/STLE International Joint Tribology Conference, 26-29 October 2003, Florida, USA.
[15] Seguineau C., Broue A., Dhennin J., Desmarres JM., Pothier A., Lafontan X., Ignat M. MEMS Characterization – A new experimental approach for measuring electrical contact resistance by using nanoindentation: proceeding of 18th Workshop on MicroMechanics Europe, 16-18 Sept 2007, Guimarães, Portugal.
[16] Holm H. Electrical contacts, Theory and applications. 4th ed., Springer-Verlag, Berlin;1967.
[17] Gilbert KW., Mall S., Leedy KD., Crawford B. A Nanoindenter Based Method for Studying MEMS Contact Switch Microcontacts: proceedings of the 54th IEEE Holm Conference on Electrical Contacts, 27-29 October 2008, Florida, USA.
[18] Broue A., Dhennin J., Seguineau C., Lafontan X., Dieppedale C., Desmarres JM., Pons P, Plana R. Methodology to Analyze Failure Mechanisms of Ohmic Contacts on MEMS Switches: proceedings of the IEEE International Symposium on Reliability Physics, 26-30 April 2009, Montreal, Quebec, Canada.
[19] Liu HK., Pan CH., Liu PP. Dimension effect on mechanical behavior of silicon micro-cantilever beams. Measurement 2008; 41(8) 885-895.
[20] McCarthy J., Pei Z., Becker M., Atteridge D. Fib micromachined submicron thickness cantilevers for the study of thin film properties. Thin solid Films 2000; 358(1) 146-151.
[21] Uchic MD., Dimiduk DM., Florando JM., Nix WD. Sample dimensions influence strength and crystal plasticity. Science 2004; 305(5686) 986-989.
[22] Frick CP., Clark BG., Orso S., Schneider AS., Arzt E. Size effect on strength and strain hardening of small-scale [111] nickel compression pillars. Materials science and engineering A 2008; 489(1-2) 319-329.
[23] Uchic MD., Dimiduk DM., Parthasarathy TA. Size-affected single-slip behavior of pure nickel microcrystals. Acta Materialia 2005; 53(15) 4065-4077.
[24] Oliver WC, Greer JR., Nix WD. Size dependence of mechanical properties of gold at the micron scale in the absence of strain gradients. Acta Materialia, 2005; 53(6)1821-1830.
[25] Greer JR., Nix WD. Nanoscale gold pillars strengthened through dislocation starvation. Physical Review B 2006; 73(24) 245410.
[26] Ng KS., Ngan AHW. Stochastic nature of plasticity of aluminum micro-pillars. Acta Materialia 2008; 56(8) 1712–1720.
[27] Wei Q., Zhang H., Schuster BE., Ramesh KT. The design of accurate micro-compression experiments. Scripta materiala 2006; 54(2) 181-186.

[28] Orso S., Frick CP., Arzt E. Loss of pseudoelasticity in nickel-titanium sub-micron compression pillars. Acta Materialia 2007; 55(11) 3845-3855.

[29] Sneddon IN. The relation between load and penetration in the axisymmetric boussinesq problem for a punch of arbitrary profile. International Journal of Engineering Science 1965; 3(1) 47-57.

[30] Saint-Patrice D, Marion F., Fendler M., Dumont G., Garrione J., Mandrillon V., Greco F., Diop M., Largeron C., Ribot H. New Reflow Soldering and Tip in Buried Box (TB2) Techniques For Ultrafine Pitch Megapixels Imaging Array: proceedings of the 58th Electronic Components and Technology Conference, 27-30 May 2008, Orlando, Florida, USA.

[31] Boutry H., Brun J., Franiatte R., Nowodzinski A., Sillon N., Dubois-Bonvalot B., Depoutot F., Brunet O., Peytavy A. Reliability characterization of Ni-based microinsert interconnections for flip chip die on wafer attachment and their evaluation in multichip simcard prototype:proceedings of the 10th IEEE Electronics Packaging Technology Conference, 9-2 December 2008, Singapore.

[32] Diop MD., Mandrillon V., Boutry H., Fortunier R., Inal K. Analysis of nickel cylindrical bump insertion into aluminium thin film for flip chip applications. Microelectronic Engineering 2010; 87(3) 522-526.

[33] Goubault de Brugière B., Marion F., Fendler M., Mandrillon V., Hazotte A., Volpert M., Ribot H. A 10μm Pitch Interconnection Technology using Micro Tube Insertion into Al-Cu for 3D Applications: proceedings of the 61st IEEE Electronic Components and Technology Conference, 31 May – 3 June 2011, Lake Buena Vista, Florida, USA.

Chapter 10

Mechanical Characterization of Black Diamond (Low-k) Structures for 3D Integrated Circuit and Packaging Applications

Vasarla Nagendra Sekhar

Additional information is available at the end of the chapter

http://dx.doi.org/10.5772/53198

1. Introduction

As Integrated Circuit (IC) technology scales down into the nanotechnology regime, it allows millions of active components to be fabricated on a single chip in accordance with the historical trend of Moore's law. Integration of all these active elements on a single IC, multilevel on-chip interconnect system must be developed in BEOL (Back End of the Line) technology [1]. Accordingly, to meet the industry requirements of scaling for improved performance, semiconductor technologies are forced to move from well-established Al/SiO_2 interconnect technology to Cu/low-k (Copper/low dielectric constant) technology [2-3]. The main objective of this transition is to reduce cross talk noise between metal lines, propagation delays and power dissipation from RC delay. Copper and low-k inter metal dielectric layers are used as multilevel interconnects to enhance the speed of logic devices. Amongst the available low-k materials, Black diamond ™ (BD, low-k, SiOC:H) has been considered as potential inter metal dielectric material for integration in ULSI (Ultra Large Scale Integration) due to its better electrical and dielectric properties. But in dielectric material processing the key issue is the trade-off between dielectric property and mechanical strength. Hence it is very important to study the mechanical properties of BD films and Cu/BD stacks.

There are four interdependent properties which are responsible for the mechanical reliability of thin film structures and they are elastic modulus, hardness, fracture toughness and interfacial adhesion [4]. Both theoretical and experimental studies recommend that hardness and elastic modulus are the two key material characteristics affecting the CMP process [5]. Generally, mechanical properties of thin films differ from those of the bulk materials. This is mainly attributed to their microstructure, interfacial mismatch stresses and molecular restructuring of the films. Parameters used to characterize the mechanical strength of BEOL interconnect materials include hardness, elastic modulus and adhesion/cohesion strength. Interfacial adhesion is the most important property to ensure the thermo-mechanical integrity of Cu/low-k stacks. Interfacial adhesive failure may occur during fabrication processes such as CMP and high temperature curing steps. In addition, delamination or cracking can also be observed during electronic packaging processes like reflow process, flip-chip bonding and back grinding due to thermo-mechanical stresses. This chapter further discusses on mechanical behavior of BD thin films of different thicknesses, single and dual dielectric stacks with Ta & TaN barriers. Nanoindentation experiments with continuous stiffness measurement (CSM) attachment have been performed on different BD film stacks to assess the multi-layer effect on the mechanical behavior of BD films. Effect of wafer backgrinding on the active side of the low-k structures have been studied using the Nanoindentation method as wafer backgrinding is one of the key technologies paving way to high performance three-dimensional (3D) microelectronic packages.

2. Low-K dielectric materials for BEOL (Back End of the Line) multilevel interconnect and 3D Integration applications

The contribution of the R-C delay by conventional multilevel interconnects scheme increases as the IC fabrication technology moving into further miniaturization. The introduction of Cu/low-k interconnect technology into BEOL, has progressively enhanced this condition when compared to the conventional Al/SiO_2 technology by reducing R-C delay in between interconnect lines. [6-7]. In addition to migrating to Cu/low-k multilevel interconnects at chip level, microelectronic industry is more focused on 3D IC integration and 3D packaging of chips at system level. In recent times, three-dimensional (3D) IC integration and packaging is gaining more attention because of its innovativeness, high performance, high functionality, and ability to reduce size of the final product. The major applications of the 3-D packaging include digital and mixed-signal electronics, wireless, electro-optical, MEMS and other integration technologies. At this juncture, the key technologies supporting the 3-D packaging, are as: through silicon vias (TSVs), wafer thinning or back grinding, precision alignment of wafer to wafer or chip to wafer, and wafer to wafer or chip to wafer bonding [8]. Among these key technologies, wafer thinning plays a vital role in the 3D packaging integration, as it allows to accommodate or stack more dies in one package and ultimately results in the reduction of package size. Besides the reduction of package size, the stacking of thin chips provides many other advantages, such as, more functionality per package and improved heat dissipation. Electronic packaging industry has to put a lot of R&D efforts and spend millions of dollars on the wafer thinning technologies as there is no manufacturing technology available for directly producing the ultra-thin wafers [9]. That is why in the past years, many wafer thinning methods, such as mechanical grinding, chemical mechanical polishing (CMP), wet etching and atmospheric downstream plasma (ADP), dry chemical etching (DCE) have been evolved. So far many researchers have extensively studied wafer thinning/back grinding processes in terms of die strength by assessing the quality of the grinded surface [10].

In the recent years, many researchers have extensively assessed the quality of back grinding process with the help of die strength evaluation. Die strength of the thinned wafer can be evaluated by using the different mechanical testing techniques such as three point bend test, four point bend test, ball on ring test, ball breaker tests and ring-on-ring tests. The mechanical tests are greatly influenced by several processes and material parameters such as surface roughness or finish, degree of thinning, stress relief process, quality of the dicing edge [11-15]. However the literature available related to the effect of grinding processes on the active side of the die/chip is limited, and this necessitates a focused study on the effect of wafer back grinding on the active side of chip. For the first time we have studied the effect of back grinding processes on the active side of low-k stack by using Nanoindentation method. Usually active side of the chip is a few microns in thickness, and it cannot be studied using those methods which are being used for conventional die strength evaluation. Therefore, we have chosen sophisticated method like nanoindentation technique for mechanical characterization of low-k films.

2.1. Different types of low-k materials

In the recent past, many low-k materials have been developed, and they can be broadly classified into Si-based and non-Si based. Si-based materials can be further classified into Si-based and silsesquioxane (SSQ) based, which include hydrogen-SSQ (HSSQ) and methyl-SSQ (MSSQ). Non-Si based low-k materials can be further divided into two groups, polymer based and amorphous carbon. Several types of low-k materials with varied low-k values and different deposition methods are being used in the IC fabrication technology. Table 1 lists the contemporary deposition techniques together with the k value of various silicon based, non-silicon based and polymer dielectric material candidates for the 0.13μm and 0.1μm technology nodes.

In silica based dielectric materials, usually tetrahedral silica is the elementary unit. Each silicon atom is at the center of the tetrahedron of oxygen atoms as shown in Fig. 1 (a). Typically Si based dielectric materials are dense structures with higher chemical and thermal stability. In silica based materials, dielectric constant (k) value can be lowered by replacing the Si-O bonds with Si-F bond, producing fluorinated silica glasses (FSG) or doping with C by introducing CH_3 groups [7].

Black diamond (BD) is one of the popular low-k materials and it is a trade mark of CVD processed dielectric material, introduced by Applied Materials Inc. [17]. It is silica based dielectric material, obtained by doping of silica with -CH_3 groups as shown in figure Fig. 1 (b) and it has chemical formula SiOC:H. It is also called hybrid dielectric material as it contains both organic (-CH_3) and inorganic (Si-O) constituents. The dielectric constant (k) value of silica based dielectrics ranges from 2.6 to 3. Typically BD thin films are fabricated by using the Chemical Vapor Deposition (CVD) method near room temperature using organosilane precursor in the presence of oxygen as oxidant. The lower density of the BD films is achieved by introducing network terminating species (-CH_3) into the Si-O matrix [19]. The density and dielectric value of the BD films can be altered by the selection of terminating groups in silica network. Empirically, a larger terminating group gives lower density, because it acts as a network terminating group only and it is not part of the Si-O network. Therefore BD films retain many of the useful thermo-mechanical properties of silicon oxide. The summary of properties of BD films is given in Table 2. BD films can achieve bulk dielectric constant of around 2.5 to 2.7, and integrated ILD stack dielectric constant of <3 [20]. The glass transition temperature of the BD is well above 450°C. The dielectric constant of the BD films can be lowered mainly by introducing constitutive porosity into the microstructure [21]. By proper selection of the compatible barriers layers, thermal and mechanical properties of BD films provide evolutionary pathway to Cu/low-k interconnect technology.

Dielectric Materials	**Fabrication Technique**	**K value**
SiO_2	CVD	3.9 – 4.5
Fluorosilicate glass (FSG)	CVD	3.2 – 4.0
Polyimides	Spin-on	3.1 – 3.4
Hydrogen silsesquioxane(HSQ)	Spin-on	2.9 – 3.2
Diamond-like Carbon (DLC)	CVD	2.7 – 3.4
Black Diamond™ (SiCOH)	CVD	2.7 – 3.3
Parylene-N	CVD	2.7
B-staged Polymers (CYCLOTENE™ and SiLK™)	Spin-on	2.6 – 2.7
Fluorinated Polyimides	Spin-on	2.5 – 2.9
Methyl silsesquioxane (MSQ)	Spin-on	2.6 – 2.8
Poly(arylene ether) (PAE)	Spin-on	2.6 – 2.8
Fluorinated DLC	CVD	2.4 – 2.8
Parylene-F	CVD	2.4 – 2.5
PTFE	Spin-on	1.9
Aerogels/Xerogels (porous silica)	Spin-on	1.1 – 2.2
Porous HSQ	Spin-on	1.7 – 2.2
Porous SiLK	Spin-on	1.5 – 2.0
Porous MSQ	Spin-on	1.8 – 2.2
Porous PAE	Spin-on	1.8 – 2.2
Air Gaps	-	1.0

Table 1. Dielectric constants of various contemporary low-k materials are the interest of 0.13μm and 0.1μm technology nodes [7, 16-18].

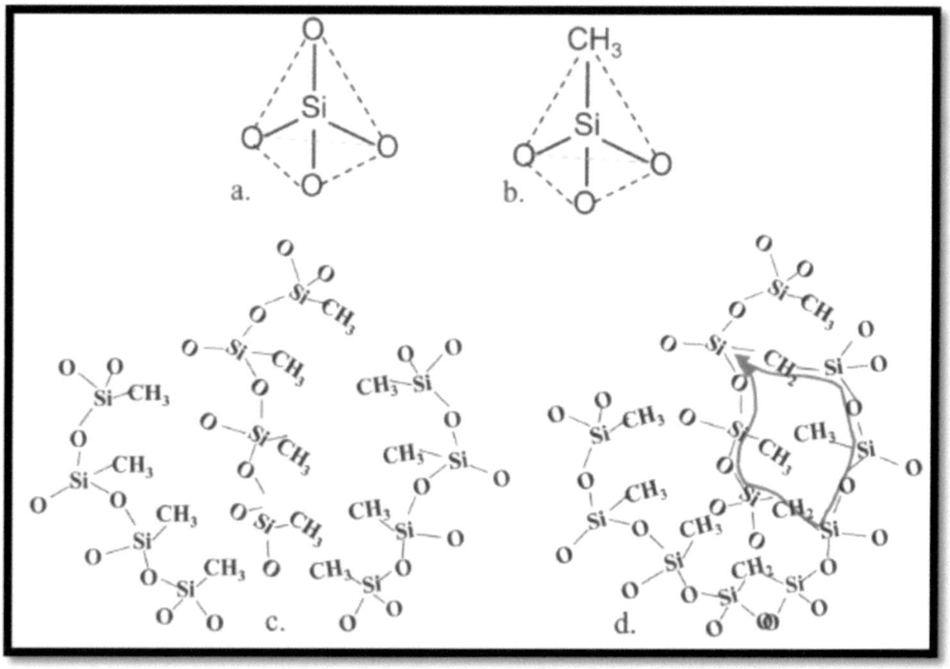

Figure 1. Elementary units of (a) SiO2 (b) carbon doped Silica, called as Black Diamond (SiOC:H) (c) C doped silica without cross linking (d) with cross linking [7]

Property	Value of the Blanket film
Dielectric Constant-Bulk film(Hg Probe)	2.5-2.7 @ 1 MHz
Uniformity (%, 1σ)	<1.5
Stress (MPa)	40-60 Tensile
Stress Hysteresis (MPa)	<20 (RT-450ºC)
Cracking Threshold (μm; blanket film)	>1.5
ASTM scratch tape test on SiN, SiON, Ta, TaN	Passed
Leakage Current (Amps/cm≤)	10-9 @ 1MV/cm
Glass transition temperature	>450ºC
Hardness (GPa)	1.5 – 3.0
Modulus (GPa)	10 - 20

Table 2. Summary of Black Diamond ™ film properties [17,21]

2.2. Required properties of low-k materials for integration

For successful integration of low-k materials into the BEOL interconnects, besides having low dielectric constant it should be chemically, mechanically and thermally stable in the system. Every IC fabrication node technology demands low-k materials with lower dielectric values and with optimum physical properties as summarized in Table 3. Choosing a new low-k material with optimal electrical, thermal and mechanical properties for current interconnects and IC fabrication technology is very crucial. As mentioned earlier, lower dielectric constants are obtained by modifying of the molecular structure of the materials, which ultimately affects the mechanical and thermal properties of the low-k materials. Good thermal stability and low coefficient of thermal expansion is needed to prevent both, damage to the film and, property changes during subsequent thermal processing. The bulk dielectric constant of the ILD stack, when low-k film is stacked with barrier layers and liners (SiC and SiN) should be less than 3.0.

Electrical isotropic k < 3 @ 1MHz	**Chemical No material change when exposed to acids, bases and strippers**	**Mechanical Thickness uniformity <10% within and <5% wafer to wafer for 8″ wafer at 3σ**	**Thermal T_g >400 °C**
Low Dissipation	Etch rate and selectivity better than oxide	Good adhesion to metal and other dielectrics	Coefficient of thermal expansion <50ppm/°C
Low leakage current	<1% moisture absorption at 100% relative humidity	Residual stress <(±)100MPa	Low thermal shrinkage
Low charge trapping	Low solubility in H_2O	High hardness	< 1% weight loss
High electric field strength	Low gas permeability	Low shrinkage	High thermal conductivity
High reliability	High purity	Crack resistance	
High dielectric breakdown voltage > 2-3 MV/cm	No metal corrosion	Tensile modulus >1GPa	
	Long shelf life	Elongation at break >5%	
	Low cost of ownership	Compatible with CMP	
	Commercially available		
	Environmentally safe		

Table 3. Summary of required properties of low-k materials [16]

2.3. Barriers and adhesion promoters for Cu/low-k structures

BEOL multilevel interconnect structures comprise varieties of materials, such as Cu, low-k materials, oxides and nitrides. Copper interconnects in BEOL technology have some challenges such as, poor adhesion to dielectric materials and diffusion of copper into silicon substrates. Copper easily reacts with silicon and forms copper silicide at low temperatures

[22-23]. Traces of copper in silicon substrate will cause adverse effect on device operation. Most of the Interconnect metals (e.g. Cu, Ag, Au and W) except Al; do not bond well to underlying substrate and ILD. Therefore, the use of diffusion barrier layers and adhesion promoters between copper and underlying silicon substrate in ICs is mandatory. An ideal diffusion barrier material should also act as good diffusion barrier and adhesion promoters. Diffusion barriers under consideration are metal nitrides, carbides and borides, and metals such as Ti, Ta, and W. Diffusion barrier should be immiscible and non-reactive with copper. As stated by the studies, sputtered Ta and TaN films act as excellent diffusion barrier layers in Cu/low-k multilevel interconnects [24-25]. In the present investigation Ta, TaN, SiC and SiN are studied as barrier layers/cap layers in Cu/low-k stacks. SiC and SiN diffusion cap layers also acts as etch stops during the BEOL processes. Silicon nitride is widely used as cap layer; first, because it acts as an excellent barrier to copper and second, because of its etch selectivity to oxides.

3. Mechanical characterization of low-k structures by using nanoindentation

Currently most of IC fabrication technology has been migrated to Cu/low-k interconnect technology, as there is an increase of RC delay associated with the conventional Al/SiO2 interconnects used in the miniaturization of IC technology. This technology transition brings several integration challenges for the Cu/low-k interconnects because of poor thermal and/or mechanical properties of low-k thin films and Cu/low-k stacks [1]. Hence, it is very important to study the mechanical properties of different Cu/low-k structures to evaluate the device reliability. In the present investigation, mechanical properties of various Cu/low-k structures are studied by using nanoindentation technique.

The desired 'k' value of the low-k films can be attained by modifying the molecular structure of the low-k film or by introduction of organic or inorganic groups into the base structure [7]. Mechanical reliability and dielectric constant of the low-k materials are mutually dependent and have inverse relation. In case of SiO_2 based low-k materials, dielectric constant strongly depends on the density of the material, which in turn also depends upon the amount of porosity introduced, as shown in Fig 2 [27]. In chemical vapor deposition (CVD) based low-k materials, dielectric constant is decreased by the introduction of terminal methyl ($-CH_3$) groups, that will break the Si-O network and create nanopores. As the concentration of Si-O bonds decreases, percentage of pores and density non-uniformity increases, ultimately increasing the probability of mechanical failure of low-k films.

In IC fabrication technology, low-k material selection and its integration into BEOL is very crucial and it should withstand chemical mechanical polishing (CMP) without any failures. In microelectronic industry many researchers have been actively working toward finding threshold values of hardness and elastic modulus that can provide Cu/low-k system the ability to withstand CMP and wire bonding processes [28-30]. Researchers at Motorola [43] have concluded that passing the CMP process of low-k material is not a simple factor of modulus, hardness, adhesion or toughness, but a combination of all of these properties. As

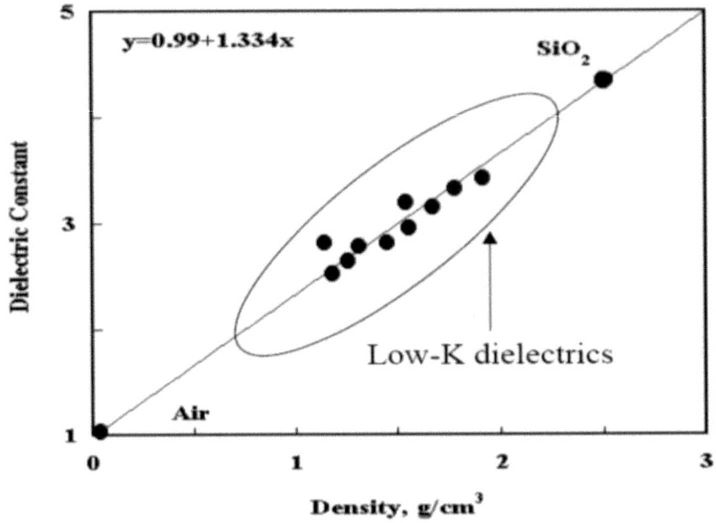

Figure 2. Dielectric constant dependence on low-k material density [27]

Dielectric thin films	Thickness (nm)	Elastic Modulus (GPa)	Hardness (GPa)	Researcher
Organo Silicate Glass (OSG)	2000	6.6-8.4	1.2-1.7	A.A. Volinsky et al [4]
USG (undoped silicate glass)	200-1000	79.06-80.66	5.65-7.52	Lu Shen et al [51]
SiLK ™	600	0.4	6.65	Lu Shen et al [55]
Porous SiLK ™	600	0.26	5.34	
MSQ –Hard	500-1000	12.5	0.936	S. Y. Chang et al [56]
MSQ-Soft	500-1000	2.7	0.19	
Low-k/barrier/Si (Anonymous)	1000	0.5	0.05	I. S. Adhihetty et al [57]
Porous low-k (carbon base)	250-540	4	0.15	Y. H. Wang et al [58]
Porous low-k (silica base)	250-540	0.35	0.45	

Table 4. Mechanical properties of various low-k materials studied by nanoindentation technique.

stated before, the mechanical properties of low-k films depends on chemical structure, amount of porosity and composition, Elastic modulus and hardness values of the different dielectric thin films varies from 2 to 14 GPa and 0.5 to 7 GPa respectively [41-45]. Volinsky et al [1] have found a linear relationship between hardness and elastic modulus for silicate low-k dielectric films in nanoindentation testing with continuous stiffness measurement (CSM) attachment. In nanoindentation testing, mechanical response of low-k films is different from the metallic films and usually exhibits little or no plasticity [27]. Hardness and elastic modulus values of various low-k films and Cu/low-k stacks were tested with nanoindentation technique by many researchers and some of those results have been summarized in table 4. Based on the extensive literature survey and the present work, it is observed that the mechanical properties of various films and Cu/low-k stacks depend on many factors, more importantly the amount of porosity (Constitutive and Subtractive), composition, molecular structure, thickness, type of stack and diffusion barrier.

3.1. Sample preparation

Materials used in the present work include BD thin films, oxide, nitride and barrier layers. Black diamond, oxide, SiC and SiN films were prepared by using PECVD technique. Sputtering technique was employed to deposit the copper seed (150 nm), Ta and TaN layers. Copper film of 1μm thickness was deposited by electroplating process. Different stacks of these layers were deposited by different experimental techniques. All thin film samples were prepared on 8″ Si (100) wafer in semiconductor fabrication plant of class 1000 clean room environment. A thin oxide layer of thickness about 5 nm is deposited on the surface of silicon substrates to improve the adhesion between the substrate and the low-k thin films.

3.2. Continuous Stiffness Measurements (CSM)

This technique was introduced in the year 1989 by Oliver and Pethica [36-37]. The CSM technique developed over the last decade by researchers offers numerous advantages. It has the unique advantage of providing mechanical properties as a function of penetration depth. Calibration and testing procedure take very less time as there is no need for multiple loading and unloading. At high frequencies it allows to avoid obscure effects of the samples like creep, viscoelasticity and thermal drift, which cause much problem in the conventional calibration method. It allows us to measure the effect of contact stiffness changes and damping changes at the point of initial contact [38]. In CSM nanoindentation technique the contact stiffness is measured during loading of the indentation test and there is no need for separate unloading cycles. This is an ideal method to determine contact stiffness and it can measure at very small penetration depths. Hence this method is unique to measure mechanical properties of thin films of few tens of nanometers. It has an additional advantage that if the specimen shows viscoelastic behavior, the phase difference between the force and displacement signals gives idea about the storage and loss modulus of the specimen [26].

In nanoindentation experiment, the CSM technique is carried out by applying a harmonic force at relatively high frequency (69.3 Hz), which is added to increasing load, P, on the loading coil of the indenter as shown in Fig. 3. The applied current that determines the nominal load of the indenter is very small, which results oscillations to the indenter with a frequency related contact area and stiffness of the sample [39].

This technique accurately measures displacements as small as 0.001 nm using frequency specific amplification. To determine the contact stiffness of the sample, the dynamic response of the nanoindenter has to be determined. A dynamic model which is used in CSM system is shown in Fig. 4. Major components of the dynamic model are the mass of the indenter, the spring constant of the leaf springs that support the indenter, the stiffness of the indenter frame and the damping constant due to the air in the gaps of the capacitor displacement sensing system.

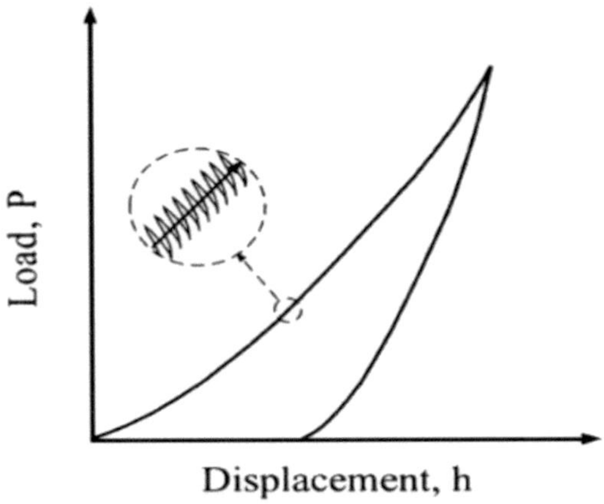

Figure 3. Schematic of nanoindentation CSM load-displacement curve [39;82].

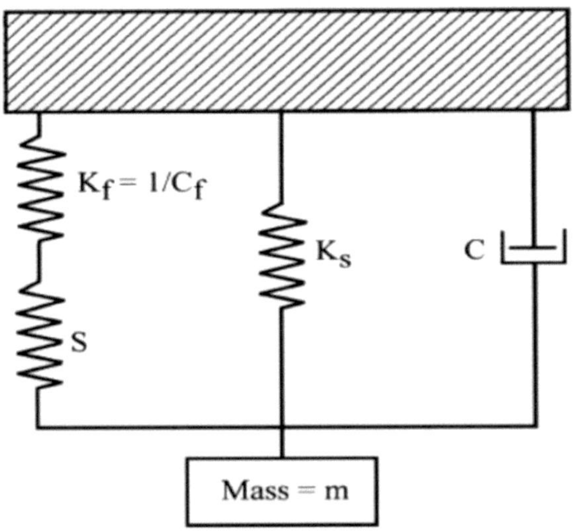

Figure 4. Schematic of components of dynamic model for the indentation CSM system [32]

By analyzing this model, the contact, S, can be calculated from the amplitude of the displacement signal from [39; 82],

$$\left|\frac{P_{OS}}{h(\omega)}\right| = \sqrt{\{(S^{-1} + C_f)^{-1} + K_S - m\omega^2\}^2 + \omega^2 D^2} \tag{1}$$

And the phase angle, ϕ between the driving force and the displacement response is

$$\tan(\phi) = \frac{\omega D}{(S^{-1} + C_f)^{-1} + K_s - m\omega^2} \quad (2)$$

Where

C_f= the compliance of the load frame (~1.13 m/MN)
K_s= the stiffness of the column support springs (~60 N/m)
D= the damping coefficient (~54 N s/m)
P_{os}= the magnitude of the force oscillation
$h(\omega)$= the magnitude of the resulting displacement oscillation
ω= frequency of the oscillation
ϕ = the phase angle between the force and displacement signals
m= mass (~4.7 gm)

4. Mechanical characterization of BD thin films with varying thickness

Thickness of the thin film structures used in the BEOL interconnects continue to decrease as the chip size decreases. Usually, thickness of the dielectric film has small effect on electrical properties [40], but it has significant influence on the mechanical properties of the film. Hence, it is very important to study the mechanical properties of these films of minute thicknesses. For this study, BD thin films of six different thicknesses, 100, 300, 500, 700, 1000 and 1200 nm have been prepared on 8" silicon substrate. Nanoindentation tests were performed on these films to study the effect of thickness on the nanomechanical behaviors. Fig.5 shows typical load-displacement curves of the BD thin films (100, 300, 500, 700, 1000 and 1200 nm) using nanoindentation CSM technique. By using this CSM technique, hardness and modulus can be determined as a function of indentation penetration depth with a single nanoindentation load-unload cycle. Berkovich indenter was employed in all experiments and a series of ten indentation tests were performed on each sample. The Poisson's ratio of the BD films of different thicknesses in nanoindentation experiment data is taken as 0.25, because the Poisson's ratio has a negligible effect on the indentation results [41]. During indentation, almost all BD films exhibit pop-in events as indicated by arrows in Fig. 5 due to failures in the films and it is evident from the optical micrographs (Fig. 6) and the corresponding pop-in events are marked on hardness and modulus vs. displacement graphs in Fig. 7. As the thickness increases, the degree of failure or the extent of damage increases in nanoindentation testing and it can be observed from the optical images shown in Fig. 6, where cracks are indicated by arrows on the micrographs. Many researchers have observed this kind of fracture behavior (pop-in event) in various types of low-k materials, bulk glasses and silica foams [42-46] Table 5 gives the information about the BD films fracture/delamination during nanoindentation testing, in terms of threshold load, threshold indentation depth and % of thickness at which film cracking occurs. For 100 nm thick film, failure is observed when the indenter tip is in the substrate and it can be seen in Fig. 5. In case of BD films with thickness 300-1200 nm, failure is observed within the films at different loads and indentation depths as shown in Fig5 and Table 5. It is found that as the BD film thickness increases the threshold load of cracking and the threshold indentation depth increases. One common trend observed in the fracture among BD films (300-1200 nm) is that, the film failure (crack or/and delmaination) occurs at around 60-65% of the film thickness in indentation testing. This data is very useful in conventional nanoindentation processes to determine the physical properties of BD films, i.e. the threshold load for the BD films without cracking, threshold loads in CMP process and to measure the fracture toughness of the BD films.

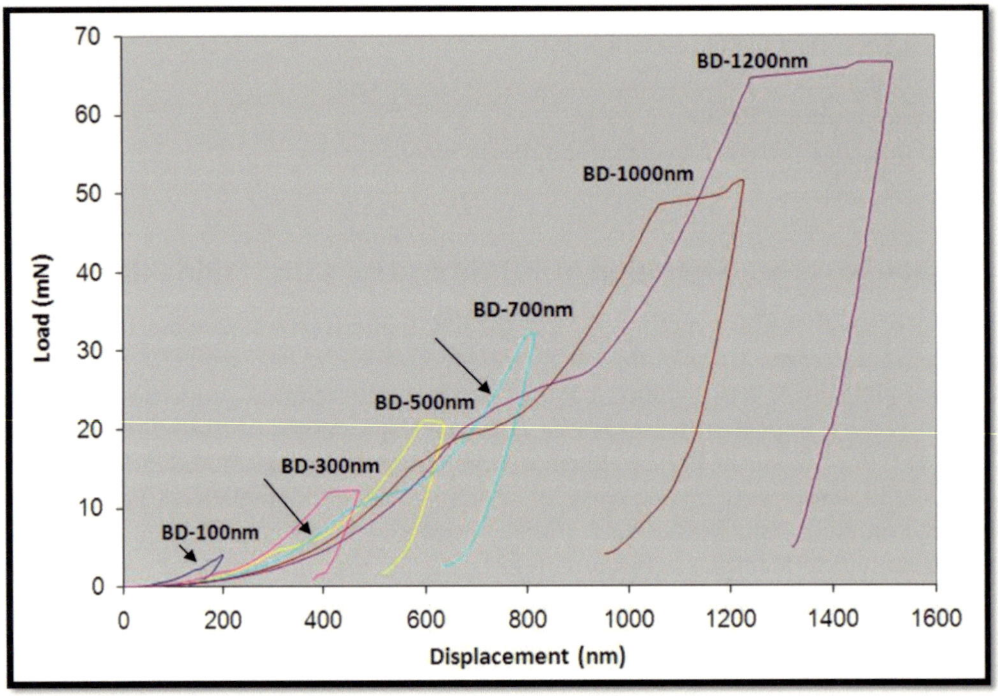

Figure 5. Load-displacement curves for Black Diamond ™ (BD) films of six different thicknesses under nanoindentation, (a) 100, (b) 300, (c) 500, (d) 700, (e) 1000 and (f) 1200 nm. Arrows on the curves indicate the pop-in events.

BD sample thickness (nm)	Threshold load of cracking (mN)	Threshold indentation depth of cracking (nm)	% of thickness at which cracking occurs (nm)
100	2.17	142	142
300	1.78	190.69	63.56
500	4.28	312.84	62.57
700	11.00	460.00	65.71
1000	18.83	644.00	64.40
1200	25.23	730.00	60.83

Table 5. Summary of fracture behavior data for all BD samples in nanoindentation testing.

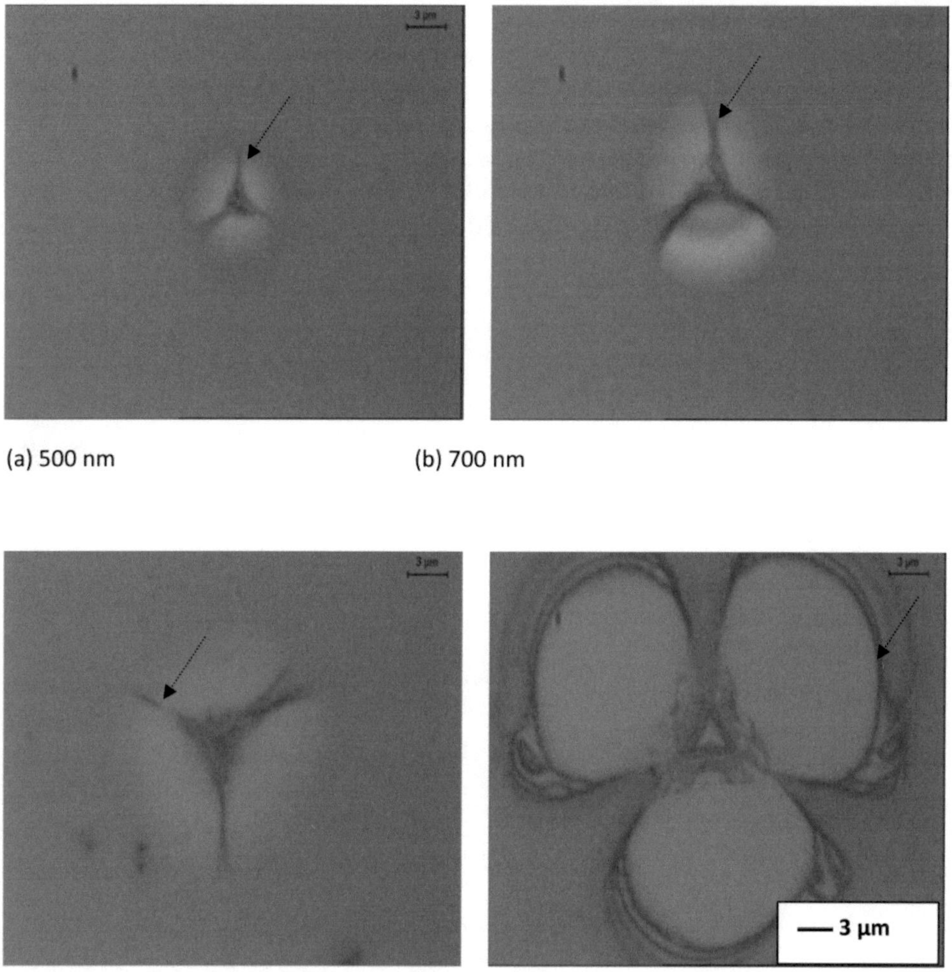

(a) 500 nm (b) 700 nm

(c) 1000 nm (d) 1200 nm

Figure 6. Optical images of residual nanoindentation impressions of (a) 500, (b) 700, (c) 1000 and 1200 nm BD thin films. Radial cracks in the films observed for all thicknesses as shown by arrows.

The hardness and elastic modulus as a function of displacement (indentation depth into the film thickness) of all BD samples are shown in Fig. 7. Results at the initial indentation depths show large fluctuations and high values of hardness and elastic modulus as the tip touches the film surface which might be due to the equipment noise and the inaccuracy of the indenter tip function at very shallow depth of indentation. As the depth of the indentation increases, both the hardness and elastic modulus reaches minima region as a property plateau and then tip starts sensing the effect of silicon substrate which results in higher hardness and elastic modulus. This phenomenon is observed only in the case of soft films on hard substrate. Usually, the initial part of the data is usually discarded in nanoindentation analysis and this kind of problem has been observed by other investigators [47-48]. Here, the averaged hardness and elastic modulus of this minima plateau region is used to define the properties of the films. This phenomenon has been validated by comparing the properties at 1/10th of the film thickness and properties at the minima property region as shown in Table 6. There are no significant differences observed in between properties of the films at 1/10th of the thickness and properties from minimum plateau region. Hence in this scenario computing properties from minimum plateau region can be more practicable. Fig. 7 compares the mechanical properties of six different thicknesses of BD samples. Hardness and elastic modulus of the BD films (100 to1200 nm) are in the range of 1.66 to 2.02 GPa and 9.27 to16.48 GPa respectively.

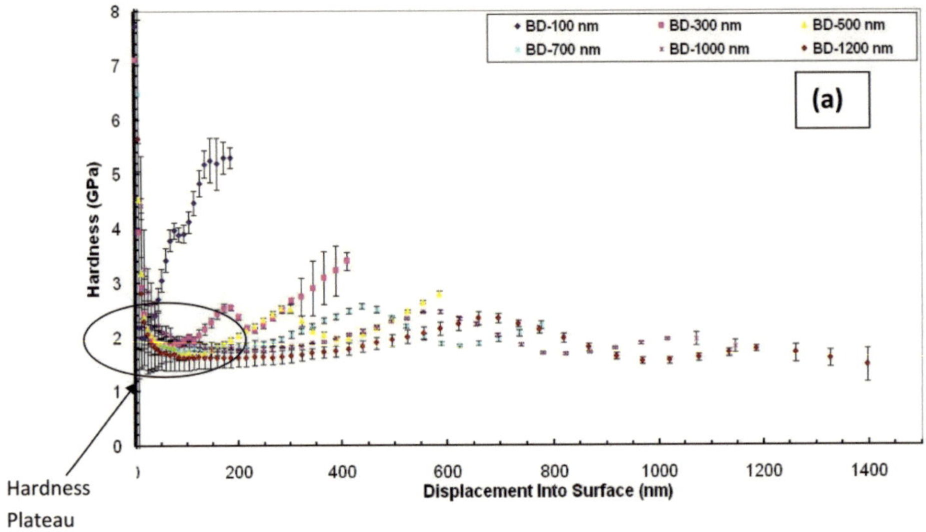

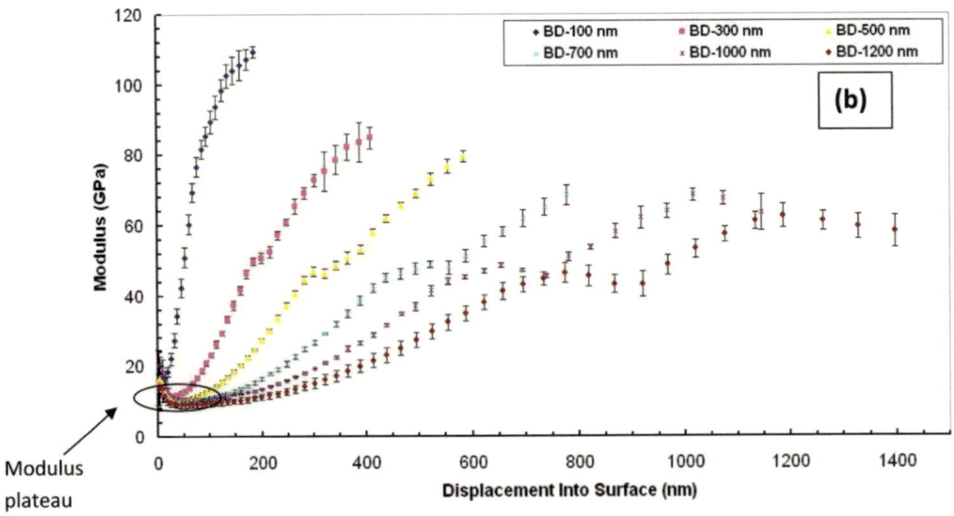

Figure 7. Hardness and (b) elastic modulus as functions of the displacement for BD films with six different thicknesses

Thickness of BD films (nm)	Property at 10% of film thickness (GPa)		Property at Plateau Region	
	H	E	H	E
100	1.99±0.25	18.53±1.5	2.02±0.36	16.48±2.1
300	1.83±0.32	13.76±1.3	1.85±0.12	11.54±1.2
500	1.63±0.14	12.91±1.4	1.70±0.05	10.58±1.5
700	1.58±0.24	11.62±1.5	1.78±0.02	9.93±1.1
1000	1.76±0.35	13.52±1.3	1.73±0.07	10.41±1.4
1200	1.62±0.84	12.72±1.5	1.66±0.09	9.27±1.7

Table 6. Summary of mechanical properties of BD films of six different thicknesses

As the indentation depth increases initially slight decrease in hardness and elastic modulus values are observed and as the indentation depth further increases these property values reach minima, then tip starts sensing the effect of the substrate which results in higher hardness and elastic modulus values. The initial decrease of properties as the indentation displacement increases is observed during nanoindentation of very thin films [49]. The averaged values of properties in minima plateau region are considered to be the real properties of the films and, in the present study, these values are used to define the properties of each film as shown table 6. The hardness values of all BD films are in the range of 1.66 to 2.02 GPa and the elastic modulus values are in the range of 9.27 to 16.48 GPa. Sharp deviations are observed in the property vs. displacement (see Fig. 7) graphs of all BD films, which corresponds to the pop-in events shown in Fig. 5, resulting from film cracking and/or delamination at the BD film-silicon substrate interface. Significant differences in mechanical properties are observed when the BD film thickness is less than 500 nm (100-500 nm), mainly in the case of elastic modulus. When the BD film thickness is greater than the 500 nm (500-1200 nm) no significant variation in mechanical properties are observed and it can be assumed that these properties are representative of the bulk properties of the BD films.

The minima plateau region for the hardness is considerably large and there is nearly no change with respect to different thicknesses, when compared with elastic modulus plateau region. The minima elastic modulus plateau region decreases as the BD film thickness decreases because the effect of substrate is more on elastic modulus for thinner films. In Fig. 7, it is observed that the sharp increase in modulus from the minima plateau region, mainly due to the effect of substrate, is much more on the elastic modulus than on the hardness of the BD films. This is because the elastic modulus is associated with the elastic deformation during nanoindentation, and in contrast, the hardness response of the material is associated with plastic deformation. Extensive simulation studies show that, the effective elastic modulus of a film experiences greater substrate effect than the hardness value [44]. BD -100 nm film shows significantly higher elastic modulus (E=16.48 GPa) when compared to higher thickness BD films and it is expected due to molecular restructuring in very thin BD films (≤100 nm), since the elastic modulus is an intrinsic material property, which largely depends on interatomic or molecular bonds [49]. Hence the higher elastic modulus of BD-100 nm film is probably expected due to stronger molecular bonding between organic ($-CH_3$) and inorganic (Si-O) constituents.

5. Effect of diffusion barriers on BD stacks integrity

For this study, four samples have been prepared with Ta and TaN barrier layers and dual stacks also have been prepared as BD/Ta/Si, BD/Ta/BD/Ta/Si, BD/TaN/Si and BD/TaN/BD/TaN/Si. Mechanical properties were assessed by nanoindentation technique. For single dielectric stack, the silicon substrate is first coated with Ta or TaN barrier layer of 25 nm thickness by using self-ionized metal plasma (SIP) technique at room temperature followed by the deposition of BD film of 1000 nm thickness by PECVD (plasma Enhanced Chemical Vapor Deposition) technique. For the deposition of dual dielectric stack this procedure was repeated one more time with BD films and barrier deposition. The application of Ta and TaN barrier layers to BD films improves stiffness in addition to other mechanical properties.

The thickness of single and dual dielectric stacks studied in the present work is 1025 and 2050 nm respectively. Fig. 8 shows the typical load-displacement curves of single and dual dielectric stacks. Both single and dual dielectric stacks show pop-in events. From Fig. 9 optical micrographs of residual nanoindentation impressions of single and double dielectric stacks reveal that massive failure of the films is more prominent in dual dielectric stacks. Dual dielectric stacks demonstrate massive failure which can be observed as more pop-in events in load-displacement curves especially at the interfaces, which are shown as dotted lines in Fig. 8. This can be confirmed by observing the residual nanoindentation marks as shown in Fig. 9. Hence, it is expected that crack formation and/or delamination may occur at the interfaces due to indenter penetration as observed by J. Vitiello [50]. In the current study, stacks having same barrier layer (Ta or TaN) are compared with respect to mechanical properties. Hardness and elastic modulus of single dielectric stacks are in the range of 1.43 to 1.91 GPa and 8.35 to 10.03 GPa respectively. No significant difference is observed in the case of double dielectric stacks. Mechanical properties of both single and dual stacks are given in Table 7.

5.1. Mechanical properties of the BD stacks with Ta barrier layer

Hardness and elastic modulus of both single and dual dielectric stack with Ta barrier layer is shown in Fig. 10. High hardness values at the film surface for both stacks are observed mainly due to the inaccuracy of indenter tip functions and surface roughness [47]. Large fluctuations in hardness values throughout the film thickness measurements are observed due to the failure of stacks during the nanoindentation. Substrate effect is more prominent in single dielectric stack when compared with double dielectric stack, this is because the thickness of both the stacks are different and there is difference in the number of interfaces. The hardness and elastic modulus values of single dielectric stack is higher than the values for dual dielectric stack and these properties are consistent throughout the thickness of the stacks as shown in Fig. 10. The averaged minima property plateau is used to define the property of each stack and these values are given in Table 7. Single dielectric stack has hardness of 1.91 GPa and modulus of 10.03 GPa, whereas for dual stack, the values are 1.38 and 7.98 GPa respectively. From Fig 10, sudden deviations in properties are observed mainly due to the film failure during nanoindentation process which corresponds to pop-in events in load-displacement curves as discussed earlier. This mechanical property data of single and dual stacks are very useful in CMP performance, for example, single dielectric stack has slower removal rate when compared to dual stack for the same CMP pressure [51-52]. This is mainly due to the higher hardness value of the single dielectric stack. This mechanical characterization data will be very helpful in deciding the CMP loads as per the dielectric stacks.

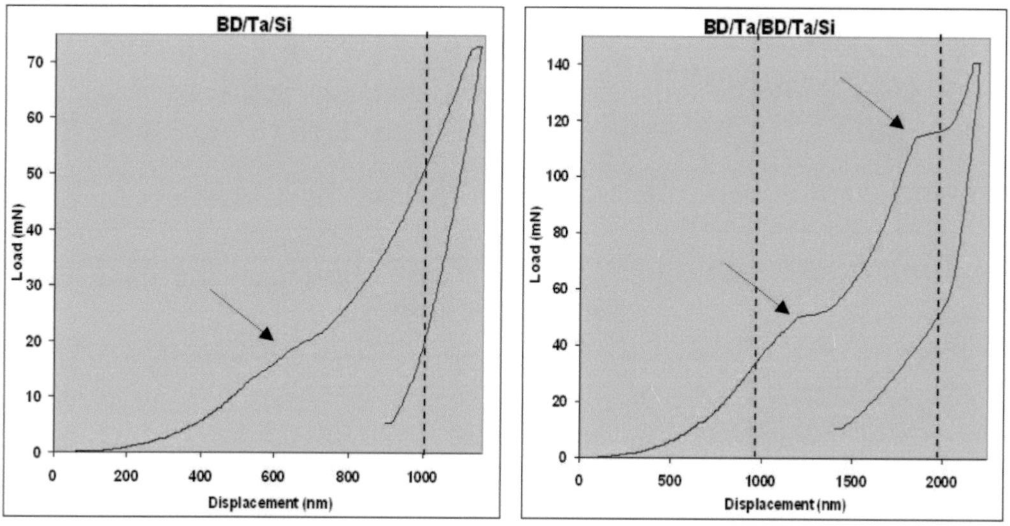

(a) (b)

Figure 8. Typical load-displacement curves of (a) BD/Ta/Si (b) BD/Ta/BD/Ta/Si, (c) BD/TaN/Si and (d) BD/TaN/BD/TaNs/Si stacks. Arrows indicate the pop-in events.

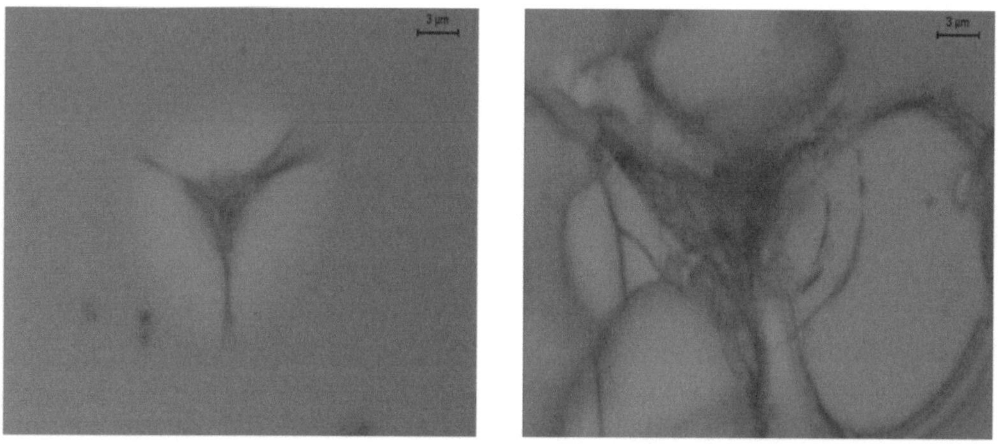

(a) BD/Ta/Si (Single dielectric stack) (b)BD/Ta/BD/Ta/Si (double dielectric stack)

Figure 9. Optical micrographs of residual nanoindentation impression on single and double dielectric stacks.

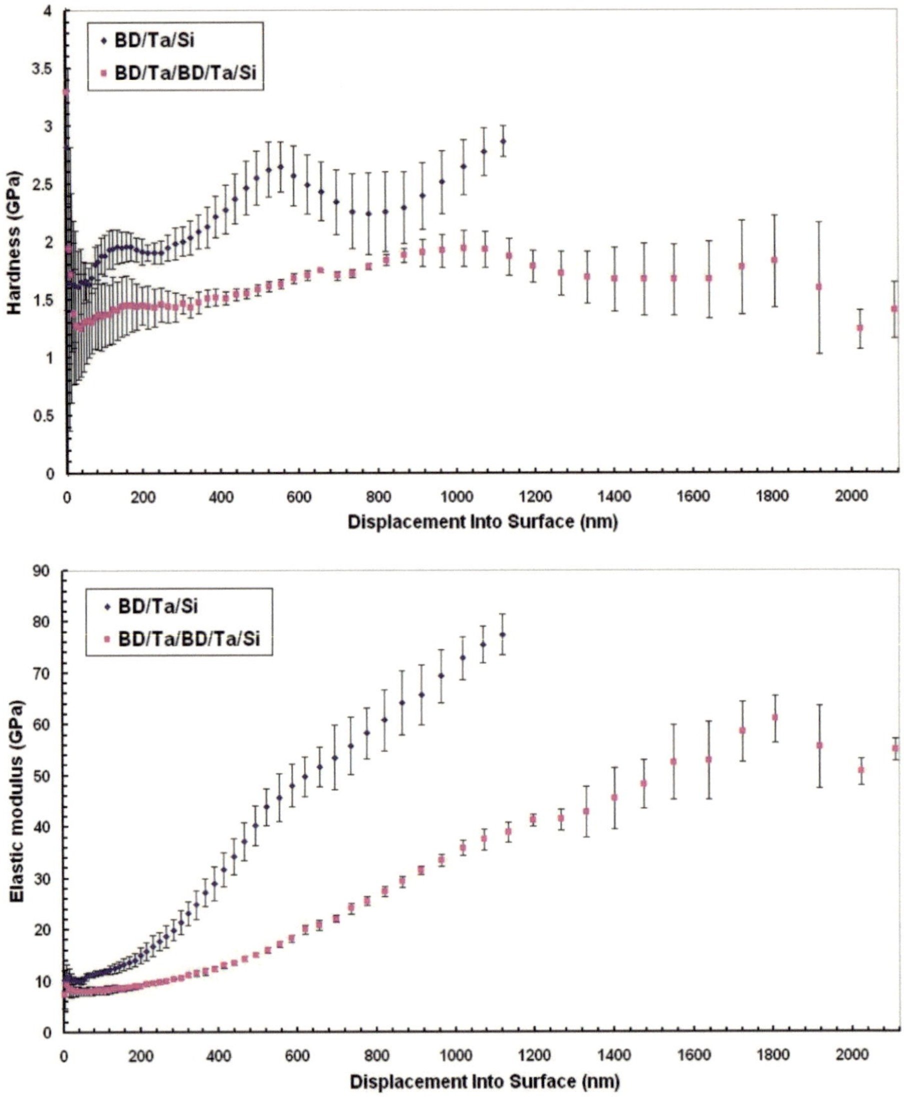

Figure 10. Hardness and elastic modulus as a function of displacement for BD/Ta/Si and BD/Ta/BD/Ta/Si stacks measured using the nanoindentation CSM technique.

5.2. Mechanical properties of the BD stacks with TaN barrier layer

The total thickness of these stacks are maintained same as in previous section, i.e. 1025 nm for single dielectric stack and 2050 nm for dual dielectric stack. Hardness and elastic modulus as a function of displacement of these stacks measured using the nanoindentation CSM technique is presented in Fig. 11. Single dielectric stacks have slightly higher mechanical properties when compared with dual dielectric stacks as summarized in table 7. In both cases, the performance of single dielectric stacks is better than that for the dual dielectric stacks. By comparing the single dielectric stacks, BD/Ta/Si has higher hardness and elastic modulus than for BD/TaN/Si stack (Table 7). Thus barrier layer greatly affects the mechanical properties of the single stack. There is no significant difference observed in the case of BD/Ta/BD/Ta/Si and BD/TaN/BD/TaN/Si dual dielectric stacks. It is anticipated that the lower mechanical properties of dual stacks is mainly due to the presence of residual stresses. Usually, dual stacks have more residual stresses when compared with single dielectric stacks as dual-stacked samples are subjected to more processing steps. Compressive stresses in the stacks result in an increased hardness values but tensile stresses cause a decrease in hardness. Therefore, dual stacks are expected to have more tensile stresses when compared with single dielectric stacks.

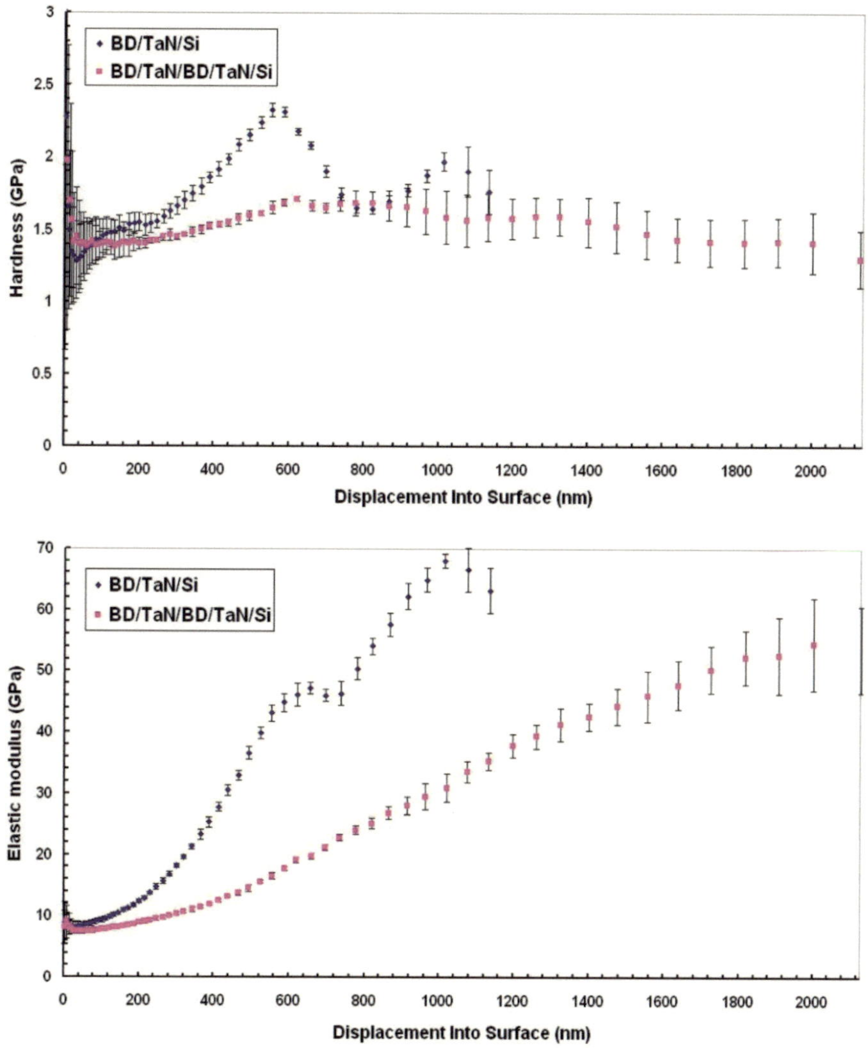

Figure 11. Hardness and elastic modulus as a function of displacement for BD/TaN/Si and BD/TaN/BD/TaN/Si stacks measured using the nanoindentation CSM technique

Multilayer Stack	Stack Thickness (nm)	Hardness (GPa)	Elastic Modulus (GPa)
BD/Ta/Si (Single dielectric stack)	1025	1.91±0.3	10.03±1.2
BD/Ta/BD/Ta/Si (double dielectric stack)	2050	1.38±0.1	7.98±1.5
BD/TaN/Si (Single dielectric stack)	1025	1.43±0.5	8.35±1.4
BD/TaN/BD/TaN/Si (dual dielectric stack)	2050	1.4±0.2	7.58±1.3

Table 7. Summary of mechanical properties of single and double dielectric stacks.

6. Effect of the wafer thinning/backgrinding on the integrity Low-k stacks

For this study, multilayer low-k stack with fifteen different thin films comprising SiN, USG, Blok (SiC), BD (Black Diamond ™, low-k), has been fabricated as shown in Fig. 12 All thin film samples were prepared on 8" Si(100) wafer in a semiconductor processing clean room of class 1000 environment. Samples fabricated in this study are exclusively designed to study the BD (low-k) integrity in the BEOL interconnects and it resembles the Cu/low-k structure of the three metallizations. So the structure has three BD low-k layers at different levels according to the BEOL (Back End of the Line) interconnect design specifications. The total thickness of the multilayer low-k stack is around 3400 nm. Therefore present study gives an outlook about the response of the low-k test structures during back grinding process. To study the backgrinding/thinning effect on the Low-k stacks, all samples have been subjected to backgrinding process by using the commercial backgrinding system. Empirically backgrinding process involves, first coarse grinding using grit #300, then fine grinding using #2000 followed by dry polishing. A total five set of samples have been prepared for this study as, normal sample (without backgrinding), Back grinded samples of four different thicknesses (BG-500, 300, 150 & 75um).

Failure loads, hardness (H) and elastic modulus (E) normal (no back grinding) and back grinded low-k stacks (BG-500, 300, 150 & 75 μm) are computed by analyzing the nanoindentation load-displacement curves as shown in Fig. 13. From Fig. 13, it is obvious that all stacks have shown pop-in event and this event can be taken as the failure load/fracture strength of the stack. These pop-in events in the nanoindentation curves are resulting from the film cracking and delamination of the stack in the form of blisters [53]. The normal stack shows pop-in event (failure load/fracture strength of the stack) at lower loads and indentation depths, as compared to the back grinded stacks. Normal stack failed at 456.25±21.22 mN load and 2422.41±58.53 nm indentation depth, whereas back grinded stacks failed in the range of 482 to 661mN load and 2405-2979 nm indentation depth. The failure load and depth values of all types of samples are summarized in the Table. 8 and. From the nanoindentation curves and optical imaging (from Fig.14), analysis it is clearly evident that the nanoindentation response of the normal and back grinded stacks is different in terms of failure load depth and fracture behavior. Normal stack and BG-500μm show extensive delamination and chipping, whereas other back grinded stacks (BG 300, 150, 75 μm) show delamination blister and this behavior is in good agreement with the nanoindentation pop-in event. BG-500μm exhibits the mixed response as it shows chipping-off during nanoindentation and moderate pop-in failure load of 482.17 mN and this might be due to the moderate degree of back grinding. In case of the other back grinded stacks (BG 300, 150, 75 μm), even higher nanoindentation loads are not able to damage/chip-off the low-k stack and cause interfacial delamination only. No significant difference in fracture strength (pop-in event) is observed among BG-300, 150, and 75 μm back grinded stacks and all these grinded stacks show higher facture strength/loads than the normal stack and BG-500μm. Accordingly the increase in failure load depends on the degree of the back grinding, but not much difference is observed when the wafers are grinded to 300, 150 and 75 μm. After back grinding, the strength of the low-k stack is enhanced, mainly in terms of nanoindentation load and indentation depth and this increase is understood mainly due to the application of mechanical pressure and thermal stresses during back grinding action. The back grinding pressure or load may improve the adhesion, especially Vander walls forces at the multilayered interfaces and cause the densification of the individual films of the stack. It is being investigated by many researchers in the packaging field that the back grinding processes are deteriorating the die strength, but this is not the same phenomenon with the active side of the chip stack.

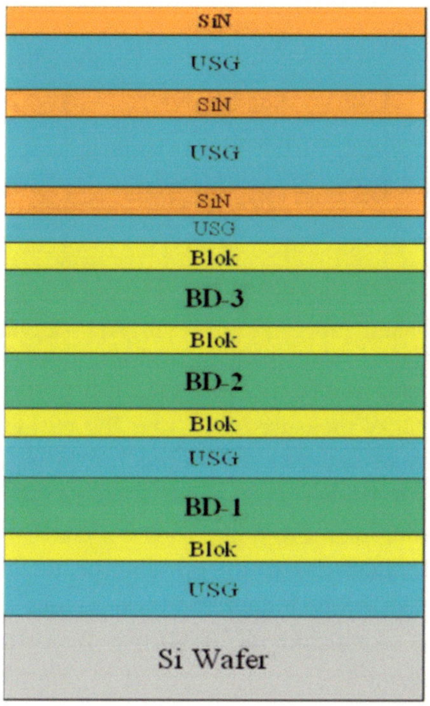

Figure 12. Low-k stack with fifteen multilayers for wafer backgrinding study

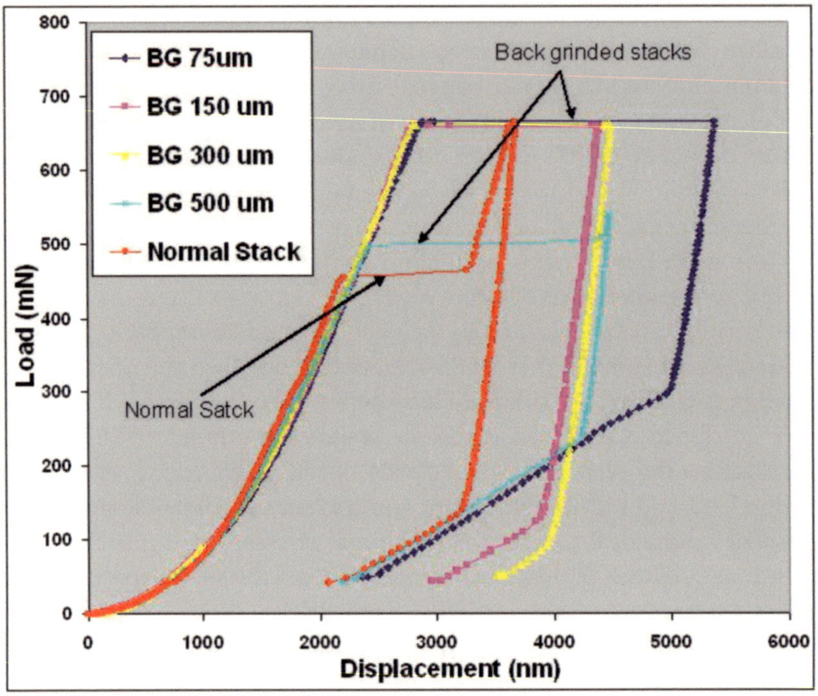

Figure 13. Typical load-displacement curves of normal and back grinded samples and its comparison.

BD samples	Fracture Behavior	
	Load (mN)	**Indentation Depth (nm)**
Normal stack	456.25±21.22	2422.41±58.53
BG-500μm	482.17±25.25	2405.86±70.47
BG-300μm	661.20±7.57	2979.79±21.58
BG-150μm	658.45±4.74	2809.01±30.60
BG-75μm	658.60±12.21	2942.52±71.20

Table 8. Summary of nanomechanical properties of normal and back grinded samples

Fig. 15 shows the hardness and elastic modulus as a function of the indentation depth for normal and back grinded stacks measured by using the nanoindentation CSM technique. Mechanical properties of all samples are not constant, but strongly influenced by the contact depth and this is mainly due to the presence of the different types of thin films with diverse physical properties and 15 interfaces. From Fig. 15, it is clear that initially all samples exhibit high hardness and modulus values due to the presence of the SiN layer on top of the stack. However, back grinded stacks show higher hardness values throughout the indentation depth and this difference is significant till ~1500 nm. Even though BG-500 μm samples shows moderate failure load values when compared to other back grinded stacks, it still shows higher mechanical properties than normal stack. There is no difference in hardness

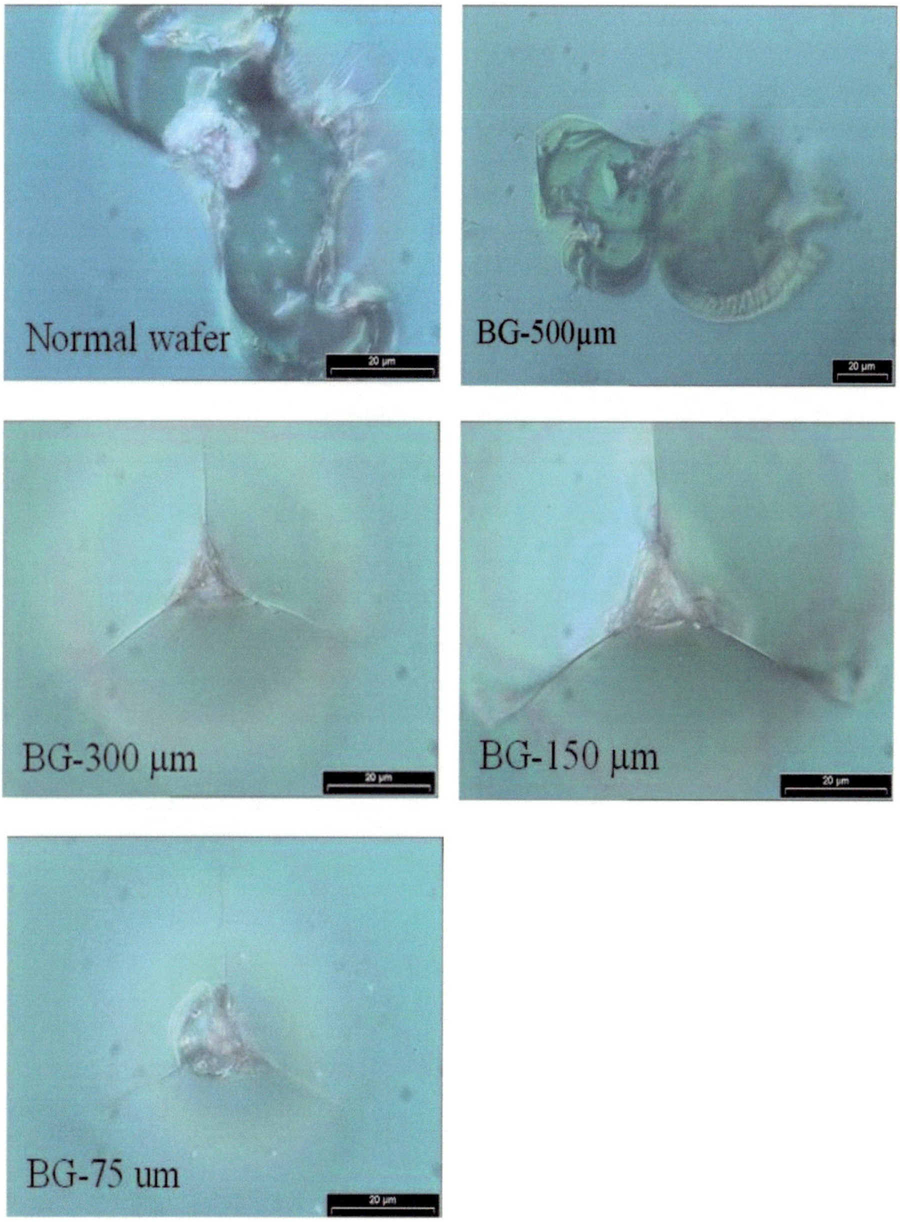

Figure 14. Optical images of residual nanoindentation impressions of normal and back grinded samples.

values among the back grinded stacks. In case of the elastic modulus, the overall trend is mixed, in which initially back grinded stacks exhibit high values and from 1000 nm depth, BG-150 µm stack follows the trend of normal stack and BG-75 & 300 µm stacks show lower modulus values. The mixed trend of modulus values of all samples is mainly due to the fact that, the elastic modulus is highly sensitive and intrinsic property. Elastic modulus is greatly influenced by layers beneath the testing films and substrate. In the case of hardness, the difference between normal stack and back grinded stacks is very clear when compared with the modulus values. As a whole, back grinded stacks exhibit the higher hardness and elastic modulus values, and this trend is quite clear in the low-k region. This might be due to back grinding loads or pressures influencing the interfaces and causing the densification of the films, especially in the low-k region. Sekhar Et al., have extensively studied the nanoscratch behavior on the backgrinded stacks by analyzing the fracture behavior by critical loads [54]. In this study the fracture or failure strength, hardness and elastic modulus of the normal and grinded stacks are analyzed and compared at overall level as the nanoindentation analysis is very complicated and not well established for the multilayered stacks.

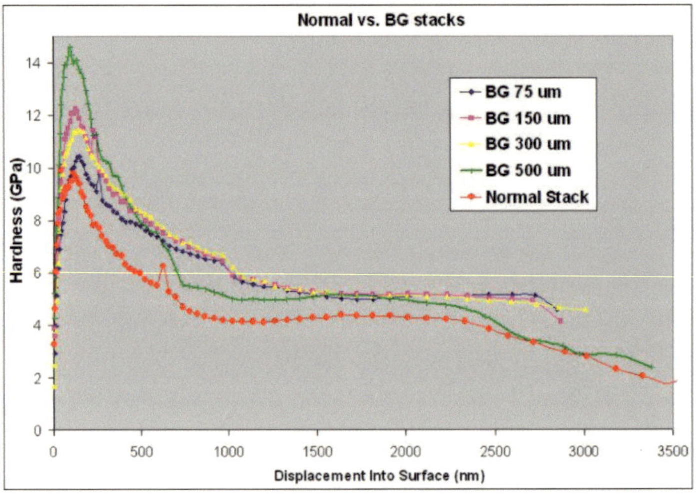

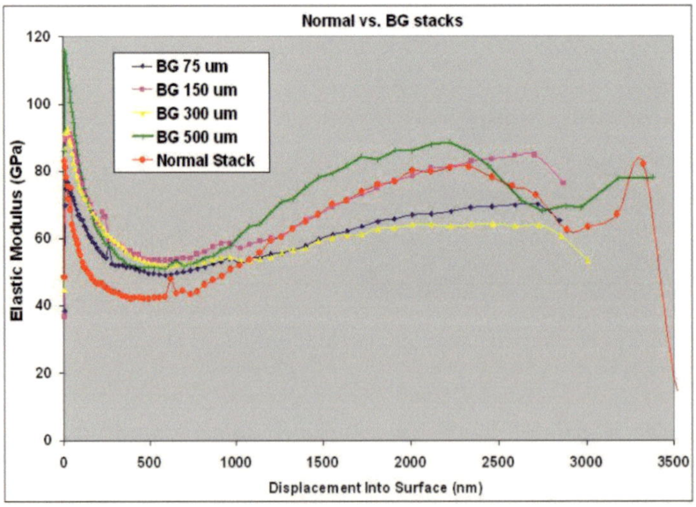

Figure 15. Hardness and elastic modulus as a function of displacement for normal and back grinded samples

7. Summary

This chapter presents the systematic mechanical characterization of the BD thin films for BEOL interconnects and 3D IC/packaging applications. For this purpose several thin film systems have been chosen which comprises, Black diamond (SiOC:H), USG (SiO_2),SiC, SiN and Ta & TaN thin films which have been prepared by using PECVD technique and sputtering technique respectively. Nanoindentation tests with continuous stiffness measurement (CSM) attachment have been performed on all samples to assess hardness (H) and elastic modulus (E) properties. The CSM attachment is preferable because it provides continuous measurement of the hardness and elastic modulus as a function of indentation depth. In this study several Low-k systems have been prepared and characterized as, mechanical characterization of low-k films with different thicknesses, influence of Ta & TaN barrier layers, and effect of wafer backgrinding/thinning on the complete low-k stack which is similar to actual BEOL with three metallization. In case of the thickness dependence of mechanical properties of BD films, hardness and elastic modulus values obtained of all BD films (100-1200 nm) are in the range of 2.02to 1.66 GPa and 16.48 to 9.27 GPa respectively. Considerable thickness dependence of the properties is observed when thickness is less than 500 nm. To study the effect of barrier layers on low-k stack, single and dual dielectric stacks have been fabricated with Ta and TaN barrier layers. In nanoindentation testing, the performance of single dielectric stacks is better than the dual dielectric stacks is mainly attributed to differences in the number of interfaces, total film thickness and residual stresses.

For the first time the effect of the back grinding process on active side of the low-k chip stack has been studied by using sophisticated nanoindentation technique. It is further investigated by analyzing the fracture or failure strength (pop-in event in nanoindentation), hardness, elastic modulus of the low-k stack for both normal and back grinded samples. After back grinding process, the overall integrity of low-k stack is enhanced and thus the back grinded stacks exhibit higher nanomechanical properties than the normal stacks. This can be attributed mainly to the straightening of the low-k interfaces and densification of the BD (low-k) layers during back grinding processes. Based on the results and their detail analysis it can be said that the thermo-mechanical stresses that applied and/or generated during wafer back grinding processes affect the interfaces and nanomechanical behavior of the low-k stack, in turn enhances the overall integrity of the back grinding stacks.

Author details

Vasarla Nagendra Sekhar
Institute of Microelectronics,
*A*STAR (Agency for Science, Technology and Research), Science Park II, Singapore*

Acknowledgement

Author would like to acknowledge the support from Dr. B. S. S. Chandra Rao for suggestions and proof reading. Many heartfelt thanks to Mrs. Sindhura for kind co-operation and encouragement.

8. References

[1] S. Wolf, 'Multilevel interconnects for ULSI', 'Silicon processing for the VLSI era': Vol 4- Deep submicron technology Chap.12, pp. 573-602, 2002, lattice press, Sunset beach CA.

[2] International Technology Roadmap for Semiconductors, 2002 Update (SIA San Jose, CA, 2003), http://public.itrs.net

[3] The international technology roadmap for semiconductors, 2000 update, (semiconductor industry association, San Jose, CA, 2000).

[4] Alex. A. Volinsky, M. B. Palacio and W. W. Gerberich, 'Incompressible pore effect on the mechanical behavior of low-k dielectric films' Mat. Res. Symp. Proc., Vol. 750 (2003).

[5] W. T. Tseng, C. W. liu, B. T. Dai and C. F. Yeh, 'Effects of mechanical characteristics on the chemical-mechanical properties of the thin films' Thin Solid Films, 290-291 (1996) 458-463.

[6] J. D. Plummer, M. Deal and P. B. Griffin, 'Silicon VLSI Technology: Fundamentals, Practice and Modeling' Prentice Hall, NJ, USA.

[7] D. Shamiryan, T. Abell, F. Lacopi and K. Maex, Jan-2004, 'Low-k dielectric materials' Materials Today, 7(1):34-39.

[8] www.semiconductor.net

[9] Reiche, Manfred & Wagner, Gerald "Wafer thinning: techniques for Ultra-thin Wafers", Advanced Packaging, March, 2003.

[10] Hoh Huey Jiun et al, "Effect of wafer thinning methods towards fracture strength and topography of silicon die" Microelectronics Reliability, 46 (2006) 836-845.

[11] Shoulung Chen et al. "The Evaluation of wafer thinning and Singulating Processes to Enhance Chip" IEEE Elelctronic Components and Technology Conference, 2005, pp. 1526-1530.

[12] Chen J, De Wolf I. " Study of damage and stress induced by backgrind in Si wafers" Semicond Sci Technol 2003; 18: 261-268.

[13] Gaulhofer E, Oyer H., "Wafer thinning and strength enhancement to meet emerging packaging requirement" IEMT Europe Symposium 200 pp.1-4.

[14] Mc Hatton C, Gumbert CM. "Eliminating backgrind defects with wet chemical etching" Solid-State Technol 1998 (November): 85-90.

[15] Blech I and Dang D "Silicon Wafer Deformation After Backside Grinding", Solid State Tech., 37(8), 1994, pp.74-76.

[16] Michael E. Clarke, Introducing low-k dielectrics into semiconductor processing: Application Notes, www.mykrolis.com

[17] www.appliedmaterials.com

[18] L. Xia, Handbook of Semiconductor Manufacturing Technology, pp 309-356.

[19] K. Maex, M. R. Baklanov, D. Shamiryan, F. lacopi, S. H. Brongersma, Z. S. Yanovitskaya, Low dielectric constant materials for microelectronics, J. Appl. Phys. 93 (11): 8793-8841.

[20] L. Xia, Handbook of Semiconductor Manufacturing Technology, pp 309-356.

[21] Ben Pang, W. Yau, P. Lee and M. Naik, A new CVD process for damascene low-k applications, Semiconductor Fabtech-10th Edition, pp 285-289.

[22] Q. Hong, C. M. Comris\e, S. W. Russel and J. W. Mayer, 1992, Phase formation in Cu-Si and Cu-Ge, J. Appl. Phys. 70: 3655-3660.

[23] J. D. McBrayer, R. M. Swanson and T. W. Sigmon, 1986, Diffusion fo metals in silicon dioxide, J. Elelctrochem. Soc. 133: 1242-1246.

[24] S. Y. jang, S. M. Lee, H. K. Baik, J. Mat. Sci: Mat in Elelctronics 7, p271 (1996)

[25] J. O. Olowolafe, C. J. Morgah, R. B. Gregory, M. Kottke, J. Appl. Phy. 72, p 4099, (1992)

[26] Anthony C. Fischer-Cripps, Nanoindentation, Second edition-2004, Springer-Verlag New York, LLC, NY.

[27] Alex A. Volinsky, Mechanical reliability and characterization of modern microelectronic interconnect structures, Nano Engineering world forum 2003.

[28] Y.L. Wang, C. Liu, S. T. Chang, M. S. Tsai, M. S. Fang, W. T. Tseng, Thin solid films, 308-309 (1997) 550.

[29] E. Hartmannsgruber, G. Zwicker, K. Beekmann, Microelectron. Eng. 50 (2000) 53.

[30] H. Yano, Y. Matsui, G. Minamihaba, N. Kawahashi, M. Hattori, Mater. Res. Soc. Proc. 671 (2001) D1.1.

[31] For more details see www.mts.com

[32] W. C. Oliver and G. M. Pharr, An improved technique for determining hardness and elastic modulus using load and displacement sensing indentation experiments, J. Mater. Res., Vol. 7, No. 6, June 1992, 1564-1583.

[33] I. N. Sneddon, (1965), 'Relation between load and penetration in the axisymmetric Boussineq problem for a punch with arbitrary profile', Int. J. Eng. Sci., 3, 47.

[34] K. L. Johnson, 1985, Contact mechanics, Cambridge, Cambridge university press.

[35] B. R. Lawn and T. R howes, (1981) Elastic recovery at hardness indnetations, J. Mater. Sci., 16, 2745.

[36] W. C. Oliver and J. B. Pethica (1989): 'Methods for continuous determination of the elastic stiffness of contact between two bodies', US patent 4,848,141, July 18.

[37] J. B. Pethica and W. C Oliver(1989) 'Thin films: Stresses and mechanical properties', MRS Symp. Proc. 130, 13-23; Pittsburgh, PA, Materials Research Society.

[38] W. C. Oliver and G. M. Pharr, 'Measurement of hardness and elastic modulus by instrumented indentation: Advances in understanding and refinements to methodology', J. Mater. Res., Vol. 19, No. 1, Jan2004.

[39] X. Li and Bharat Bhusan, 'A review of nanoindentation continuous stiffness measurement technique and its applications' Material Characterization 48 (2002) 11-36.

[40] E. Wu, J. Sune, W. Lai, E. Nowak, J. McKenna, A. Vayshenker and D. Harmon, Solid State Electron. 46, (11) (2002) 1787.

[41] Shaohua Chen, Lei Liu, Tzuchiang Wang, "Investigation of the mechanical properties of thin films by nanoindentation, considering the effects of thickness and different coating-substrate combinations", Surface & Coatings Technology 191 (2005) 25-32.

[42] R. J. Nay, O. L. Warren, D. Yang, and T. J. Wyrobek, 'Mechanical characterization of low-k dielectric materials using nanoindentation', Microelectronic Engineering, 75 (2004) 103-110.

[43] J. B. Vella, Alex A. Volinsky, Indira S. Adhihetty, N. V. Edwards, and William W. Gerberich, 'Nanoindentation of silicate low-k dielectric thin films' Mat. Res. Sym. Proc., Vol. 716, Materiual Research Society, B12.13.1-6.

[44] Lugen Wang, M. Ganor, and S. I. Rokhlin, 'Nanoindentation analysis of mechanical properties of low to ultralow dielectric constant SiOCH films', J. Mater. Res., Vol.20, No.8, Aug 2005, 2080-2093.

[45] D. J. Morris, S. B. Myers, and R. F. Cook, 'Sharp probes of varying acuity: Instrumented indentation and fracture behavior' J. Mater. Res., 19, 165 (2004).

[46] Y. Toivola, A. Stein, and R. F. Cook, 'Depth-sensing indentation response of ordered silica foam', J. Mater. Res., 19, 260 (2004).

[47] Wu Tang, Lu Shen and Kewei Xu, 'Hardness and elastic modulus of Au/NiCr/Ta multilayers on Al2O3 substrate by nanoindentation continuous stiffness measurement technique' Thin solid films (2005), Article in press.

[48] Kaiyang Zeng, Zhi-Kuan Chen and Bin Liu, 'Study of mechanical properties of light-emitting polymer films by nanoindentation technique' Thin Solid Films 477 (2005) 111-118.

[49] J. Wang, F. G. Shi, T. G. Nieh, B. Zhao, M. R. Brongo, S. Qu and T. Rosenmayer, 'Thickness dependence of elastic modulus and hardness of on-wafer low-k ultrathin polytetrafluoroethylene films' Scripta mater. 42 (2000) 687-694.

[50] J. Vitiello, A. Fuchsmann, L. L. Chapelon, V. Arnal, D. Barbier and J. Torres, 'Imapct of dielectric stack and interfaces adhesion on mechanical properties of porous ultra low-k' Microelectronic Engineering, 2005, Article in press.

[51] Lu Shen, Kaiyang Zeng, Yihua Wang, Babu Narayanan, Rakesh Kumar, Dtermination of the hardness and elastic modulus of low-k thin films and their barrier layer for microelectronic applications, Microelectronic Engineering 70 (2003) 115-124.

[52] V. N. Sekhar, T. C. Chai, S. Balakumar, Lu Shen, S. K. Sinha, A. A. O. Tay and Seung Wook Yoon, " Influence of Thickness on Nanomechanical Behavior of Black Diamond TM Low Dielectric Thin Films For Interconnect and Packaging Applications" J Mater Sci: Mater Electron (2009) 20:74-86

[53] Lugeb Wang et al, "Nanoindentation analysis of mechanical properties of low to ultraz dielectric constant SiOCH films" J. Mater. Res., Vol. 20, No.8, Aug 2005, 2080-2093.

[54] V. N. Sekhar, Lu Shen, Aditya Kumar, Tai Chong Chai, Xiaowu Zhang, C. S. Premchandran, Vaidyanathan Kripersh, Seung Wook Yoon and John H. Lau, " Study on the Effect of Wafer Back Grinding Process on Nanomehcanical Behavior of Multilayered Low-k Stack" IEEE Transaction on CPMT, Vol.2, No. 1, Jan-2012

[55] Lu Shen, Kaiyang Zeng, Comparison of mechanical properties of porous and non-porous low-k dielectric films, Microelectronic Engineering 71(2004) 221-228.

[56] S. Y. Chang, H.L. Chang, Y. C. Lu, S. M. Jang, S. J. Lin, M. S. Liang, Mechanical property analyses of porous low-dielectric-constant films for stability evaluation of multilevel-interconnect structures, Thin Solid Films 460 (2004) 167-174.

[57] I.S.Adhihetty, J.B. Vella, A. A. Volinsky, C. Goldberg, W. W. Gerberich, Mechanical properties, adhesion and fracture toughness of low-k dielectric thin films for microelectronic applications, Proceedings of the 10th international congress on fracture, Honululu, USA, DEC-2001.

[58] Y. H. Wang, M. R. Moitreyee, R. Kumar, S. Y. Wu, J. L. Xie, P. Yew, B. Subramanian, L. Shen, K. Y. Zeng, The mechanical properties of ultra-low dielectric constant films, Thin Solid Films 462-463 (2004) 227-230.

Section 5

Nanoindentation on Biological Materials

Chapter 11

Nanoindentation of Human Trabecular Bone – Tissue Mechanical Properties Compared to Standard Engineering Test Methods

Ondřej Jiroušek

Additional information is available at the end of the chapter

http://dx.doi.org/10.5772/50152

1. Introduction

There are generally two types of bone tissue: trabecular and cortical. Cortical bone is the more dense tissue found on the surface of bones and trabecular bone is a highly porous structure that fills the proximal and distal ends of all long bones (e.g. femur or tibia) and is also present as a filler in other bones (e.g. in vertebral bodies). While at the molecular level both the cortical and trabecular bone are made of the same constituents, their overall mechanical properties are quite different.

Material properties of trabecular bone bone are influenced not only by the architecture and connectivity of individual trabeculae, but also by the properties at the molecular level. At this level one can consider bone to be a composite mineral consisting of organic and inorganic constituents. To relate the overall mechanical properties (strength, stiffness, yield properties) to its microstructure, it is necessary to measure the properties of individual trabeculae. One has to bear in mind the very small dimensions of single trabecula. Although the trabecular microstructure is dependent on the anatomical site, the typical length of trabecula is 1-2 mm with diameter around 100 microns. This makes assessment of mechanical properties at the level of single trabeculae quite a challenging task.

To measure the mechanical properties of trabecular bone at the tissue level (at the level of individual trabeculae) five main methods has been designed and used: (i) tensile or three-(four-) point bending tests, (ii) buckling studies, (iii) acoustic methods, (iv) back-calculation from finite element simulations, (v) nanoindentation. There is a significant scatter in the material properties obtained by any of these methods, even when the same method is used by different authors. The published mechanical properties of human trabecular bone vary between 1 GPa and 15 GPa. The cause of this broad discrepancy in results might be in sample preparation, different testing protocols or anisotropy and asymmetry of the micro-samples. In this chapter, two most favorite methods (nanoindentation and micromechanical testing) will be compared.

2. Composition and structure of bone

At the *molecular* level, bone is a composite material made of collagen matrix stiffened by crystalline salts composed primarily of calcium and phosphate. Collagen is a fibrous protein found in flesh and connective tissues. It is a soft organic material and the main structural protein in the human body. The collagen provides the bone with toughness, while the rigidity and stiffness of the bone is provided by inorganic salts. Apart from collagen, other proteins are present in the bone. These include glycoproteins and protein-polysacharides ("proteoglycans") which create an amorphous mixture of extracellular material.

These proteins comprise about 30–50% of the bone volume. In addition to this protein constituent of bone, there is the inorganic constituent – a mineral very similar to hydroxyapatite, which is a naturally occurring mineral form of calcium apatite and can be described chemically as $Ca_{10}(PO_4)_6(OH)_2$. Hydroxyapatite in bone includes calcium phosphate, calcium carbonate, calcium fluoride, calcium hydroxide and citrate. In bone, the mineral is present in form of crystals with the shape of plates or rods, with thickness about 8 to 15 Å, width 20 to 40 Å and length 200 to 400 Å.

To summarize, on molecular level bone is a *hard* or *mineralized* or *calcified* tissue, consisting from organic matrix impregnated with the inorganic bone mineral. About 70% of the bone weight is given by the inorganic mineral. Generally, bone mineral shows positive correlation with bone strength, however, in metabolic diseases, such osteoporosis it fails to predict the bone strength correctly because these metabolic disorder results in weaker bones in presence of greater mineral density.

Going from the *molecular* level up, one can see that the collagen molecules and crystals of hydroxyapatite are assembled into microfibrils. These fibrils are again assembled into fibers with thickness about 3 to 5 μm. This level is often called *ultrastructural* level. The next level is important from the material properties point of view. In this level, the fibers are assembled either randomly into *woven* bone or are organized into lamellae forming *lamellar* bone. These lamellae can be either in concentric groups, called osteons or can form linear lammelar groups, called *plexiform* bone. This level is called *microstructural* level.

At the same level, the bone is different not only in terms of lamellar organization, but also in terms of its architectural organization. There are two types of architectural structure present in all types of bones. On the surface of the bones, there is a thin layer of dense bone, called *cortical* or *compact* bone. Under this dense layer the ends of all weight-bearing bones are filled with less dense type of bone, called *trabecular* or *spongional* bone (see Fig. 1). The primary function of the cortical bone is to support the whole body, to provide for the movement and to protect the soft inner organs. Secondary function of the cortical bone is to store and release calcium.

The trabecular bone gives supporting strength to the ends of long bones, vertebral bodies and other bones providing structural support and flexibility. The inner structure of trabecular bone is a result of structural optimization provided by remodeling processes. Result of these processes is a strong, but lightweight structure with superior mechanical properties without the weight of compact bone.

From the above mentioned arises the importance of the bone composition and structure at molecular, ultrastructural and microstructural levels for the resulting mechanical properties

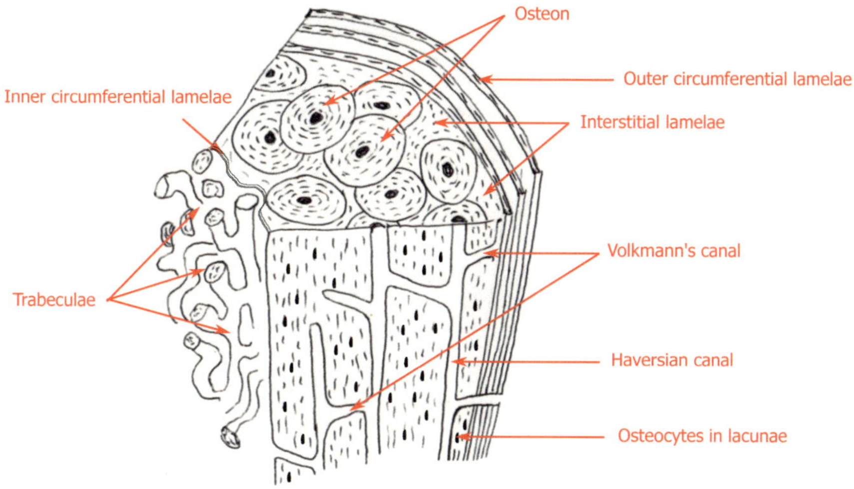

Figure 1. A schematic picture of bone structure showing trabecular, lamellar and cortical bone. Interstitial and circumferential (inner and outer) lamelae are distinguished.

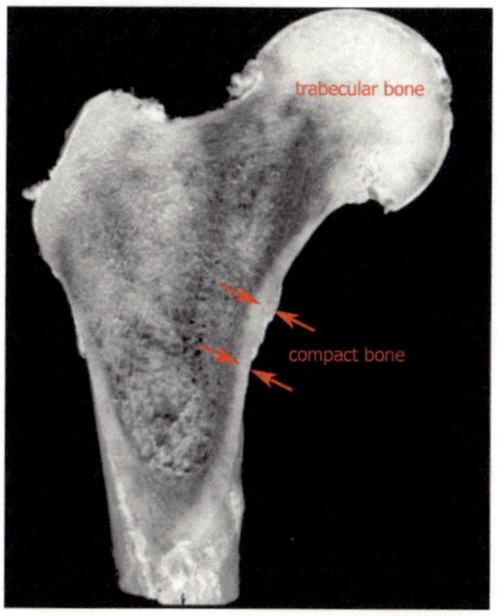

Figure 2. Longitudinal cut through proximal part of human femur. Red arrows indicate the thickness of cortical bone layer.

of both cortical and trabecular bone. And because in the living organism the skeleton is a metabolically active organ undergoing continuous remodeling process one has to distinguish between the newly deposited bone and fully calcified (old) bone. Bone remodeling is a lifelong process composed of *resorption* (old bone tissue removal) and *formation* (aposition of new bone). For the trabecular bone, most of the remodeling process takes place on the surface of the trabecular structure, the interstitial bone (bone tissue in the middle part of trabeculae) is often excluded from remodeling [1, 2]. This phenomenon has a consequence in highly nonuniform distribution of mineral content [3]. In the new bone tissue, most of the mineral is deposited in a couple of days (approximately 70%), then the process slows down dramatically and the remaining mineral is deposited in the following years [4]. Thus, the interstitial bone has a larger mineral content compared to the bone on the surface.

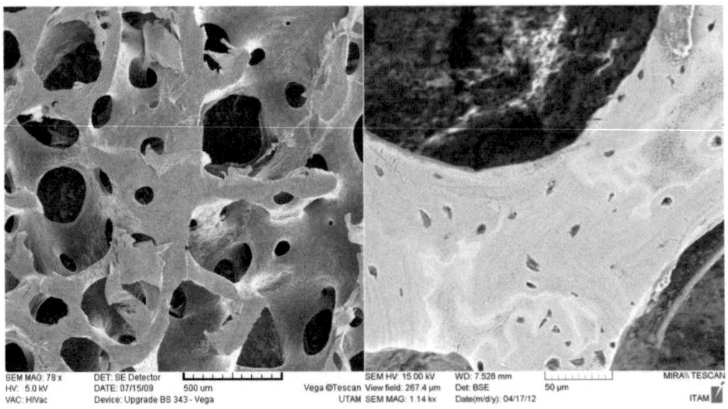

Figure 3. Trabecular bone as seen in scanning electron microscope (SEM). Sample of healthy bone from human proximal femur. The dark dots in the enlarged image are lacunae that contain osteocytes (most numerous bone cell).

2.1. Bone quality assessment

Traditionally, bone quality is described by bone mass or Bone Mineral Density (BMD) measured by dual energy X-ray absorptiometry (DXA, previously DEXA) [5], [6]. Today, BMD is used as a predictor of fracture risk in both healthy and osteoporotic bone. In DXA measurements, two X-ray beams with different energy levels are used and the absorption of X-rays by the tissues is measured. The two distinct energy levels allows to differentiate the bone tissue from the soft tissues (one X-ray beam is absorbed mainly by the soft tissues). The BMD is evaluated after the X-ray absorption by the soft tissues is eliminated.

One of the reasons for DXA failing to assess the risk of osteoporotic fracture properly [7] is the fact that bone mechanical properties correlate not only to the bone mineral density, but more importantly to its three-dimensional inner structure [8, 9]. Moreover, recent studies, e.g. [10] show the inability of DXA to account for the large variability in skeletal size and body composition in growing children.

Proper, precise and reliable assessment of morphometric and mechanical properties of trabecular bone have both biological and clinical importance. Fracture risk assessment at early time is essential for better treatment of bone diseases such as osteoporosis. To improve the reliability of fracture risk assessment, it is necessary to improve the diagnostic possibilities of modern imaging systems, such as microfocus computed tomography (micro-CT) or high-resolution peripheral quantitative computed tomography (HR-pQCT) for assessment of the 3D microstructure of cortical and trabecular bone in vivo and to improve measurement techniques to assess the material properties of trabecular (and cortical) bone at the microstructural level.

2.2. Measurement of bone mechanical properties

Mechanical properties of engineering materials are usually measured experimentally by performing mechanical tests. For engineering materials a sample of the material of known dimensions is prepared and exposed to mechanical load to measure its strength, elastic constants or other material properties under various loading and environmental conditions. For engineering materials there is a set of standardized test methods such as those developed by the American Society for Testing and Materials (ASTM) and the International Organization for Standardization (ISO). However, testing of biological materials is more complicated – often it is not possible to prepare a sample with specific dimensions and, more importantly, because the capability of the tissue to retain its natural material properties ex vivo is limited, environmental conditions, namely temperature and hydration must be considered during the tests.

Most of the tissues exhibit viscoelastic or viscoplastic behavior. This is highlighted in soft tissues, such as ligaments and tendons, but to some extent it is characteristic for all tissues. Mechanical properties of trabecular bone are inhomogeneous or site-specific, i.e. they depend on position in the material [11]. Moreover, they are dependent on anatomic location [12], e.g. Young's modulus of trabecular bone from femoral head is different from vertebral body. They are non-linearly viscoelastic [13] and display anisotropic behavior, which is usually described as orthotropic [14]. The orthotropic behavior of trabecular bone is given by the microstructural arrangement of the trabecular network; at the level of single trabecula the properties can be considered isotropic. The only study that studies the anisotropy in elastic properties of trabecular bone is [15] in which the authors used microindentation with depth equal to 2.5 μm in six different regions of interest – three on axial and three on transverse sections through trabeculae. In axial direction the bone exhibit significantly higher indentation modulus than in transverse direction, suggesting the transverse anisotropy for vertebral trabecular bone. However, there is no other study proving the anisotropy of indentation modulus for trabecular bone. This is in contrary to the compact bone where the anisotropy is present at different level of detail, i.e. single osteon is anisotropic [16] with a higher stiffness value aligned along the axial direction.

As already mentioned, experimental conditions, especially temperature and water content are influencing the results of bone mechanical testing [17]. Generally, bone stiffness increases slightly with decreasing temperature. Higher temperature has a consequence in collagen denaturation which leads to a significant decrease in the toughness of bone, but the effect on the stiffness of bone is low [18]. Effects of hydration were studied both at the microlevel, for instance in [19] nanoindentation creep tests were used for quantification of elastic and viscoelastic properties as a function of water content, or at the macrolevel, e.g. in [20] the authors studied the effect of dehydration and rehydration on the flexural properties of whole bones (mouse femora). Generally speaking, hydration increase is associated with significant (up to $\sim$40%) decrease in bone stiffness [21] and elastic properties are regained at different rates. The change in bone properties due to the dehydration can be overcome by some methods that are able to prevent the loss of hydration. Usually, the samples are tested in a phosphate-buffered saline (PBS) bath or the PBS is dripped on its surface during the experiment.

From the aforementioned complexity of material properties of trabecular bone it is evident that great care must be paid to proper sample preparation, handling during the experiment and keeping stable environmental conditions. There is a great potential of incorrectness in existing testing procedures and one must be aware of all possible error sources. Both methods, that will be discussed – nanoindentation and micromechanical testing – have their advantages and disadvantages. While micromechanical tests are more straightforward, the sample extraction, handling and measurement of the deformation is very demanding. On the other hand, nanoindentaion of a block of trabecular bone has its own limitations, not only due to difficulties in preserving the bone natural properties, but due to other effects which will be described in the following section.

3. Nanoindentation of trabecular bone

Nanoindentation measures the Young's modulus of small-volume samples with the help of a diamond tip pressed in the polished surface of the sample while the applied force and indentation depth are measured. The depth sensing indentation methods have been developed because of the difficulty in precise measurement of the contact area between the sample and the diamond tip. These methods record continuously the displacement of the tip and the contact area is calculated from the known geometry of the tip and the indentation depth.

Today, the most used model for nanoindentation is a model developed by Oliver and Pharr [22, 23] which calculates elastic properties from the unloading part of the indentation curve. This approach assumes that the unloading is purely elastic and the contact can be therefore treated as Hertzian. It has been proved that the technique is valid for axisymmetric indenters with infinitely smooth profile [22, 24] though it assumes at least four simplifying assumptions: i) perfect geometry of the indenter, ii) zero adhesive and frictional forces, iii) specimen is treated as infinite half-space and iv) material is linear elastic and incompressible.

Since in the Oliver and Pharr method an elastic contact analysis is used, only elastic properties of the material can be directly obtained. Another approach is to use finite element (FE) simulation of the contact in which various aspects of the contact, e.g. the three-dimensional nature, nonlinear material properties, time effects, can be treated. First FE studies of nanoindentation were those of Dumas [25] and Hardy [26], later studies focused on various aspects of the indentation problem, e.g. on plasticity [27]. Parameters of advanced material models can be ascertained by fitting the experimental load–displacement curves to curves obtained from the FE simulations [28, 29].

Early studies on bone nanoindentation showed great variability of measured elastic properties obtained from specimens from different anatomical locations [30, 31], although there were also few studies concluding that at the tissue level, elastic properties of trabecular and cortical bone are similar, e.g. [32]. Statistically significant difference in elastic properties and hardness of microstructural components of cortical bone, individual trabeculae and interstitial lamellae has been shown by Rho in [30]. It was shown by Zysset in [31] that hardness and elastic modulus differ substantially among lamellar types, anatomical sites and individuals. Authors suggested that tissue heterogeneity plays an important role in bone fragility and adaptation. Dependence of the elastic moduli on the direction was shown also by Rho et al. [33] resulting in larger elastic moduli in the longitudinal direction than in the transverse. Variability in bone mechanical properties dependent on anatomical site has been proven for diaphyseal and metaphyseal parts, showing greater elastic modulus and hardness for diaphyseal than metaphyseal tissues [34] confirming that tissue properties vary with anatomical location and may reflect differences in the average tissue age or mineral and collagen organization. This heterogeneity in elastic moduli of human bone at the lamellar level was observed also by other authors [35]. For compact bone, it is therefore important to distinguish between the osteonal, interstitial, and lamellar tissue which all have higher elastic moduli than trabecular bone from the same anatomical location.

These effects can be explained by a characteristic bone mineralization density distribution (BMDD) which describes local mineral content (calcium concentrations) in the (heterogeneously mineralized) bone matrix [36]. BMDD is a measure of bone matrix mineralization and compared to BMD is a more local measure for the amount of bone mineral. Combined with nanoindentation, BMDD measurements can provide important information about the structure-function relationship and explain the above-mentioned great variability in local mechanical properties of bone.

It is evident, that precise measurement of local mechanical properties is of key interest for proper description of bone mechanics. For cortical bone, it is required to measure the elastic modulus and hardness at the level of individual osteons, whereas for trabecular bone it is important to distinguish between individual trabeculae and to account for the cross-sectional difference. The only study that uses nanoindentation of individual trabeculae in their cross-sections is the study by Brennan el al. [39], however, the properties were measured only in three distinct areas (core, middle, outer) and the authors used quasi-static nanoindentation. Another possibility to measure variations in material properties in the cross-section of trabecula is to use modulus mapping (MM) [40]. Both techniques will be discussed in the following paragraphs.

3.1. Sample preparation techniques

Results of nanoindentation are highly influenced by the sample preparation procedure. This is especially true for biological tissue samples due to the difficulty of sample fixation. It is well-known both for compact and trabecular bone, that hardness and elastic modulae are dependent on the water content, resulting in up to 40% difference in measured indentation modulus [21]. For wet conditions the indentation modulus decreases.

Because the intention of this study was to compare different techniques to assess the bone properties and because the compared methods (nanoindentation and micromechanical testing) are time-consuming all the samples were tested in dry conditions. The water promotes enzymatic degradation of the bone collagen matrix and because the nanoindentation experiments with large set of different parameters take hours it would be very difficult to maintain stable conditions during the experiment. For this reason all tested samples were dried in stable conditions (48 hours at 40°C) prior the experiments. The author is aware of the fact, that revealed stiffness is higher than stiffness of the bone tested under wet conditions.

Optimizing of the sample preparation process was presented by Dudikova et al in [41]. Different approaches are described and results of the selected surface preparation procedures compared. Effects of the grain size, load and duration time of grinding on surface roughness are analyzed using confocal laser scanning microscopy. Monitoring and optimization of roughness reduction procedure used for preparation of samples for nanoindentation tests was compared to evaluate the optimal forces and times of grinding. The most suitable procedure with respect to time and cost was proposed.

3.2. Quasi-static nanoindentation of trabecular bone

The standard compliance method by Oliver and Pharr assumes elastic, isotropic materials with negligible adhesion. Bone and other tissues exhibit viscoelastic or time-dependent behavior. The viscoelastic behavior has a consequence in the nanoindentation tip sinking in the material's surface under constant load (creep). This creep behavior is observed in the force–displacement curve as a 'nose' in the unloading part. To avoid this effect (slope of the unloading part is used for calculation of the Young's modulus in the compliance method) a hold period when the maximum load is kept constant for 3-120 s is introduced to allow the creep to diminish prior unloading. The issue of removing the creep effect from the contact-depth and contact-area measurement using the trapezoidal load function (see Fig. 5) during nanoindentation is addressed in [42]. Introducing holding period to the loading function for the minimization of the creep effects on evaluated elastic properties is widely used also for other types of viscoelastic materials ranging from polymers to cementitious composites, see e.g. [43–45].

To illustrate the quasi-static nanoindentation of trabecular bone a detailed experimental procedure will be described. To measure material properties of trabecular bone in human proximal femur a set of different indentation experiments was undertaken using a small cubic sample of bone tissue obtained from embalmed cadaver (male, 72 year) using a diamond blade saw (Isomet 2100, Buehler Ltd., USA). The fat and marrow was removed from the sample using a soft water jet followed by repetitive ultrasonic cleaning. The sample was fixed in a low shrinkage epoxy resin and polished with diamond discs of grain size 35 and 15 µm. The surface was finished with monocrystalline diamond suspension of grain size 9, 3 and 1 µm. For the final polishing aluminum-oxide Al_2O_3 suspension with grain size 0.05 µm on a soft cloth was used.

Prior the mechanical testing the surface roughness of the sample was measured in a confocal laser scanning microscope (Lext OLS3000, Olympus America Inc., USA). The peak roughness R_p (the highest peak in the roughness profile) of the finished surface was 15 nm. The sample was then fixed in nanoindenter and indented using two different peak forces, 10 mN and 20 mN. For both peak forces a grid of 20 indents was performed with different set of parameters. Apart from the two peak forces, three different loading rates were used (20, 120, 240 mN/min) and three different holding times (10, 20, 40 s).

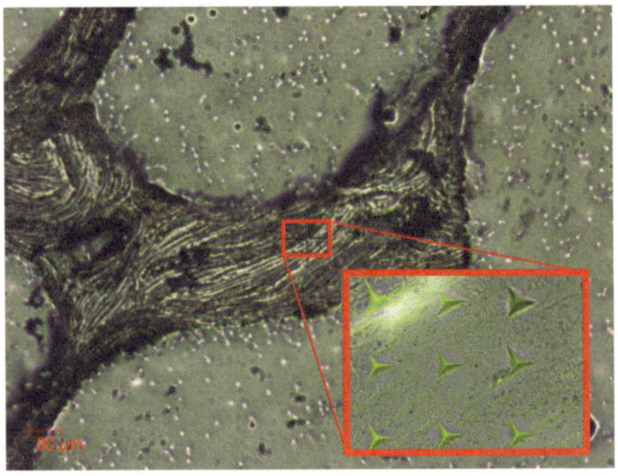

Figure 4. Sample trabecula with places of indents (Berkovich indenter)

Berkovich tip (a three sided pyramid) was used in the experiments. The indents were made with 10 µm grid size (see Fig. 4). For each indent force-depth curves were plotted and hardness and modulae of elasticity were calculated for each nanoindentation curve with the Oliver-Pharr method [22]:

$$\frac{1}{E_r} = \frac{1-\nu^2}{E} + \frac{1-\nu_i^2}{E_i} \tag{1}$$

where E_i , ν_i is Young's modulus and Poisson's ratio of the diamond tip. E, ν is Young's modulus and Poisson's ratio of the tested material (bone). In this study, Poisson's ration of the

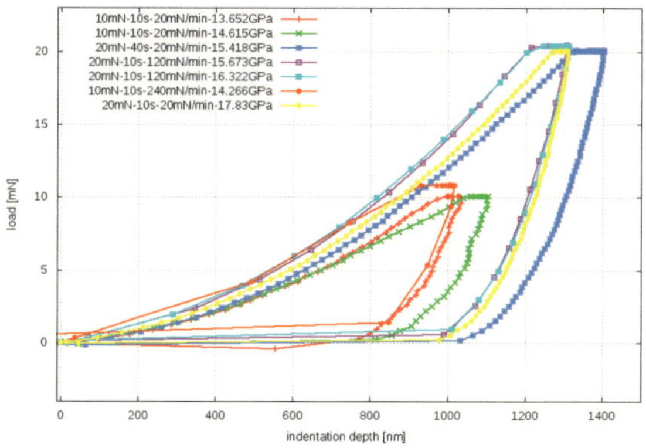

Figure 5. Set of nanoindentation curves measured with different peak forces and velocities.

	Dwell time [s]				
peak force [mN]	1.25	2.5	5	10	30
10	13.04±0.63	10.58±0.48	11.80±0.60	11.43±0.99	13.23±0.39
20	15.34±1.71	14.22±0.94	15.75±1.19	14.46±1.30	14.21±1.67

Table 1. Young's modulus from nanoindentation using different hold times for two maximum indentation forces.

sample ν was taken from our tensile experiments with single trabeculae. Reduced modulus E_r is calculated from following equation introduced by Sneddon [46]:

$$E_r = \frac{\sqrt{\pi}}{2}\frac{S}{\sqrt{A}}, \; S = \frac{dP}{dh} \tag{2}$$

where A is the projected area of elastic contact, S is the contact stiffness (experimentally measured from the unloading data).

In average, there were 17 successful indents in each set of parameters. Obtained values of Young's modulus for selected indentation curves (see Fig. 5) for different holding times that were later used in the inverse FE calculations are shown in the Tab. 1.

3.3. Inverse calculation of material parameters from FE model of nanoindentation

The previously presented derivation of material properties from nanoindetation experiment is based on linear elastic assumptions, i.e. while the loading stage is elasto-plastic, the unloading stage is purely elastic. Young's modulus of elasticity of the indented material was derived from the unloading curve. Using the Oliver-Pharr method only elastic material constants can be derived directly from the measurements. However, two nonlinear phenomena are present in the nanoindentation problem: i) contact with friction between the indenter and material's surface, and ii) elasto-plasticity of the tested material. Because it is not possible to derive analytical solution for this problem a numerical modeling using approximate solution of the complex problem must be used. Usually, finite element method is used to obtain the approximate solution. This part shortly describes a "numerical experiment" performed to establish the parameters of a material model for trabecular bone.

The indentation problem was modeled as rotationally axisymmetric problem in which the Bercovich pyramidal indenter was replaced with equivalent cone. The sharp tip of the cone was rounded due to the use of nonlinear contact between indenter and specimen. For better numerical convergence the sharp tip of the cone is usually rounded 100 $\sim$ 300 nm. In the presented model, radius r =200 nm was chosen. The FE model (see Fig. 6) was composed from 13,806 2-D structural solid elements (6,997 nodes) with linear shape functions. Between the indenter and the surface of the material frictionless contact was modeled.

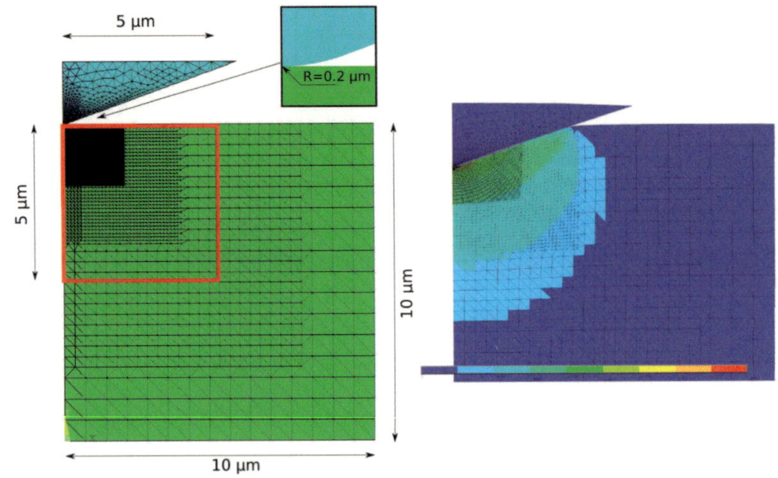

Figure 6. Axisymmetric FE model of nanoindentation experiment used for back-calculation of parameters of material model (elasto-plastic with kinematic hardening)

For the diamond nanoindenter elastic material model (E_i=1140 GPa, μ_i=0.2) was used. The trabecular bone was modelled using elasto-plastic material model with two different yield criteria – von Mises yield criterion and pressure-dependent Drucker-Prager yield condition. Bilinear isotropic hardening was chosen for both considered models. In case of von Mises plasticity with kinematic hardening rule, four material constants are needed for complete description, Young's modulus E, Poisson's ratio ν, yield stress σ_y and tangent modulus E_{tan}. Since the Drucker-Prager model is a smooth version of the Mohr–Coulomb yield surface, it is usually expressed in terms of the cohesion d, angle of internal friction φ and dilatation angle. Therefore the model is given by five constants: Young's modulus E, Poisson's ratio ν, cohesion d, friction angle φ and dilatation angle θ.

In both considered models, Young's modulus and Poisson's ratio were taken from the nanoindentation experiment. Remaining material constants ($\langle \sigma_y, E_{tan} \rangle$ or $\langle d, \varphi, \theta \rangle$) were evaluated by fitting the nanoindentation curves.

The set of nanoindentation curves with different load speeds, holding times and maximal forces was sampled using linear approximation. Values of force and penetration depth at approximation points were calculated for each nanoindentation curve. The indenter was loaded incrementally with force values in each load step of the FE simulation (450 load steps per simulation). A least-squares approach was used to compare the experimental force-penetration depth curves with curves obtained from each FE simulation with one set of material parameters. Flowchart of the fitting procedure is schematically shown in Fig. 7.

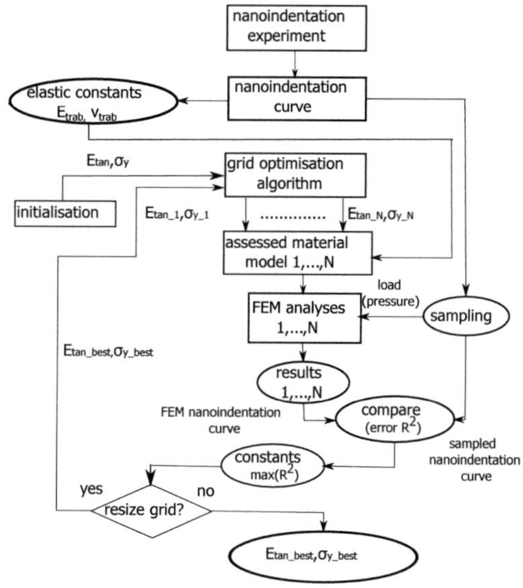

Figure 7. Flowchart of the procedure used to establish the parameters of elasto-plastic material model by fitting the nanoindentation curves from FEA to experimental curves.

Axisymmetric FE model was loaded according to the experiment using prescribed boundary conditions. A convergence study was performed to ensure the sufficient mesh density. Results from the least square fitting obtained both for the von Mises and Drucker-Prager plasticity criterion for one selected case (peak force 10 mN, holding time 5 s) are shown in Tab. 2. The fitting procedure was undertaken for every set of indentation parameters (maximal force, loading rate, dwell time).

Model	E [GPa]	ν	σ_y [MPa]	E_{tan} [MPa]	d [MPa]	β[°]	ψ[°]
von Mises	11.4	0.2	92	1420	–	–	–
D-P	11.4	0.2	–	–	23	31	4

Table 2. Best fit of the material constants for von Mises and Drucker-Prager (D-P) material models

An extension of the presented material model is visco-elasto/plastic material model with damage, described in total by 10 material constants; two elastic constants: Young's modulus E and Poisson's ratio ν, two constants describing the von Mises yield criterion with bilinear isotropic hardening: yield stress σ_y, tangent modulus E_{tan}, four constants for implicit creep C_1, C_2, C_3, C_4 with time hardening according to the equation:

$$\dot{\varepsilon}_{cr} = C_1 \sigma^{C_2} t^{C_3} e^{-C_4/T} \tag{3}$$

where $\dot{\varepsilon}_{cr}$ is the change in equivalent creep strain with respect to time, σ is the equivalent stress, t is the time at end of a substep and T is the temperature, and finally, two constants D_1, D_2 for damage model published in [37]:

$$E_{new} = (1 - d_c)E_0 \tag{4}$$

in which isotropic damage parameter d_c is defined as:

$$d_c = D_1(1 - e^{-D_2 \varepsilon_{eqv}^{pl}}) \tag{5}$$

where E_{new} is the degraded Young's modulus which is calculated at the end of each loadstep, E_0 is the initial Young's modulus and ε_{pl}^{eqv} is the accumulated equivalent plastic strain at the end of loadstep. Table 3 gives overview on the material constants fitted using the whole set of nanoindentation curves for different loading rates, dwell times and maximal indentation forces.

	mean value	standard deviation
Young's modulus [GPa]	15.39	1.4
Poisson's ratio [-]	0.2[1)]	–
yield stress σ_y [MPa]	180	43
tangent modulus E_{tan} [MPa]	1854	336
C_1 [-]	3.1×10^{-18}	4.1×10^{-18}
C_2 [-]	6.1	0.42
C_3 [-]	0.88	0.71
C_4 [-]	0	–[2)]
D_1 [-]	0.73	0.037
D_2 [-]	25.3	7.48

Table 3. Best fit of the material constants for visco-elasto/plastic material model with damage. Value indicated by [1)] was expertly determined and value indicated by [2)] was not varied.

For more detailed description of fitting this material model to the whole set of nanoindentation curves as well as application of the material to modeling of deformation behavior of single trabecula see [38].

3.4. Use of modulus mapping to map mechanical properties over cross-section of individual trabeculae

As it has been pointed out in Section 2 material properties of trabecular bone can vary in the cross-section of individual trabecula. This is caused by the highly uneven distribution of mineral content as a consequence on non-uniform deposition of new bone. Nanoindentation, as a measurement tool for very local mechanical properties is able to distinguish this variation in mineral content (interstitial bone has a larger mineral content compared to the bone on the trabecular surface).

Common approach in nanoindentation of trabecular bone is to cut a block sample of the tissue and prepare a larger area to be indented. This results in uncertainty whether interstitial or superficial bone is indented and can cause problems in evaluation of the results, even when tissue from one anatomical location is used. The only study that uses nanoindentation of individual trabeculae in their cross-sections is the study by Brennan el al. [39], however, the properties were measured only in three distinct areas (core, middle, outer) and the authors used quasi-static nanoindentation. Another option is to use a quantitative technique for mapping the elastic modulus in a larger area. Modulus mapping technique can provide highly valuable information about the elastic properties in larger area, since it is equivalent to performing a dynamic indentation test in a matrix of 256×256 points.

In their pioneering work Asif et al. [47] used modulus mapping to measure elastic properties of a carbon fiber epoxy composite. MM has been used to measure the nanoscale elastic properties of the collagen fibers, fibrils and mineral deposits in extrafibrillar space [48] in order to evaluate properties of nanocomposite films to mimic the hierarchy of natural bone. Recently the technique was used to measure the local variations in dentin and enamel in human teeth [49] but no verification with other experimental method has been done. In this part, MM was used to evaluate mechanical properties in a large area of single trabecula cross-section. Average material properties obtained with MM are compared to elastic modulae measured by quasi-static nanoindentation.

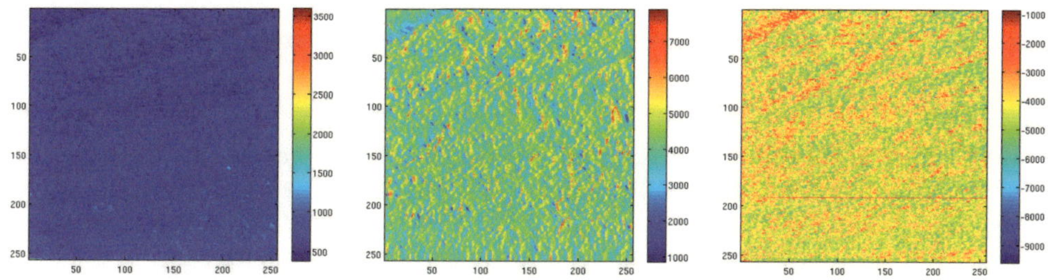

Figure 8. Exemplar image maps of (a) complex modulus, (b) loss modulus and (c) storage modulus for one indented cross-section of human trabecula

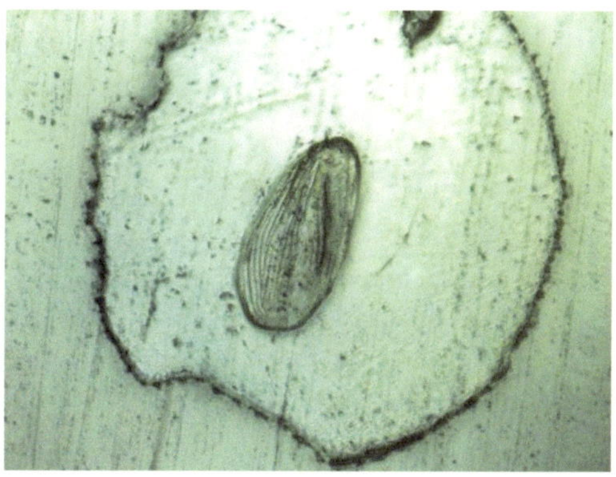

Figure 9. Cross-section of trabecula embedded in epoxy resin. Image acquired by optical microscopy.

To measure elastic properties in a larger area (35×35 μm) of trabecula's cross-section, modulus mapping technique (combination of nanoDMA and in-situ SPM) was used. In this process, the probe is sinusoidally oscillating over the polished surface with a given frequency and load. From the recorded displacement amplitude and phase lag storage and loss modulae are determined. During MM a small sinusoidal force is superimposed on top of a larger quasi-static force. Motion of the vibrating system of indenter and the surface sample can

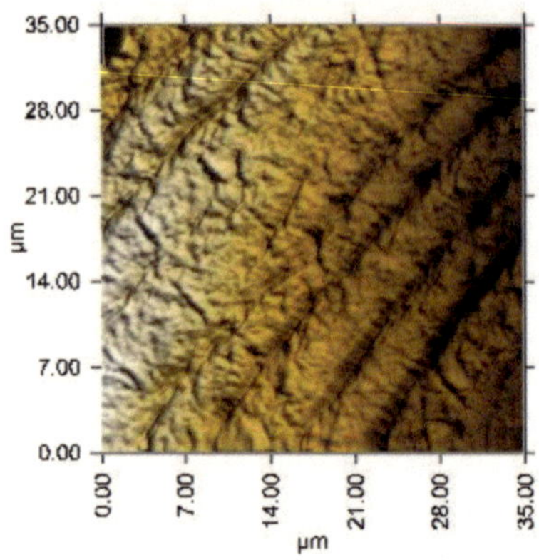

Figure 10. Typical topography of the surface sample acquired by in-situ SPM (dimensions of scanned area 35×35 µm.)

be described by equation of motion for one degree of freedom. Harmonic equation describing the motion is:

$$F_0 \sin(\omega t) = m\ddot{x} + c\dot{x} + kx \tag{6}$$

in which F_0 is the magnitude of the harmonic force, ω is the circular frequency of the system, c is the damping coefficient and k is the stiffness of the system. The system is here assumed to be linear viscoelastic. Denoting C_i stiffness of the indenter, C_s stiffness of the sample and A_0 amplitude of the system's response, we can write following equation for the time evolution of the dynamic response:

$$A_0 = \frac{F_0}{\sqrt{(k - m\omega^2)^2 + [(C_i + C_s)\,\omega]^2}} \tag{7}$$

Denoting $k = k_s + k_i$ (k_s is stiffness of the sample, k_i is stiffness of the indenter and k is the total spring stiffness) we can calculate the phase difference φ between the force and displacement from:

$$\tan\phi = \frac{(C_i + C_s)\,\omega}{k - m\omega^2} \tag{8}$$

Prior the measurement, a dynamic calibration of the system is performed to establish three parameters of the system (indenter mass m, damping coefficient of the capacitive displacement sensor C_i, stiffness of the indenter k_i), leaving only stiffness k_s and damping coefficient C_s of the sample as unknown values. In indentation, the contact stiffness k_s is proportional to the projected contact area A_c:

$$k_s = 2E^*\sqrt{\frac{A_c}{\pi}} \tag{9}$$

Using this, storage modulus $E^{'}$, loss modulus $E^{''}$ and phase shift between the force and displacement δ can be calculated using following equations:

$$E^{'} = \frac{k_s\sqrt{\pi}}{2\sqrt{A_c}},\ E^{''} = \frac{\omega C_s\sqrt{\pi}}{2\sqrt{A_c}},\ \tan\delta = \frac{\omega C_s}{k_s} \tag{10}$$

From the storage and loss modulae, complex modulus E^* can be computed using:

$$E^* = E^{'} + iE^{''} \tag{11}$$

As a result of the MM technique, the stiffness is continuously measured in the pixel matrix and the modulus includes the real and imaginary part providing the storage ($E^{'}$) and loss ($E^{''}$) modulae of the material.

3.5. Comparison with quasi-static nanoindentation

In our experiments, the 256×256 square matrix represented physical area 35×35 µm. In each point of the matrix storage and loss modulae were evaluated. To compare results from MM technique with quasi-static indentation, each sample was indented with a set of 9 indents in the center of the area used for MM. In quasi-static indentation, maximal force 1000 µN was applied in 5 s loading part, which was followed by 5 s holding part, finished with 5 s unloading part.

Modulus mapping shows the trend of larger stiffness in core, smaller values are measured in superficial areas. Both quasi-static nanoindentation and MM can be used to measure the elastic properties of extracted trabeculae, however, to identify material constants for more complex material model (e.g. von Mises plasticity with kinematic hardening) it is necessary to use different experimental program, e.g. nanoindentation with various strain rates or micromechanical testing as will be described in the following chapter.

4. Micromechanical testing of isolated trabeculae

Apart from nanoindentation, delicate tensile or bending tests can be performed with extracted trabeculae. The most important advantage of micromechanical testing is that it is well-developed and straightforwad technique. Moreover, using mechanical testing it is possible not only to measure the Young's modulus, but also yield properties or properties at different strain rates can be assessed. In this section a method for measuring elastic modulus and yield stress using micromechanical testing of individual trabeculae will be described in detail.

4.1. Shape from silhouettes

The difficulty of performing mechanical tests with samples of such small dimensions (diameter of an average human trabecula from proximal femur is about 100 µm) lies not only in complicated manipulation with the samples, but also in precision of the measurement. The dimensions of the samples must be measured with micrometer precision and due to the irregularity in the samples' shape it is necessary to measure the real shape of each sample. To assess the precise shape of the trabeculae, every sample can be scanned in a micro-CT device (resolution of 1-5 µm is needed) and its geometry reconstructed with filtered backprojection for the cone beam scanning geometry.

Another option is to reconstruct the 3-D shape of the sample from its silhouettes. This is a method commonly used to obtain 3-D shapes from silhouettes captured by multiple cameras with the volume intersection method. The problem is stated as follows: Given a set of calibrated images of an object taken from different angles can we reconstruct its 3-D model? The simplest method to perform this reconstruction is to use the so called *visual hull* which is the shape maximally consistent with the silhouettes. The visual hull is the intersection of the silhouette cones produced by images taken from different viewpoints (see Fig. 11). It can be easily shown that the visual hull cannot capture concavities not visible in the silhouettes which can lead to reconstruction artifacts such as erroneous additional connected components. In the reconstruction, the foreground object is separated from the background by binary thresholding and silhouette image is this binary foreground/background image.

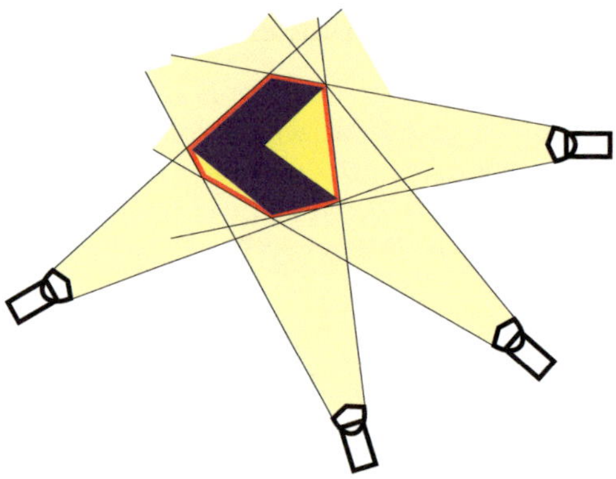

Figure 11. The principle of visual hull. It is the intersection of silhouette cones (green), which is a bounding geometry of the object

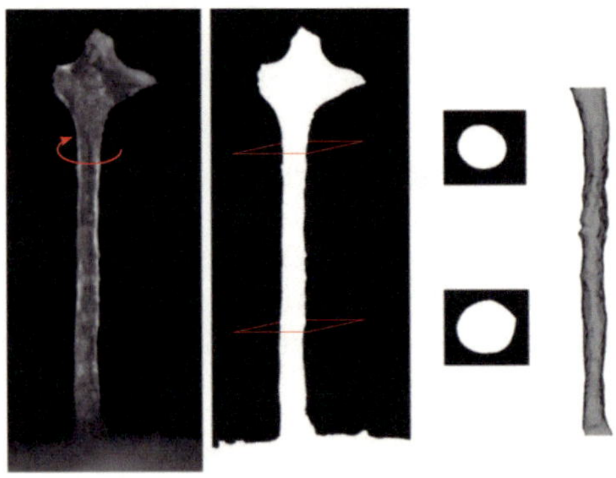

Figure 12. Reconstruction of the geometry of extracted trabeculae. Only middle part of the sample (between the supports) is needed.

It can be shown that for convex objects the reconstruction from images converges to the true shape as more view angles are added. This technique has been used in our experiments with isolated trabeculae. Prior the testing, each sample was mounted on a rotational stage driven by stepper motor and its shade was captured with a CCD camera attached to a microscope. The sample was rotated by 360° in 1° increments and its silhouettes captured with the camera. From the shade images the real geometry of each sample was reconstructed with error less than 4%.

The geometry of the trabecula is represented by its surface given as a set of connected triangles (see Fig. 12). The volume of the sample is discretized with tetrahedral elements using a constrained Delaunay approach [50]. After shape optimization of the tetrahedral elements, mid-side nodes are added for more accurately calculated gradients in deflections, strains and stresses in all FE calculations.

4.2. Displacement tracking using DIC

To measure the applied load and displacements with high accuracy a novel experimental device (Fig. 13) has been developed which enables to control the displacement of the loading tip with sub-micron positioning capability by means of differential micrometer with 0.1 μm sensitivity. Positioning of the supports and the sample is provided by precision double-row ball bearing linear stages with 1 μm sensitivity. This very high accuracy in positioning is necessary for perfect loading and exact geometry of the experimental setup (see the close-up view in Fig. 15a and Fig. 15b).

Figure 13. Experimental setup for the optical measurement of extracted trabecula in three-point bending.

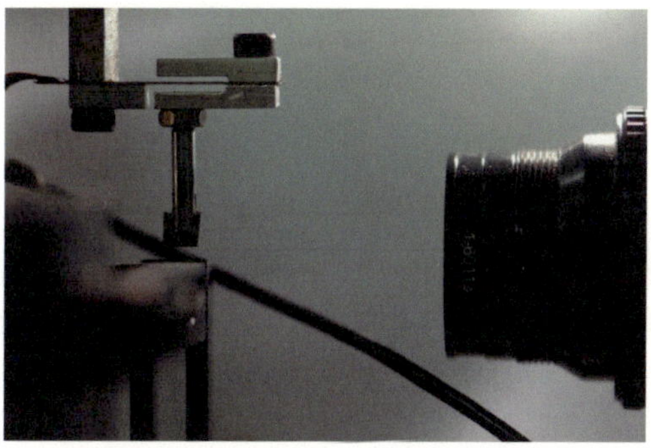

Figure 14. Close-up view of the sample showing the loading tip attached to the miniature load cell. High magnification (12×) lens on right.

The deformation of the sample is measured optically using a high-resolution CCD camera. The strain field is evaluated with an image correlation algorithm [51] applied to the part of the sample between the supports.

DIC is based on the maximization of a correlation coefficient that is computed in subsets of the image surrounding the measurement points. Correlation between the pixel intensity in the image subsets between two corresponding images (at time t_i and t_{i+1}) is computed for every measured point with a non-linear optimization technique (inverse compositional algorithm). Coefficients of the affine transformation between the deformed and original position are used to compute the deformation gradient tensor F. Green-Lagrange deformation tensor is then computed from: $E = \frac{1}{2}(F^T F - I)$.

In the bending tests, only displacements are evaluated with DIC. The displacements are used to find the best fit between the experimental results and results computed using FE simulation of the bending test with real geometry of the sample. This procedure is described in the following text.

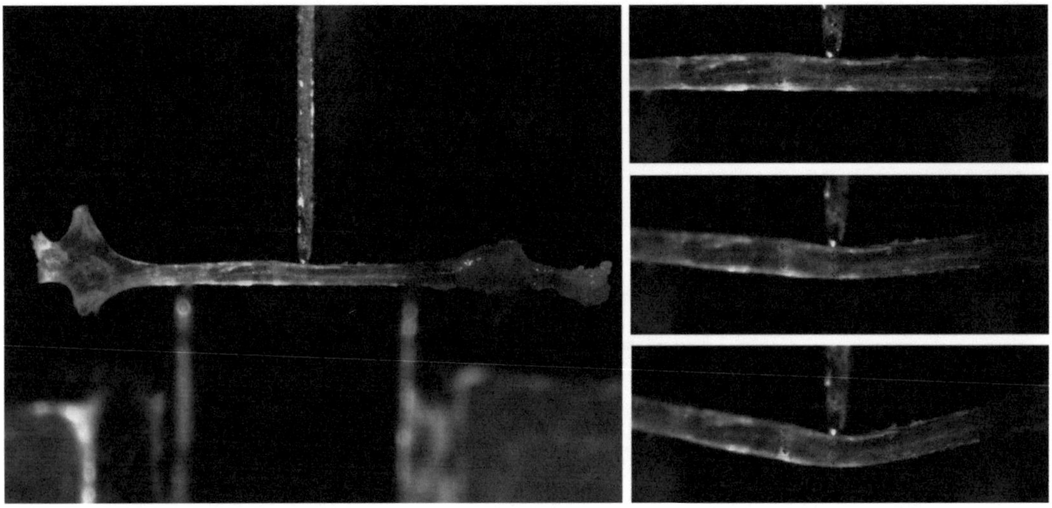

a) *b)*

Figure 15. Three-point bending of isolated trabecula. Straight sample resting on supports with span length equal to 1 mm (a). Principle of tracking "markers" on the sample using DIC (b).

4.3. Results from 3-point bending

To account for the real geometry of each irregularly shaped specimen, FE model reflecting the true geometry is developed using the shape-from-silhouettes approach. Exact position of the supports and the loading tip is established from the high-resolution images. Boundary conditions and loading of the FE model are prescribed to match exactly the experimental conditions for each sample. Using similar fitting procedure as described for the FE model of nanoindentation a set of material parameters is determined to find the best fit between the experimental force–displacement curves and those determined from the numerical simulation.

Because the displacements are actually measured in three distinct locations (using the image correlation), the same positions are determined in the FE model and displacements in each load increment are calculated in these points. Values from the FE simulation are compared to the measured values. Example of the best-fit for selected "markers" in the FE model are depicted (together with displacement field in the model) in Fig. 16.

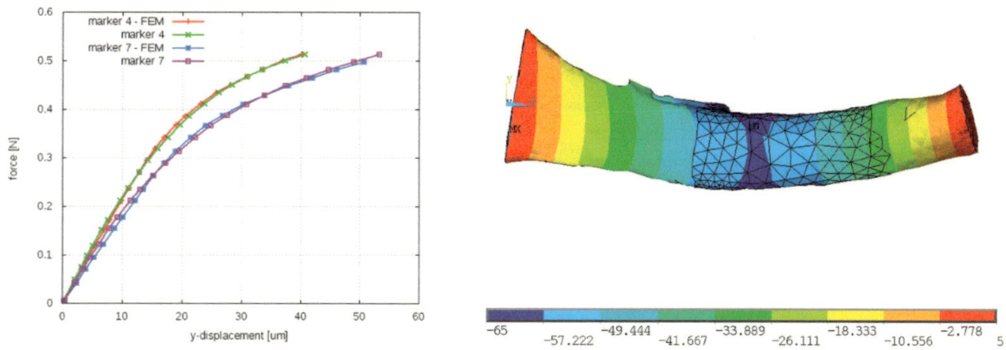

Figure 16. Example of fitting the force-deflection curves calculated by the FE model to experimentally determined ones (left). Only part of the sample between the supports is considered by the FE model of the specimen's real geometry(right).

4.4. Results from micromechanical testing compared to nanoindentation

To compare nanoindentation with micromechanical testing, only the elastic properties were determined by the two different methods. For the nanoindentation, only parts of the trabeculae outside the supports were used. In this region, the bending moment is zero and this part of the sample can be considered undamaged. After the bending test, each sample was embedded in the upright position in the low shrinkage epoxy resin and polished to reveal the cross-section close to one of the ends of trabecula. After surface polishing the nanoindentation experiments described in section 3 were undertaken. Elastic properties obtained from nanoindentation and those assessed by the micromechanical bending tests are compared in Tab. 4.

sample No.	quasi-static nanoindentation [GPa]	bending test [GPa]
1	12.04	11.07
2	15.93	13.26
3	16.22	13.87
4	17.82	14.52
5	17.91	14.70
6	12.07	10.51

Table 4. Average values of elastic modulae obtained by quasi-static indentation in the cross-section of each trabecula compared to the results of micromechanical tests with single trabecula (three-point bending)

From Tab. 4 it is clear, that both methods give very similar results in terms of elastic properties. The values of elastic modulus measured by three-point bending are 10% - 20% lower that those assessed by nanoindentation. This can be explained by the fact that nanoindentation is not performed in the very outer part of the cross-section, thus giving more information about the core hardness, which is stiffer. Another reason might be in overestimation of the cross-sectional area by the shape-from-silhouettes method. The overestimation of the cross-sectional area is well under 3%, however, the calculation of bending stiffness can lead to overestimation over 5%.

5. Application to constitutive modeling of trabecular bone tissue

The importance of assessment of mechanical properties at the tissue level, i.e. at the level of single trabeculae can be demonstrated using indirect determination of trabecular bone mechanical properties from high-resolution images of its internal architecture. These micro-FE models, i.e. models properly describing the geometry of the trabecular bone architecture are nowadays a common tool for assessment of mechanical properties of trabecular bone [52]. The high-resolution images can be obtained from micro-CT images, peripheral quantitative computed tomography (pQCT) or, more recently with high-resolution magnetic resonance (MR) imaging.

These models can be used to study the relation between the microstructure, material properties at the tissue level and overall mechanical properties. It can be used to predict evolution of mechanical properties with osteoporotic changes to the microstructure and thus to assess evolution of osteoporotic fracture risk. Moreover, combined with experimental tests with small-scale samples of trabecular bone, these models can be used to verify the tissue constitutive models. This can be achieved by comparing the response of micro-FE model with the response of the real sample.

Additionally, microtomographic scanning of a sample can be performed in a time-lapse fashion, i.e. a sequence of a time-lapse micro-CT [53] of the specimen under gradually incremented load is performed and the deformed microstructure is recorded. A custom, laboratory micro-radiographic system composed of micro-focus X-ray tube and a large flat panel detector [54] can be used for such an experiment. Reconstruction of the internal structure is provided using backprojection algorithm for equiangular cone-beam projection data [55] for each load increment.

To compare the strains in the loaded sample to the values calculated by the FE analysis a Digital Volume Correlation (DVC) [56, 57] method is used. DVC can be seen as a natural extension of existing digital image correlation techniques [59] to three dimensions. Example of a comparison between the experimentally measured strain distribution in a loaded sample and numerical simulation can be found in [58].

The bone microstructure can be discretized using a tetrahedral mesh. In the nodal points of the mesh the displacement vector is computed using X-ray digital volumetric correlation technique. For each of the tetrahedral element the Green-Lagrange strain tensor is calculated from the displacements of its vertices. The overlaid tetrahedral mesh serves not only for the calculation of the deformation tensor, but also for easy visualization of the vector and tensor fields and for fast and direct comparison with results of numerical simulations. The numerical simulations can use the existing tetrahedral mesh or the mesh can be easily refined/coarsened if needed. Illustrative example of the FE model and displacement field computed with DIC is shown in Fig. 18.

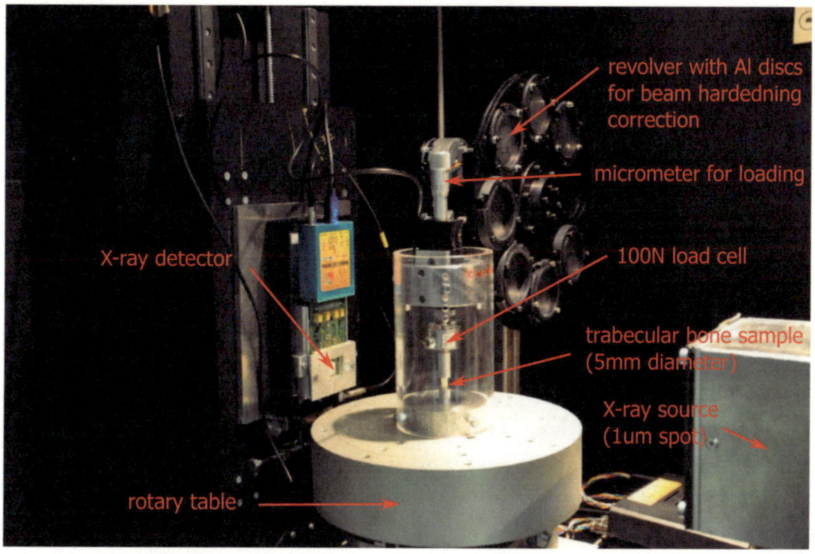

Figure 17. Experimental setup for time-lapse X-ray rentgenography. Sample is tested in three-point bending and X-ray projections of the deformed state are captured using exposures with very small time (250 ms)

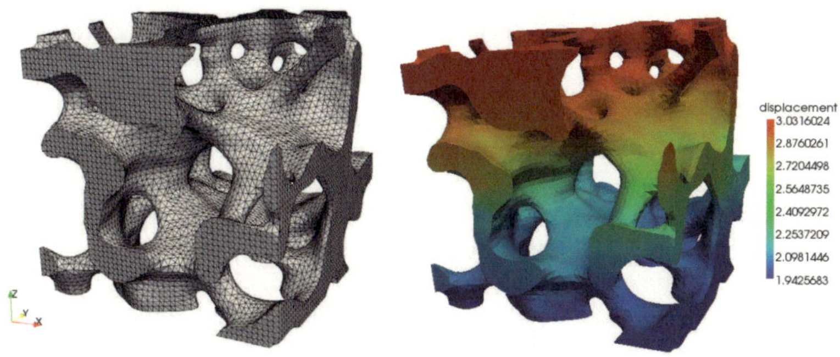

Figure 18. Small part of the micro-FE model of trabecular bone. FE mesh of the model (a) and displacement field (b). Illustrative image.

6. Conclusions

An overview of techniques used to determine material properties of trabecular bone at different microstructural levels was presented in the chapter. The focus was paid on two important methods for measurement of mechanical properties at the tissue level: i) nanoindentation and ii) micromechanical testing. In the part focused on nanoindentation two methods, i.e. quasi-static indentation and modulus mapping were compared. Apart from measurement of the elastic properties, a method based on fitting the response of FE model of indentation to experimental curves was introduced.

Apart from nanoindentation, another approach to measure material properties of trabeculae was described. This approach is based on precise micromechanical tests performed with individual trabeculae. A methodology how to account for the imperfect shape of the small-scale specimens was introduced. This methodology is based on developing exact FE model of each tested specimen using the shape-from-silhouettes approach. Discussion of precision of the measurements as well as precision of the geometry reconstruction was shortly given.

In the last part of the chapter, application of the tissue material properties to modeling of trabecular bone was given. Short description of the development process of micro-FE models, i.e. FE models of the trabecular micro-architecture was described and utilization of microfocus Computed Tomography or other high-resolution imaging methods was outlined.

Precise and reliable measurement of tissue mechanical properties of trabecular bone as a key factor for the overall bone strength is important not only for reliable assessment of osteoporotic fracture risk assessment but also for precise evaluation of the effects of treatment procedures, e.g. in animal studies. Although the term bone quality is poorly defined, it is clear that current diagnostic techniques for measuring the bone quality in vivo do not yield reliable results. To improve the current diagnostic methods all the influencing factors, i.e. tissue properties, geometric, microarchitectural and connectivity factors, must be taken into account and a combination of all described methods must be used.

Acknowledgements

Support of the Czech Science Foundation (P105/10/2305) is gratefully acknowledged.

Author details

Ondřej Jiroušek
Institute of Theoretical and Applied Mechanics
Academy of Sciences of the Czech Republic, v.v.i.,
Prague, Czech Republic

7. References

[1] Birkenhager-Frenkel DH, Nigg AL, Hens CJJ, Birkenhager JC. Changes of interstitial bone thickness with age in men and women. Bone 1993; 14(3) 211-216.
[2] Boyde A, Elliott JC, Jones SJ. Stereology and histogram analysis of backscattered electron images: Age changes in bone. Bone 1993; 14(3) 205-210.
[3] van der Linden JC, Birkenhager-Frenkel DH, Verhaar JAN, Weinans H. Trabecular bone's mechanical properties are affected by its non-uniform mineral distribution. Journal of Biomechanics 2001; 34(12) 1573-1580.
[4] Parfitt AM. The composition, structure and remodeling of bone a basis for the interpretation of bonemineral measurements. In: Dequeker J, Geusens P, Wahner HW. (eds.) Bone Mineral Measurements by Photon Absorptiometry: Methodological Problems, 1988; Leuven University Press, Leuven, 9-28.
[5] Petersen MM, Gehrchen PM, Nielsen PK, Lund B. Loss of bone mineral of the hip assessed by DEXA following tibial shaft fractures. Bone 1997; 20(5), 491-495.

[6] Pafumi C, Chiarenza M, Zizza G. Role of DEXA and ultrasonometry in the evaluation of osteoporotic risk in postmenopausal women. Maturitas 2002; 42 113-117.
[7] Cummings SR, Bates D, Black DM. Clinical use of bone densitometry: scientific review. JAMA 2002; 288 1889-1897.
[8] Silva MJ, Gibson LJ. Modeling the mechanical behavior of vertebral trabecular bone: Effects of age-related changes in microstructure. Bone 1997; 21(2) 191-199.
[9] Sievanen H, Weynand LS, Wacker WK, Simonelli C, Burke PK, Ragi S, Del Rio L. A Novel DXA-Based Hip Failure Index Captures Hip Fragility Independent of BMD. Journal of Clinical Densitometry 2008; 11(3) 367-372.
[10] Wren TAL, Liu X, Pitukcheewanont P, Gilsanz V. Bone Densitometry in Pediatric Populations: Discrepancies in the Diagnosis of Osteoporosis by DXA and CT. The Journal of Pediatrics 2005; 146(6) 776-779.
[11] Carte DR, Hayes WC. The compressive behavior of bone as a two-phase porous structure. Journal of Bone and Joint Surgery - Series A 1977; 59 (7) 954-962.
[12] Goldstein SA. The mechanical properties of trabecular bone: Dependence on anatomic location and function. Journal of Biomechanics 1987; 20 (11-12) 1055-1061.
[13] Deligianni DD, Maris A, Missirlis YF. Stress relaxation behaviour of trabecular bone specimens. Journal of Biomechanics 1994; 27 (12) 1469-1476.
[14] van Rietbergen B, Odgaard A, Kabel J, Huiskes R. Direct mechanics assessment of elastic symmetries and properties of trabecular bone architecture. Journal of Biomechanics 1996; 29 (12) 1653-1657.
[15] Wolfram U, Wilkea H, Zysset PK. Transverse isotropic elastic properties of vertebral trabecular bone matrix measured using microindentation (effects of age, gender and vertebral level). Bone 2009; 44 S392–S393.
[16] Reisinger AG, Pahr DH, Zysset PK. Principal stiffness orientation and degree of anisotropy of human osteons based on nanoindentation in three distinct planes Journal of the Mechanical Behavior of Biomedical Materials 2011; 4 (8) 2113-2127.
[17] Yamashita J, Li X, Furman BR, Rawls HR, Wang X, Agrawal CM. Collagen and bone viscoelasticity: A dynamic mechanical analysis. Journal of Biomedical Materials Research 2002; 63 (1) 31-36.
[18] Wang X, Bank RA, Tekoppele JM and Agrawal CM. The role of collagen in determining bone mechanical properties. Journal of Orthopaedic Research 2001; 19 1021–1026.
[19] Bembey AK, Bushby AJ, Boyde A, Ferguson VL, Oyen ML. Hydration effects on the micro-mechanical properties of bone. Journal of Materials Research 2006; 21(8) 1962-1968.
[20] Broz JJ, Simske SJ, Greenberg AR, Luttges MW. Effects of rehydration state on the flexural properties of whole mouse long bones. Journal of Biomechanical Engineering 1993; 115 (4 A) 447-449.
[21] Bembey AK, Oyen ML, Bushby AJ, Boyde A. Viscoelastic properties of bone as a function of hydration state determined by nanoindentation. Philosophical Magazine 2006; 86 (33-35) 5691-5703.
[22] Oliver WC and Pharr GM. An improved technique for determining hardness and elastic modulus using load and displacement sensing indentation experiments. Journal of Materials Research 1992; 7(6) 1564-1583.

[23] Pharr GM. Measurement of mechanical properties by ultra-low load indentation. Materials Science and Engineering A 1998; 253(1-2) 151-159.

[24] Pharr GM, Oliver WC, Brotzen FR. On the generality of the relationship among contact stiffness, contact area, and the elastic modulus during indentation. Journal of Materials Research 1992; 7(3) 613–617.

[25] Dumas G, Baronet CN. Elastoplastic indentation of a half-space by an infinitely long rigid circular cylinder. International Journal of Mechanical Sciences 1971; 13 (6) 519-530.

[26] Hardy C, Baronet CN, Tordion GV. Indentation of an elastic-perfectly-plastic half-space by a hard sphere. Journal of Basic Engineering 1972; 94(1) 251-253.

[27] Cheng YT, Cheng CM. Scaling Relationships in Conical Indentation of Elastic Perfectly Plastic Solids. International Journal of Solids Structures 1999; 36 1231–1243.

[28] Wang X, Allen MR, Burr DB, Lavernia EJ, Jeremic B, Fyhrie DP. Identification of material parameters based on Mohr–Coulomb failure criterion for bisphosphonate treated canine vertebral cancellous bone. Bone 2008; 43 (4) 775-780.

[29] Jiroušek, O., Nemecek, J., Kytýr, D., Kunecký, J., Zlámal, P., Doktor, T., Nanoindentation of trabecular bone-comparison with uniaxial testing of single trabecula. Chemicke Listy 2011; 105 (17) s668-s671.

[30] Rho JY, Tsui TY, Pharr GM. Elastic properties of human cortical and trabecular lamellar bone measured by nanoindentation. Biomaterials 1997; 18(20) 1325-1330.

[31] Zysset PK, Guo XE, Hoffler CE, Moore KE, Goldstein SA. Elastic modulus and hardness of cortical and trabecular bone lamellae measured by nanoindentation in the human femur. Journal of Biomechanics 1999; 32(10) 1005-1012.

[32] Turner CH, Rho J, Takano Y, Tsui TY, Pharr GM. The elastic properties of trabecular and cortical bone tissues are similar: Results from two microscopic measurement techniques. Journal of Biomechanics 1999; 32(4) 437-441.

[33] Rho JY, Roy ME 2nd, Tsui TY, Pharr GM. Elastic properties of microstructural components of human bone tissue as measured by nanoindentation. Journal of Biomedical Materials Research 1999; 45(1) 48-54.

[34] Hoffler CE, Moore KE, Kozloff K, Zysset PK, Brown MB, Goldstein SA. Heterogeneity of bone lamellar-level elastic moduli. Bone 2000; 26(6) 603-609.

[35] Hengsberger S, Kulik A, Zysset P. Nanoindentation discriminates the elastic properties of individual human bone lamellae under dry and physiological conditions. Bone 2002; 30(1) 178-184.

[36] Roschger P, Paschalis EP, Fratzl P, Klaushofer K. Bone mineralization density distribution in health and disease. Bone 2008; 42(3) 456-466.

[37] Zhang J, Michalenko MM, Kuhl E. Characterization of indentation response and stiffness reduction of bone using a continuum damage model. Journal of the mechanical behavior of biomedical materials 2010; 3(2), 189-202.

[38] Zlámal P, Jiroušek O, Kytýr D, Doktor T. Indirect determination of material modelparameters for single trabecula based o nanoindentation and three-point bending test. Proceedings of the 18th international conference Engineering Mechanics 2012; 1 394-395.

[39] Brennan O, Kennedy OD, Lee TC, Rackard SM, O'Brien FO Biomechanical properties across trabeculae from the proximal femur of normal and ovariectomised sheep. Journal of Biomechanics 2009; 42(4) 498-503.

[40] Jiroušek O, Kytýr D, Zlámal P, Doktor T, Šepitka J, Lukeš J. Use of modulus mapping technique to investigate cross-sectional material properties of extracted single human trabeculae. Chemické Listy 2012; accepted.
[41] Dudíková M, Kytýr D, Doktor T, Jiroušek O. Monitoring of material surface polishing procedure using confocal microscope. Chemické Listy 2011; 105(17) 790-791.
[42] Tang B and Ngan AHW. Accurate measurement of tip–sample contact size during nanoindentation of viscoelastic materials. Journal of Materials Research 2003; 18 1141-1148.
[43] Oyen ML, Cook RF. A practical guide for analysis of nanoindentation data. Journal of the Mechanical Behavior of Biomedical Materials 2009; 2(4) 396-407.
[44] Nemecek J. Creep effects in nanoindentation of hydrated phases of cement pastes. Materials Characterization 2009; 60(9) 1028–1034.
[45] Mencík J, He LH, Nemecek J. Characterization of viscoelastic-plastic properties of solid polymers by instrumented indentation. Polymer Testing 2011; 30 101–109.
[46] Sneddon IN. The relation between load and penetration in the axisymmetric Boussinesq problem for a punch of arbitrary profile. International Journal of Engineering Science 1965; 3(1) 47-57.
[47] Asif SA, Wahl KJ, Colton RJ, Warren OL. Quantitative imaging of nanoscale mechanical properties using hybrid nanoindentation and force modulation. Journal of Applied Physics 2001; 90(3) 1192-1200.
[48] Khanna R, Katti SK, Katti DR. Bone Nodules on Chitosan-Polygalacturonic Acid-Hydroxyapatite Nanocomposite Films Mimic Hierarchy of Natural Bone. Acta Biomaterialia 2011; 7 1173-1183.
[49] Balooch G, Marshall GW, Marshall SJ, Warren OL, Asif SAS, Balooch M. Evaluation of a new modulus mapping technique to investigate microstructural features of human teeth. Journal of Biomechanics 2004; 37 1223-1232.
[50] Edelsbrunner H and Shah NR. Incremental Topological Flipping Works for Regular Triangulations. Algorithmica 1996; 15 223-241.
[51] Jandejsek I, Valach J, Vavrík D. Optimization and Calibration of Digital Image Correlation Method. In Experimentální analýza napetí 2010. Olomouc: Univerzita Palackého v Olomouci, 121-126.
[52] Verhulp E, van Rietbergen B, Müller R, Huiskes R. Indirect determination of trabecular bone effective tissue failure properties using micro-FE simulations. Journal of Biomechanics 2008; 41(7) 1479-1485.
[53] Nazarian A. and Müller R. Time-lapsed microstructural imaging of bone failure behavior. Journal of Biomechanics 2004; 37(1) 55-65.
[54] Jakubek J, Holy T, Jakubek M, Vavrik D, Vykydal Z. Experimental system for high resolution X-ray transmission radiography. Nuclear Instruments and Methods in Physics Research Section A: Accelerators, Spectrometers, Detectors and Associated Equipment 2006; 563(1), 278-281.
[55] Vavrík D, Soukup P. Metal grain structure resolved with table-top micro-tomographic system. Journal of Instrumentation 2011; 6 (11), art. no. C11034.
[56] Jiroušek O, Jandejsek I, Vavrík D. Evaluation of strain field in microstructures using micro-CT and digital volume correlation. Journal of Instrumentation 2011; 6 (1), art. no. C01039.

[57] Jandejsek I, Jiroušek O, Vavrík D. Precise strain measurement in complex materials using digital volumetric correlation and time lapse micro-CT data. Procedia Engineering 2011; 10(1) 1730-1735.

[58] Jiroušek O, Zlámal P, Kytýr D, Kroupa M. Strain analysis of trabecular bone using time-resolved X-ray microtomography. Nuclear Instruments and Methods in Physics Research, Section A: Accelerators, Spectrometers, Detectors and Associated Equipment 2011; 633 (SUPPL. 1) S148-S151.

[59] Lucas BD and Kanade T. An iterative image registration technique with an application to stereo vision. Proceedings of the 1981 DARPA Image Understanding Workshop, April, 1981 121-130.

Section 6

Design of New Nanoindentation Devices

Chapter 12

Design, Analysis and Experiments of a Novel *in situ* SEM Indentation Device

Hongwei Zhao, Hu Huang, Zunqiang Fan, Zhaojun Yang and Zhichao Ma

Additional information is available at the end of the chapter

http://dx.doi.org/10.5772/48304

1. Introduction

Instrumented indentation as a powerful tool to determine mechanical properties of materials has been widely used in fields of materials science, biomechanics, surface engineering, semiconductor, MEMS/NEMS, biomedicine and so on (Fischer-Cripps, 2004) . Up to now, there are some commercially available indentation instruments. However, most of them can only carry out ex-situ indentation tests because of their complex structures and big volumes. In previous literatures, many phenomena such as pile-up and sink-in (Huang et al., 2005; Keryvin et al., 2010), which are difficult to explain but significantly affect the load-depth (*P-h*) curve and the determination of the contact area are observed. In order to reveal mechanical behavior of materials under the indentation load in detail and explain discontinuities during the initial loading segment better, more direct observation methods during indentation tests should be developed (Ghisleni et al., 2009).

In recent years, based on the transmission electron microscope (TEM) and scanning electron microscope (SEM), in situ indentation technique is presented by researchers (Gane & Bowden, 1968; Minor et al., 2001; Zhou et al., 2006). In situ TEM indentation has the capability to visually observe microstructure variations beneath the indenter such as phase transformation, dislocation formation and propagation (Minor et al., 2006), and dislocation-grain boundary interaction (De Hosson et al., 2006). But it also has disadvantages. For example, the specimens are very small and need complex and laborious preparation, and the specimens need to be electron transparent. In addition, the method that uses the mechanical properties of specimens with limited scale to evaluate the bulk materials is questionable (Ruffell et al., 2007). Via in situ SEM indentation, it is possible to observe the surface deformation during the whole indentation process, which is helpful to correct the reduced elastic modulus and hardness obtained by the Oliver and Pharr method (Oliver & Pharr, 2004) by accounting for pile-up and sink-in phenomena. In addition, mechanism of deformation, crack formation and propagation, shear band formation, damage, pile-up, sink-in of materials will be studied deeply (Rzepiejewska-Malyska et al., 2009; Nowak et al., 2009). Compared with in situ TEM indentation, some physical phenomena such as phase transformation of materials, or nucleation and dislocation corresponding to high pressure applied by the indenter, are hardly observed by in situ SEM indentation. However, in situ SEM indentation is still a potential and attractive method because of its large field of view, simple specimen preparation and compatibility of materials with different dimensions from millimeter to micro/nanometer.

In this chapter, emphasis is put on in situ SEM indentation. Design of the indentation device compatible with the SEM has a few challenges (Huang et al., 2012), such as the small volume of the SEM chamber, short working distance, electromagnetic compatibility, the vacuum environment and vibration compatibility. Several studies of indentation inside the SEM have been done by researchers (Bangert et al., 1982; Bangert & Wagendristel, 1985; Motoki, 2006). The most representative are the device developed by Rabe in 2004 (Rabe et al., 2004) and the product — PI 85 SEM PicoIndenter manufactured by Hysitron Inc (Hysitron Incorporated). The load resolution of Rabe's device is 100 μN for a range up to 1.5 N and the maximum available load is about 500 mN due to using the stick-slip actuator. Because of integrating a built-in strain gauge, the closed-loop displacement resolution is a little low about 50 nm with an indentation displacement range of 20 μm. The PI 85 SEM PicoIndenter has high load and displacement resolution about 3 nN and 0.02 nm respectively. However, the maximum indentation load is limited to be 10 mN and the maximum indentation displacement is 5 μm, which limits its more wide applications. In addition, this product is very expensive and up to now, there are very few scientists and researchers who can use this advanced equipment. So, in situ SEM indentation devices with large ranges, high precision, compact structures and low cost are still required.

In this chapter, a novel in situ indentation device with dimensions of 103 mm × 74 mm × 60 mm is developed and it is compatible with the SEM—Quanta 250. Integrating the stepper motor, the piezoelectric actuator and the flexure hinge, the coarse positioner and the precision driven unit were designed respectively, which can be used to realize coarse adjustment of the specimen and precision loading and unloading process of the indenter. A novel indenter holder was designed to ensure vertical penetration of the indenter. Closed-loop control of the indentation process was established to solve the problem of nonlinearity of the piezoelectric actuator and to enrich the control modes. Experiments were carried out to evaluate performances and verify the feasibility of the developed device.

2. Principle of in situ indentation tests inside the SEM

The schematic diagram of the developed indentation device for in situ indentation tests inside the SEM is shown in Fig. 1. Here, the diamond indenter is fixed and the specimen can move toward the indenter. This design considers the work distance of the SEM and makes the dimension of one side of the device small, which is helpful for in situ observation of the indentation region. The z axis coarse positioner driven by a stepper motor is designed to realize coarse adjustment of the specimen. The precision driven unit consisting of the piezoelectric actuator and the flexure hinge can realize precise loading and unloading of the indenter. During the indentation test, load and displacement are measured by the load sensor and the displacement sensor respectively. The measured load is the real penetration load but the measured displacement is not the real penetration depth considering instrument compliance and thermal drift. The base has a tilt angle of 20 degrees, providing a good observation angle.

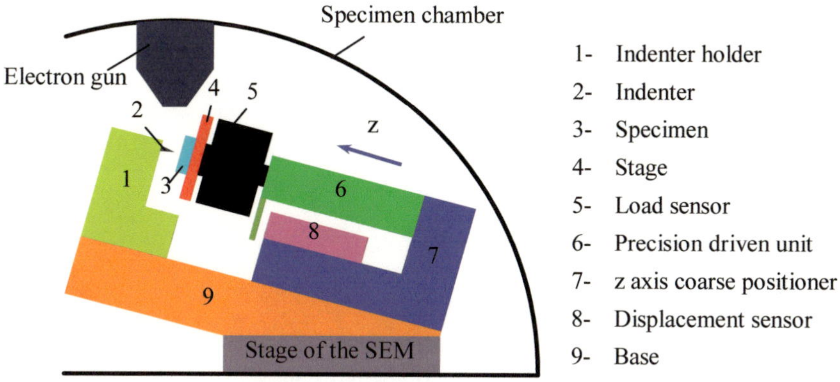

Figure 1. The schematic diagram of the proposed device for in situ indentation tests inside the SEM

3. Description of the developed device

The developed in situ indentation device is shown in Fig. 2. The output shaft of the stepper motor is connected to the knob of the z axis coarse positioner by a shaft coupling. But the installation is different from the usual. The shaft coupling is fixed with the knob but it can make relative slip with the output shaft of the stepper motor. The torque coming from the stepper motor drives the knob of the z axis coarse positioner to rotate. Just like the principle of the spiral micrometer or the ball screws, the z axis coarse positioner can realize linear motion. Though clearance exists, there is no effect on the indentation test because the reverse movement is just a process that the specimen is away from the indenter after the indentation process. The piezoelectric actuator and the flexure hinge are used to drive the specimen toward the indenter and realize precise loading and unloading process. The load sensor and the displacement sensor are used to measure the load and displacement during the indentation test respectively.

3.1. The precision driven unit

After the specimen is positioned at a suitable location by the z axis coarse positioner, the precision driven unit begins to finish the indentation process. So, the precision driven unit is very important for the total device. It determines many performances of the device, such as position accuracy, the penetration displacement resolution, the maximum penetration load and depth, dynamic performances and so on. Most of commercial indentation instruments use electromagnetic and electrostatic drivers to realize precision motion of indenters. Due to complex structures and control of electromagnetic and electrostatic drivers, sizes of these indentation instruments are usually large. In addition, electromagnetic interference coming from these two kinds of drivers will affect the image of the SEM. So, they are not suitable for the design of in situ indentation devices.

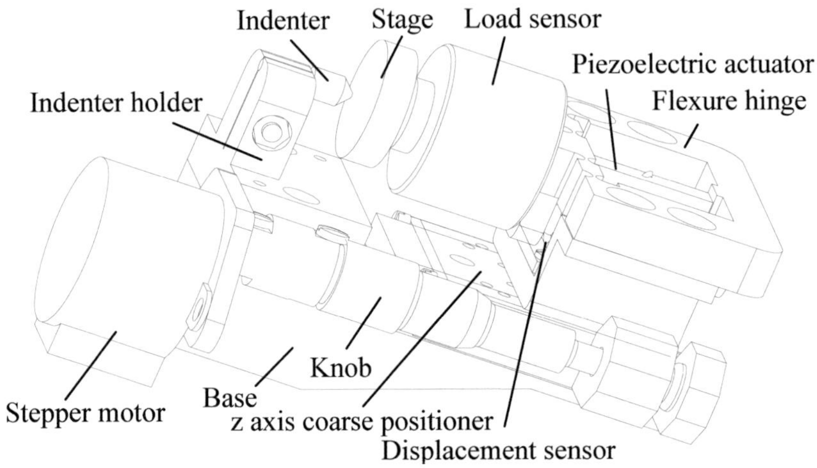

Figure 2. Model of the developed in situ indentation device

Here, combining the piezoelectric actuator with the flexure hinge, a compact precision driven unit was designed as shown in Fig. 3. The piezoelectric actuator takes advantages of small size, unlimited resolution, large force generation, fast response, low power consumption and no wear (Huang et al., 2011; Huang et al., 2012). Also the flexure hinge can overcome shortcomings such as friction, lubrication and backlash which usually exist in the conventional mechanisms with sliding and rolling bearings (Kang et al., 2005). So, purpose of precision driving and miniaturization can be achieved easily by the combination of the piezoelectric actuator and the flexure hinge.

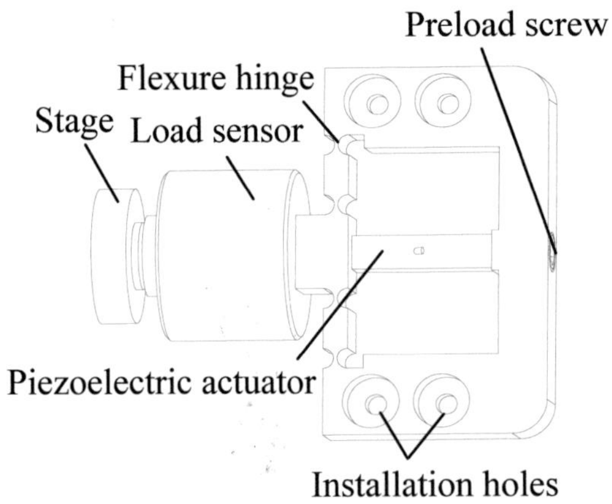

Figure 3. The precision driven unit

As shown in Fig. 3, the precision driven unit has a very simple structure which is helpful for the miniaturization of the device and can reduce the assembling. Also the simple structure is beneficial to the cleaning and it can reduce the import of pollution into the chamber of the SEM. The preload screw provides a preload force for the piezoelectric actuator to ensure suitable stiffness and displacement of the precision driven unit. The circular hinge was designed to reduce stress concentration of the flexure hinge. In order to ensure that the hinge has enough strength, structural static analysis of the flexure hinge was carried out by the software ANSYS 10.0. Material of spring steel 65Mn was selected to fabricate the hinge and its parameters are as follows. The Young's modulus is 206 GPa. The Poisson ratio is 0.288. As shown in Fig. 4(a), a displacement load of 10 μm was applied on the surface where the piezoelectric actuator was located. All degrees of freedom of the four installation holes were constrained. The analysis result is shown in Fig. 4(b). The maximum Von Mises stress is about 47.808 MPa which is less than the yield strength of 65Mn being 432 MPa. From the result, conclusion can be deduced that the flexure hinge is safety during the work process. In order to analyze the stress distribution of the hinge in depth, stress values of the hinge from point A to point B are extracted and drawn in Fig. 5. It is obvious that larger stress appears near the circular hinges. But the stress value is accepted.

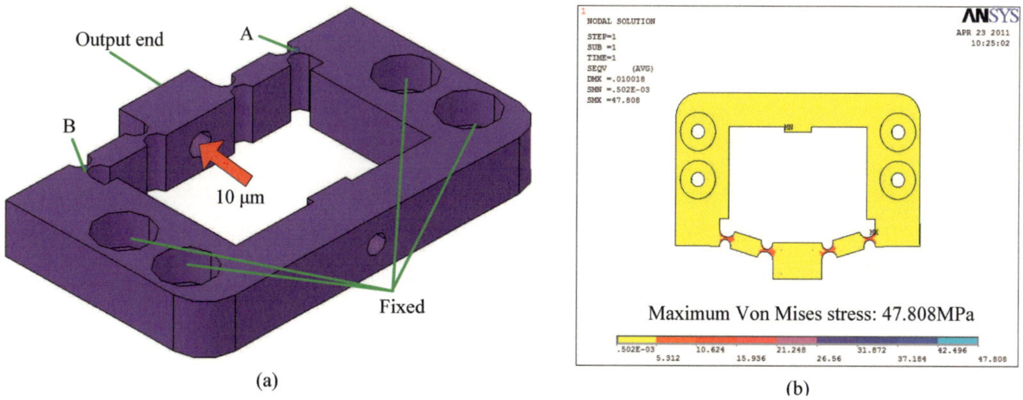

Figure 4. Model and structural static analysis result of the flexure hinge. (a) shows the loading and boundary conditions in detail; (b) gives the maximum Von Mises stress under the analysis conditions

3.2. The indenter holder

During the indentation test, the indenter should penetrate into and withdraw from the surface of the specimen vertically, which should be guaranteed during the design. So, design of the indenter holder is important and an ingenious structure should be developed. In this paper, a novel indenter holder is proposed as shown in Fig. 6. Fig. 6 (a) is the assembly drawing of the indenter unit. Fig. 6 (b) is a local view of the indenter holder which was fabricated by wire-cutting. In the center of the indenter holder, a long and thin groove is designed. A hole located at the upper of the groove is used for installation of the indenter. The axis of the hole is parallel to the installation surface. The indenter uses the hole and the positioning surface to realize precision positioning and then it is clamped by the lock nut. The positioning holes are designed to guarantee the position of the indenter holder and the installation holes are used to realize connection with the base.

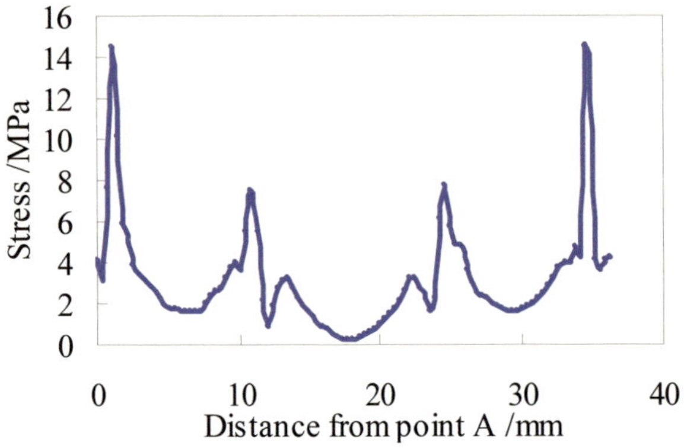

Figure 5. Stress distribution of the flexure hinge from point A to point B

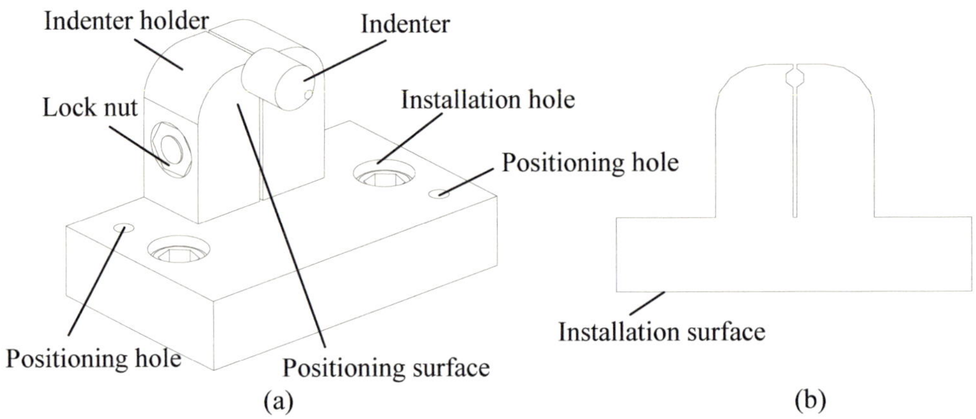

Figure 6. The designed indenter holder. (a) The assembly drawing of the indenter unit; (b) A local view of the indenter holder

4. Performance evaluation of the indentation device

The prototype of the developed in situ indentation device with dimensions of 103 mm × 74 mm × 60 mm was fabricated and assembled as shown in Fig. 7. Output performances of the device were tested.

4.1. Output performances of the z axis coarse positioner

The z axis coarse positioner is mainly used to realize coarse adjustment of the specimen and solve the problem that output displacement of the piezoelectric actuator is very limited. So, larger range with certain accuracy is required for the z axis coarse positioner. Output performances of the z axis coarse positioner were tested with the established test system as shown in Fig. 8. The stepper motor drives the stage to output displacement which is measured by the laser displacement sensor with the resolution of 10 nm. The measured signal is collected and sent to the computer for further processing. Define the parameter n as pulses per step and it can be given different value for the stepper motor. Fig. 9 is the output displacement curves when n=2048, 100 and 10 respectively. Obvious difference is observed. When n=2048, large and continuous output displacement about 239 μm/step is obtained, which can be used for coarse adjustment of the specimen when it is far away from the indenter. When n=100 and 10, small output displacement per step, about 10 μm/step for n=100 and 0.333 μm/step for n=10, is obtained. The case that n is given a small value can be used when the specimen is near the indenter. Analyzing the experimental results, we can get another difference. That is, for different n, the average output displacement per pulse is different. Maybe this is caused by the crawl phenomenon between the guide rail and the slider of the z axis coarse positioner. This difference has no effect on the indentation process because the z axis coarse positioner is only used for coarse adjustment rather than precision loading and unloading.

Figure 7. The prototype of the developed in situ indentation device

Figure 8. The established system for output performance tests of the z axis coarse positioner

4.2. Output performances of the precision driven unit with open-loop control

Output performances of the precision driven unit with open-loop control were tested. Use the power to supply voltage signal to the piezoelectric actuator manually and the output displacement is measured by the embedded displacement sensor. The experimental results are shown in Fig. 10 and Fig. 11. Fig. 10 indicates that the manual displacement resolution is about 20 nm. Fig. 11 gives the maximum output displacement of the precision driven unit about 11.44 μm when the maximum applied voltage is 100 V. With open-loop control, the precision driven unit has certain accuracy. But nonlinearity mainly caused by hysteresis of the piezoelectric actuator is also existed as shown in Fig. 11, which will affect the indentation process. So, measures should be taken to ensure linear output of the precision driven unit.

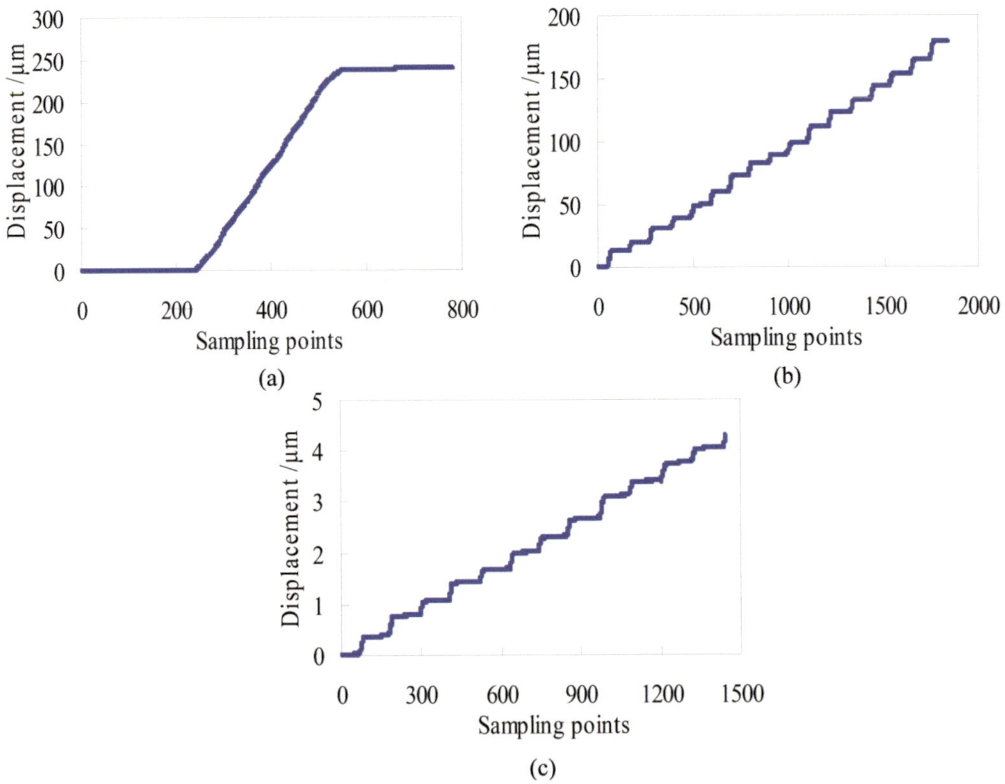

Figure 9. Output displacement curves of the z axis coarse positioner with different n. (a) n=2048; (b) n=100; (c) n=10

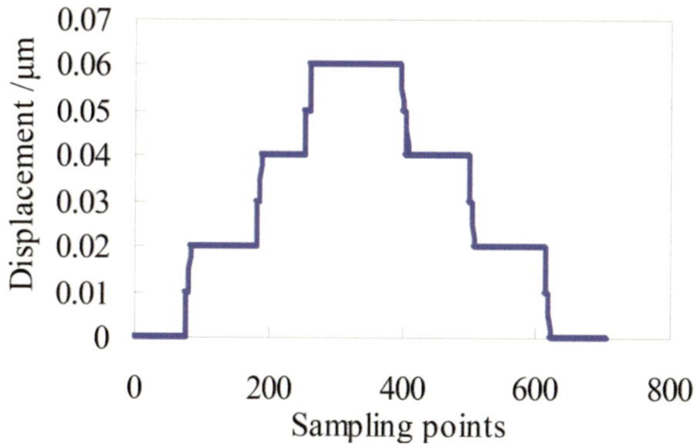

Figure 10. Resolution plot of the precision driven unit with manual operation

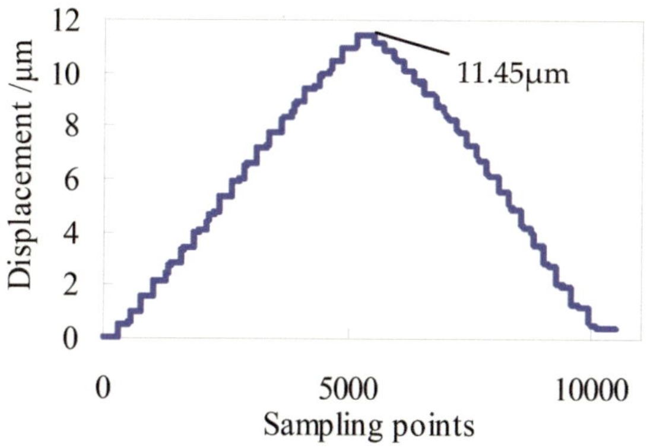

Figure 11. The output displacement curve of the precision driven unit with the maximum applied voltage of 100 V

4.3. Closed-loop control of the indentation process

In order to solve the problem of nonlinearity and ensure the testing stability, closed-loop control of the indentation process is established. Fig. 12 is the control schematic. Two control modes are developed, the displacement control mode and the load control mode. Here takes the displacement control mode as an example. Control parameters such as the maximum output displacement and the loading time, are set by the software. Control orders are sent to the power via parallel communication. The power applies voltage signal to the piezoelectric actuator and then the actuator outputs precise displacement which is measured by the embedded displacement sensor. The measured displacement signal is collected by the A/D card and sent to the PC. Comparing the measured value with the setting value, error value is obtained and then sent to the power. Repeat the process above until the measured value is equal to the setting value.

Under ambient temperature and general testing circumstance without constant temperature control and vibration isolation, three different maximum displacements and three different loads are selected to verify the feasibility of closed-loop control. Typical displacement control results and load control results are shown in Fig. 13 and Fig. 14 respectively. Via closed-loop control, output is continues and linear. Obviously, when the output displacement and load are small, fluctuations appear. In order to quantify fluctuations, a constant output displacement of 2 μm and a constant load of 20 mN were kept for 100 second respectively under the above testing circumstance. The fluctuation curves are shown in Fig. 15. Amplitude of the displacement fluctuation is less than 20 nm and amplitude of the load fluctuation is about 140 μN without constant temperature control and vibration isolation. Compared with open-loop control, closed-loop control can ensure good linearity output. Also, abundant loading and unloading modes can be developed via closed-loop control.

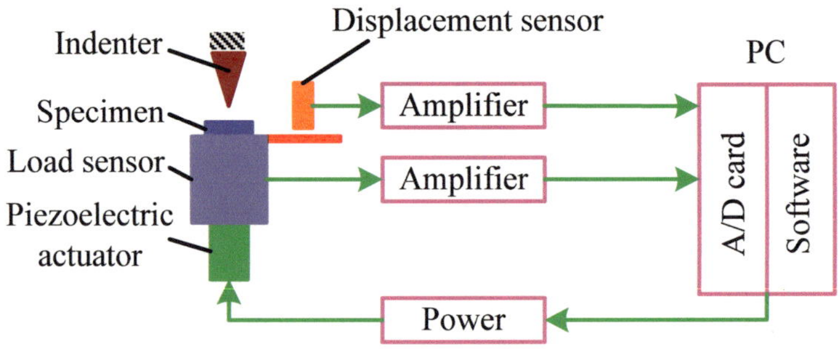

Figure 12. Closed-loop control schematic of the indentation process

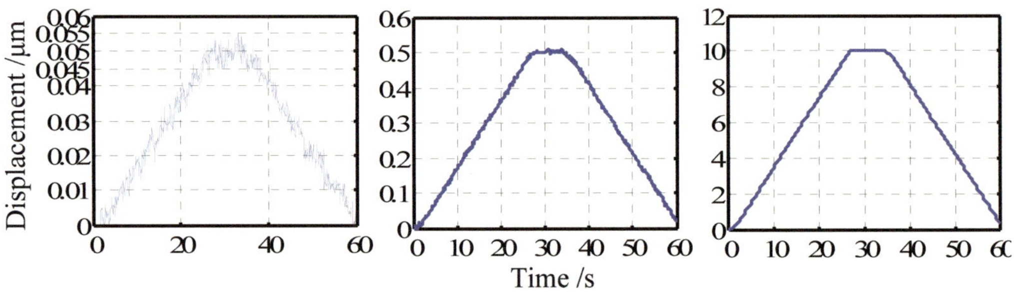

Figure 13. Typical displacement control results

4.4. Repeatability of the developed device

Comparing the load-displacement curves for different maximum loads is a good method to evaluate the repeatability of the indentation instrument. Fig. 16 shows curves between the load and the measured displacement resulting from 10 ex-situ indentation tests on fused quartz under ambient temperature and general testing circumstance without constant temperature control and vibration isolation. The loading curves agree well with each other and the unloading curves distinguish with each other because of different maximum indentation loads. The results indicate that the developed device has good repeatability, which is the premise for further calibration.

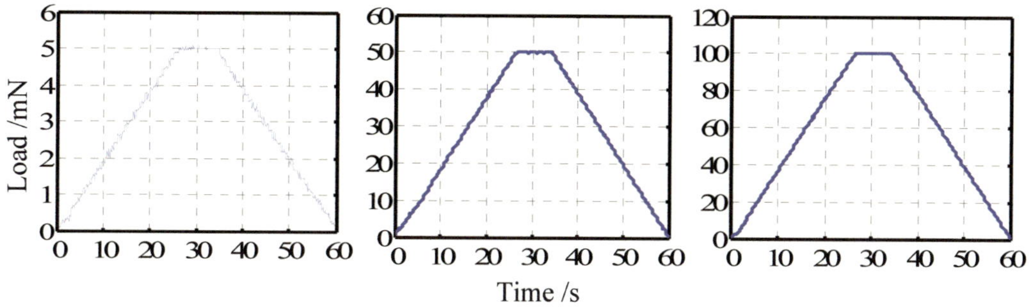

Figure 14. Typical load control results

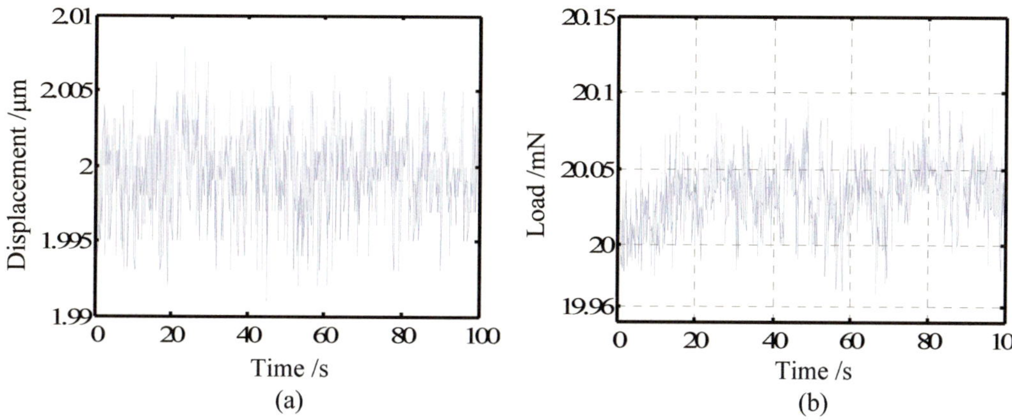

Figure 15. Output results for a constant output displacement of 2 μm (a) and a constant load of 20 mN (b)

4.5. Thermal drift

Two kinds of drift behaviors, creep within the specimen as a result of plastic flow and thermal drift due to thermal expansion or contraction of the apparatus are observed in indentation tests. Creep is time-dependent deformation under the constant load or stress and it is related with viscoelastic or viscoplastic deformation of materials. Thermal drift comes from the instrument itself and the circumstance, and it will cause the measured displacement h_t larger or smaller than the actual indentation depth h. So measures should be taken to correct the thermal drift. Assuming that the thermal drift for an indentation test is constant, the developed device allows for a hold series of data points to be accumulated at the end of the unloading from the maximum load. During the hold period at the end of the unloading, creep is less likely to occur because the load is very low. Then fitting the load and measured displacement, the thermal drift rate will be obtained. As shown in Fig. 17, thermal drift rate for the indentation test of fused quartz is about -0.5 nm/s. The corrected and uncorrected curves are shown in Fig. 18.

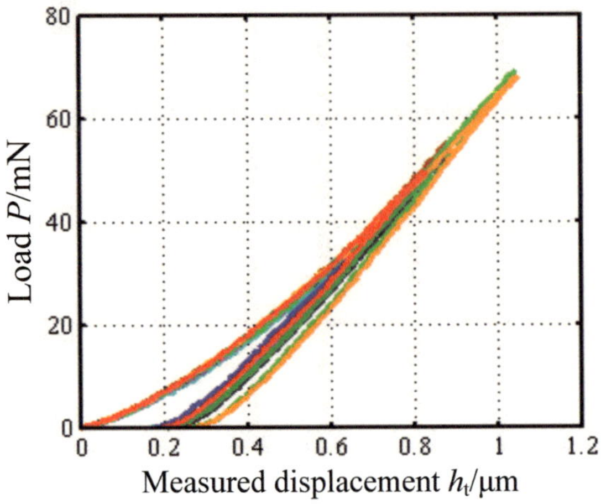

Figure 16. Curves between the load and the measured displacement resulting from 10 ex-situ indentation tests on fused quartz

4.6. Determination of instrument compliance C_m

Instrument compliance C_m is the most important factor that affects the load-depth curve, which also makes the measured displacement h_t larger than the actual indentation depth h. Especially, when the instruments are miniaturization, this effect will be even more obvious, making calibration of instrument compliance more difficult. Commercial ex-situ indentation instruments exhibit C_m in the range of 0.1 nm/mN to 1 nm/mN. But in situ indentation devices developed by previous researchers obviously have larger compliance because of their miniaturization structures. Instrument compliance of Rabe's device is about 5 nm/mN while instrument compliance of the PicoIndenter is not determined in literature (Rzepiejewska-Malyska et al., 2008; Ghisleni et al., 2009). Although in situ indentation pays more attention to visually and dynamically observe the deformation and damage process of materials (Rzepiejewska-Malyska et al., 2008) and properties of materials can be obtained by the ex-situ indentation tests, accurate and quantitative load-depth curves are also required which are helpful to make the connection between the surface deformation and the applied indentation load.

Considering instrument compliance C_m, relationship between the actual indentation depth h and the measured displacement h_t can be given as (Oliver & Pharr, 1992)

$$h = h_t - C_m P \tag{1}$$

where P is the indentation load.

There are a few methods proposed by previous researchers to calibrate instrument compliance (Doerner & Nix, 1986; Van Vliet et al., 2004; Nurot & Sun, 2005), and the most representative method is the iterative procedure developed by Oliver and Pharr (Oliver & Pharr, 2004) based on the assumption that Young's modulus of the sample does not vary with indentation depth, but their method is mathematically and time intensive (Costa et al., 2004). Here, like the literature (Huang et al., 2011), a reference material Indium-Tin Oxide (ITO) and a commercial indentation instrument-CSM's Nanoindentation Tester are used to calibrate instrument compliance of the developed device.

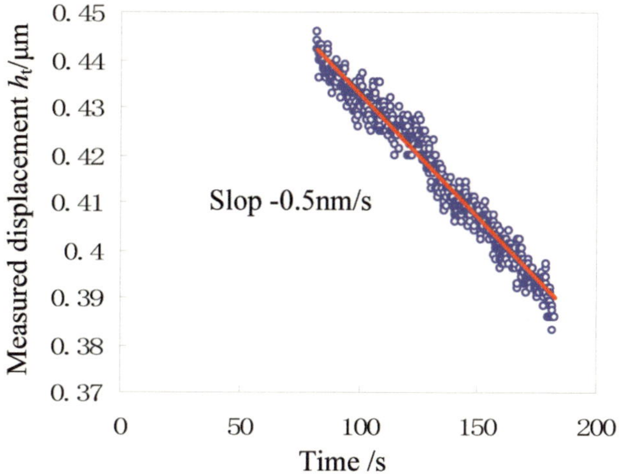

Figure 17. Relationship between the measured displacement h_t and time for a constant load

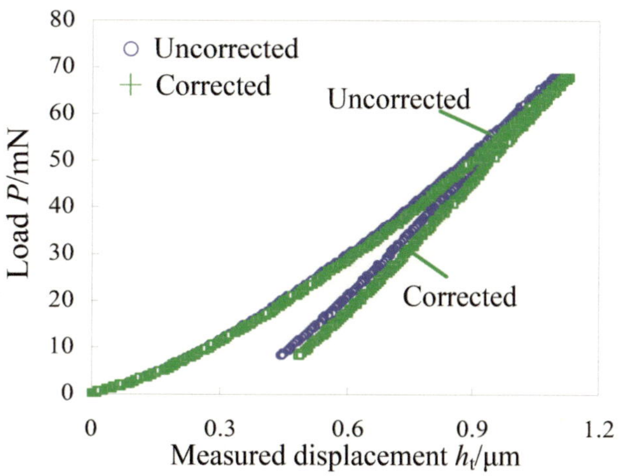

Figure 18. Comparison of the uncorrected curve and the corrected curve considering of the thermal drift

Indentation tests of ITO with maximum indentation loads of 30 mN, 50 mN, 100 mN, 150 mN, 250 mN and 500 mN were carried out with the Berkovich indenter using the CSM's Nanoindentation Tester and the experimental curves are used as standard load-depth curves. Then, with the same experimental conditions, indentation tests were carried out using the developed in situ SEM indentation device. By comparing curves obtained from the commercial device and the designed device, six values of instrument compliance can be obtained respectively, and the average value of 4.5 nm/mN is obtained as instrument compliance C_m, which is also larger than values of commercial ex-situ indentation instruments. By equation (1), the corrected curve with the maximum indentation load of 500 mN can be obtained as shown in Fig. 19. The slope of the corrected curve is obviously larger than that of the raw curve while the unloading point remains the same because the load is zero. In order to verify the feasibility of the calibration result, the corrected curve and the curve obtained from the commercial device are drawn in the same figure shown in Fig. 20, from which we can see that these two curves coincide with each other well. According to the Oliver and Pharr method (Oliver & Pharr, 1992), the contact depth between the indenter and the sample is about 1.709 μm and hardness of ITO is about 6.97 GPa. Indentation tests of single crystal copper and bulk metallic glass in the following section will verify generality of instrument compliance.

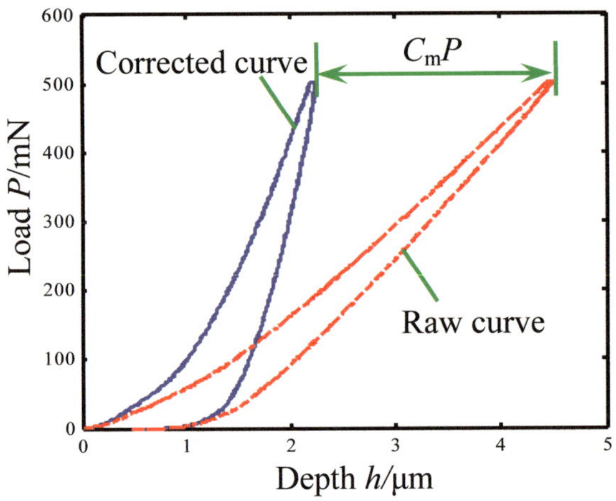

Figure 19. The corrected curve and the raw curve

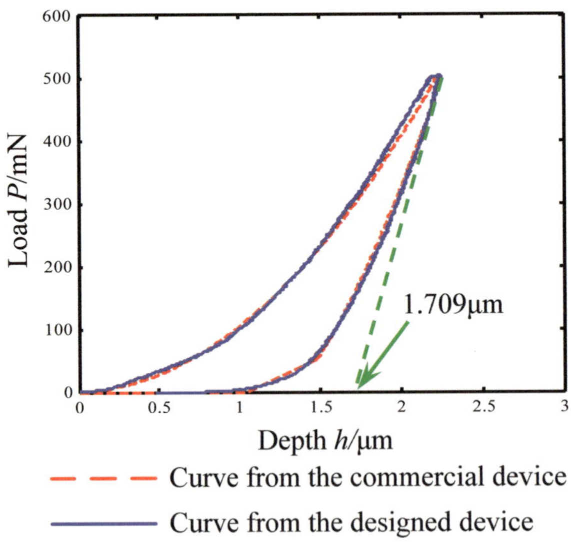

Figure 20. Comparison between curves from the commercial device and the designed device

4.7. Parameter comparison with the existing product or device

Up to now, work related with in situ indentation is concentrated on Rabe's device (Rabe et al., 2004), the PI 85 SEM PicoIndenter (Hysitron Incorporated) and the self-made device (Huang et al., 2012). Here, main parameters of these three devices are summarized in Table 1. Comparing these parameters, the self-made in situ indentation device has the advantage in high price-performance ratio.

	Self-made device	Rabe's device	PI 85 SEM PicoIndenter
Maximum load	2.5 N	500 mN	10 mN
Maximum displacement	15 μm	20 μm	5 μm
Load noise floor	140 μN	Unknown	400 nN
Displacement noise floor	20 nm	Unknown	1 nm
Load resolution	50 μN	100 μN	3 nN
Displacement resolution	5 nm	50 nm	0.02 nm
Compliance	4.5 nm/mN	5 nm/mN	Unknown
x positioning range	12 μm (Addible)	10 mm	5 mm
x positioning resolution	10 nm	150 nm	50 nm
y positioning range	12 μm (Addible)	20 μm	5 mm
y positioning resolution	10 nm	7 nm	50 nm
Price (include tax)	$100, 000.00	$100, 000.00	$ 250, 000.00

Table 1. Summarized parameters of three kinds of main in situ indentation devices

5. Indentation tests

5.1. Ex-situ indentation tests

The developed indentation device has the miniaturized structure but still it has the function to carry out ex-situ indentation tests. Fig. 21 is the load-depth curve of single crystal copper via the ex-situ indentation test under ambient temperature and general testing circumstance without constant temperature control and vibration isolation. Fig. 22 is the fitting curve of the unloading portion.

According to the literature (Oliver & Pharr, 1992), the unloading curves are distinctly curved and usually well approximated by the power law relation:

$$P = \alpha(h - h_{\mathrm{f}})^{m} \tag{2}$$

where α and m are power law fitting constants, and h_f is the depth of the residual impression. By fitting portion of the unloading data with the least squares fitting procedure, the constants α and m can be obtained. The fitting results are given in Fig. 22 and the fitting equation is expressed as

$$P = 804.709(h - 0.757)^{1.2450} \tag{3}$$

According to the Oliver and Pharr method (Oliver & Pharr, 1992), hardness of single crystal copper is 0.95 GPa and the Young's modulus is 107.42 GPa.

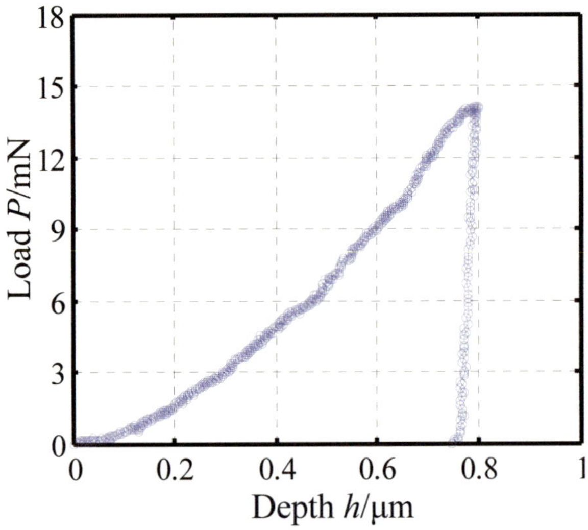

Figure 21. The load-depth curve of single crystal copper

5.2. In situ indentation test inside the SEM

Due to unique mechanical properties such as high strength, modulus, hardness and elastic limit (Inoue et al., 2004; Scully et al., 2007), bulk metallic glasses (BMGs) are considered as emerging structural materials since their discovery in 1960 (Klement et al., 1960). Mechanical properties and deformation mechanism of BMGs are given more and more attention in recent years. Because of its high spatial and temporal resolution, indentation remains an important tool to study mechanical properties of BMGs (Schuh, 2006). Unique phenomena such as the serrated flow and shear band formation underneath indenters have been reported during the ex-situ indentation tests (Wang et al., 2000; Schuh & Nieh, 2003; Li et al., 2009). However, most of previous researchers can only study shear band formation of BMGs after the indentation tests, limiting further research on the deformation mechanism. Here, in situ indentation of a Zr-based bulk metallic glass using the Berkovich indenter was carried out via the developed device.

The device was installed on the stage of the SEM—Quanta 250. The experimental results are shown in Fig. 23. The load-depth curve is shown in Fig. 23 (a). Fig. 23 (b) is the status that there is a distance between the indenter and the specimen. Fig. 23 (c) is the status that the indenter is penetrating into the specimen. Fig. 23 (d) is the indentation morphology. From Fig. 23 (c) and (d), pile-up and shear bands are observed obviously. This is only an example of in situ indentation tests inside the SEM. Via in situ observation of a whole indentation process, more direct inspection of deformation and damage of materials can be realized, bringing the possibility of further studying and revealing deformation and damage mechanism of materials.

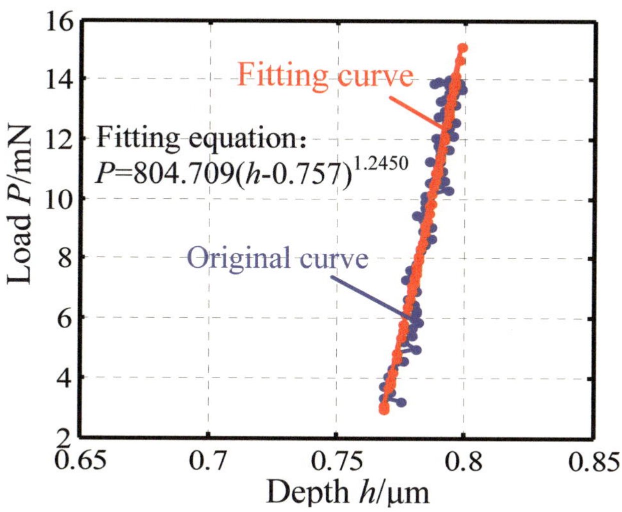

Figure 22. The fitting results of portion of the unloading data

Of course, in situ indentation techniques can be used for other research. Taking advantage of high positioning accuracy, it can be used to study effect of initial contact and surface roughness on experimental results. Using a special indenter such as the flat-ended indenter, in situ uniaxial compression of micro- and nano- structures for example carbon nanotubes and micropillars can be carried out (Buzzi et al., 2009; Niederberger et al., 2010). Also, using a suitable clamping way, in situ tensile tests are feasible (Lucca et al., 2010). Adding heating or precise electrical measuring equipments, the device can be used to study multi-physical field coupled performance of materials (Schuh et al., 2005; Nowak et al., 2009) which will be an exciting and significant work. So, the in situ indentation test inside the SEM is a promising method to study mechanics of materials in depth.

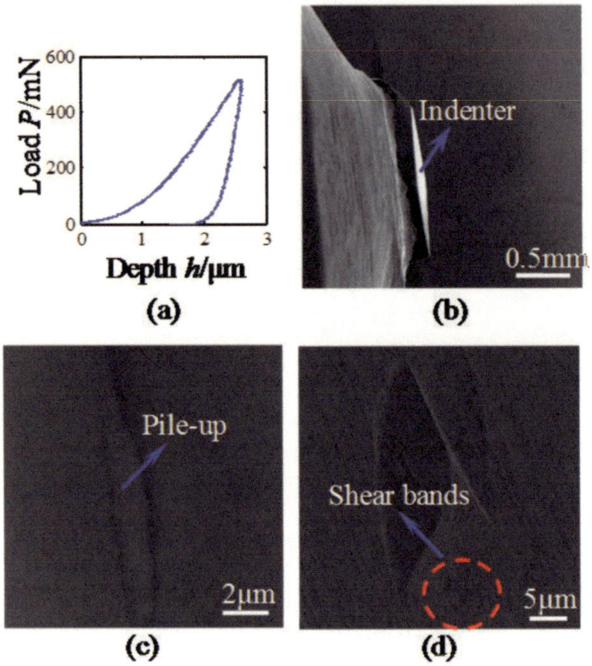

Figure 23. Experimental results of the in situ indentation test of a Zr-based bulk metallic glass. Fig. 23 (a) is the load-depth curve. Fig. 23 (b) is the status that there is a distance between the indenter and the specimen. Fig. 23 (c) is the status that the indenter is penetrating into the specimen. Fig. 23 (d) is the indentation morphology

6. Conclusions

In this chapter, a novel and miniaturization in situ indentation device was presented. Integrating the stepper motor, the piezoelectric actuator and the flexure hinge, the coarse positioner and the precision driven unit were designed respectively, which can be used to realize coarse adjustment of the specimen and precision loading and unloading process of the indenter. A novel indenter holder was designed to ensure vertical penetration of the indenter. Closed-loop control of the indentation process was established to solve the problem of nonlinearity of the piezoelectric actuator and to enrich the control modes.

Performances of the developed device were evaluated, and ex-indentation and in situ indentation tests of materials were carried out to verify the feasibility of the device. The device has the indentation load range of about 2.5 N with the load resolution of 50 μN and indentation depth range of about 15 μm with the displacement resolution of 5 nm. Compliance of the device was calibrated and it is about 4.5 nm/mN. Also thermal drift was corrected. Experimental results indicate that the developed device has good repeatability and it can be used as an ex-situ indentation device to characterize properties of materials. What's more, it can be installed on the stage of the SEM to realize in situ indentation tests of materials, which is a promising method to study deformation and damage mechanism of materials in depth. Compared with other two in situ indentation device or product, the developed in situ indentation device has the advantage in high price-performance ratio. By in situ indentation tests and other extended applications, it is hoped that the developed device will be a significant tool for advanced materials research.

Author details

Hongwei Zhao, Hu Huang, Zunqiang Fan, Zhaojun Yang and Zhichao Ma
Jilin University, China

Acknowledgement

This research is funded by the National Natural Science Foundation of China (Grant No. 50905073, 51105163), National Hi-tech Research and Development Program of China(863 Program) (863 Program) (Grant No. 2012AA041206), Key Projects of Science and Technology Development Plan of Jilin Province (Grant No. 20110307) and International Science and Technology Cooperation Program of China (Grant No. 2010DFA72000). We also gratefully acknowledge Institute of Microstructure and Properties of Advanced Materials, Beijing University of Technology for the support of the SEM.

7. References

Bangert, H., Wagendristel, A. & Aschinger, H. (1982). Ultramicrohardness measurement on thin films, fibers and finely structured surfaces. *Vacuum. Tech.*, Vol.31, pp. 200–203

Bangert, H. & Wagendristel, A. (1985). Ultralow-load hardness tester for use in a scanning electron microscope. *Rev. Sci. Instrum.*, Vol.56, pp. 1568-1572

Buzzi, S., Dietiker, M., Kunze, K., Spolenak, R. & Löffler, J.F. (2009) Deformation behavior of silver submicrometer-pillars prepared by nanoimprinting. *Philos. Mag.*, Vol.89, pp. 869-884

Costa, A.L.M., Shuman, D.J., Machado, R.R. & Andrade, M.S. (2004). Determination of the Compliance of an Instrumented Indentation Testing Machine, In: *Proc. of HARDMEKO*, Washington D.C., USA

De Hosson, J.T.M., Soer, W.A., Minor, A.M., Shan, Z., Stach, E.A., Asif, S.A. & Warren, O.L. (2006). In situ TEM nanoindentation and dislocation-grain boundary interactions: a tribute to David Brandon. *J. Mech. Sci.*, Vol.41, pp. 7704-7719

Doerner, M.F. & Nix, W.D. (1986). A method for interpreting the data from depth-sensing indentation instruments. *J. Mater. Res.*, Vol.1, pp.601-609

Fischer-Cripps, A.C. (2004). *Nanoindentation*. Springer-Verlag, New York

Gane, N. & Bowden, F.P. (1968). Microdeformation of Solids. *J. Appl. Phys.*, Vol.39, pp. 1432-1435

Ghisleni, R., Rzepiejewska-Malyska, K., Philippe, L., Schwaller, P. & Michler, J. (2009). In situ SEM indentation experiments: Instruments, methodology, and applications. *Microsc. Res. Techniq.*, Vol.72, pp. 242-249

Huang, W.M., Su, J.F., Hong, M.H. & Yang, B. (2005). Pile-up and sink-in in micro-indentation of a NiTi shape-memory alloy. *Scripta Mater.*, Vol.53, pp. 1055–1057

Huang, H., Zhao, H., Yang, J., Wan, S., Mi, J., Shi, C., Yuan, Y., Ma, Z. & Yang, Z. (2011). Design and analysis of a miniaturization nanoindentation and scratch device. *Advanced Materials Research*, Vol.314-316, pp. 1792-1795

Huang, H., Zhao, H., Mi, J., Yang, J., Wan, S., Yang, Z., Yan, J., Ma, Z. & Geng, C. (2011). Experimental research on a modular miniaturization nanoindentation device. *Rev. Sci. Instrum.*, Vol.82, pp. 095101 (1-6)

Huang, H., Zhao, H., Ma, Z., Hu, L., Yang, J., Shi, G., Ni, C., Pei, Z. (2012) Design and analysis of the precision-driven unit for nano-indentation and scratch test. *J. Manuf. Syst.*, Vol.31, pp. 76–81

Huang, H., Zhao, H., Mi, J., Yang, J., Wan, S., Xu, L. & Ma, Z. (2012). A novel and compact nanoindentation device for in situ nanoindentation tests inside the scanning electron microscope. *AIP Advances*, Vol.2, 012104 (1-10)

Hysitron Incorporated. Cited 2011; Available from: http://www.hysitron.com/

Inoue, A., Shen, B.L., Koshiba, H., Kato, H. & Yavari, A.R. (2004). Ultra-high strength above 5000 MPa and soft magnetic properties of Co–Fe–Ta–B bulk glassy alloys. *Acta Mater.*, Vol.52, pp. 1631–1637

Kang, D., Kim, K., Choi, Y., Gweon, D., Lee, S. & Lee, M. (2005). Design and control of flexure based XYθz stage. *J. Mech. Sci. Technol.*, Vol.19, pp. 2157-2164

Keryvin, V., Vu, X.D., Hoang, V.H. & Shen, J. (2010). On the deformation morphology of bulk metallic glasses underneath a Vickers indentation. *J. Alloys Compd.*, Vol.504S, pp. S41-S44

Klement, W., Wilens, R.H. & Duwez, P. (1960). Non-crystalline Structure in Solidified Gold–Silicon Alloys. *Nature*, Vol.187, pp. 869-870

Li, N., Liu, L., Chan, K.C., Chen, Q. & Pan, J. (2009). Deformation behavior and indentation size effect of $Au_{49}Ag_{5.5}Pd_{2.3}Cu_{26.9}Si_{16.3}$ bulk metallic glass at elevated temperature. *Intermetallics*, Vol.17, pp. 227-230

Lucca, D.A., Herrmann, K. & Klopfstein, M.J. (2010). Nanoindentation: Measuring methods and applications. *CIRP Ann.-Manuf. Techn.*, Vol.59, pp. 803-819

Minor, A.M., Morris, J.W. & Stach, E.A. (2001). Quantitative in situ nanoindentation in an electron microscope. *Appl. Phys. Lett.*, Vol.79, pp. 1625-1627

Minor, A.M., Asif, S.A., Shan, Z., Stach, E.A., Cyrankowski, E., Wyrobek, T.J. & Warren, O.L. (2006). A new view of the onset of plasticity during the nanoindentation of aluminum. *Nat. Mater.*, Vol.5, pp. 697-702

Motoki, T., Gao, W., Kiyono, S. & Ono, T. (2006). A nanoindentation instrument for mechanical property measurement of 3D micro/nano-structured surfaces. *Meas. Sci. Technol.* Vol.17, pp. 495-499

Niederberger, C., Mook, W.M., Maeder, X. & Michler, J. (2010). In situ electron backscatter diffraction (EBSD) during the compression of micropillars. *Mater. Sci. Eng. A*, Vol.527, pp. 4306–4311

Nowak, R., Chrobak, D., Nagao, S., Vodnick, D., Berg, M., Tukiainen, A. & Pessa, M. (2009). An electric current spike linked to nanoscale plasticity. *Nat. Nanotechnol.*, Vol.4, pp. 287-291

Nowak, J.D., Rzepiejewska-Malyska, K., Major, R.C., Warren, O.L. & Michler, J. (2009) In-situ nanoindentation in the SEM. *Mater. Today*, Vol.12, pp. 44-45

Nurot, P. & Sun, Y. (2005). Improved method to determine the hardness and elastic moduli using nano-indentation. *KMITL Sci. J.*, Vol.5, pp. 483-492

Oliver, W.C. & Pharr, G.M. (2004). Measurement of hardness and elastic modulus by instrumented indentation: Advances in understanding and refinements to methodology. *J. Mater. Res.*, Vol.19, pp. 3–20

Oliver, W.C. & Pharr, G.M. (1992). An improved technique for determining hardness and elastic modulus using load and displacement sensing indentation measurements. *J. Mater. Res.*, Vol.7, pp.1564-1583

Rabe, R., Breguet, J.-M., Schwaller, P., Stauss, S., Haug, F.-J., Patscheider, J. & Michler J. (2004). Observation of fracture and plastic deformation during indentation and scratching inside the scanning electron microscope. *Thin Solid Films*, Vol.469/470, pp. 206–213

Ruffell, S., Bradby, J.E., Williams, J.S. & Munroe, P. (2007). Formation and growth of nanoindentation-induced high pressure phases in crystalline and amorphous silicon. *J. Appl. Phys.*, Vol.102, pp. 063521 (1-8)

Rzepiejewska-Malyska, K., Buerki, G., Michler, J., Major, R.C., Cyrankowski, E., Asif, S.A.S. & Warren O.L. (2008). In situ mechanical observations during nanoindentation inside a high-resolution scanning electron microscope. *J. Mater. Res.*, Vol.23, pp. 1973-1979

Rzepiejewska-Malyska, K., Parlinska-Wojtan, M., Wasmer, K., Hejduk, K. & Michler, J. (2009). In-situ SEM indentation studies of the deformation mechanisms in TiN, CrN and TiN/CrN. *Micron*, Vol.40, pp. 22-27

Schuh, C.A. & Nieh, T.G. (2003). A nanoindentation study of serrated flow in bulk metallic glasses. *Acta Mater.*, Vol.51, pp. 87-99

Schuh, C.A., Mason, J.K. & Lund, A.C. (2005). Quantitative insight into dislocation nucleation from high-temperature nanoindentation experiments. *Nat. Mater.*, Vol.4, pp. 617-621

Schuh, C.A. (2006). Nanoindentation studies of materials. *Mater. Today*, Vol.9, pp. 32-40

Scully, J.R., Gebert, A. & Payer, J.H. (2007). Corrosion and related mechanical properties of bulk metallic glasses. *J. Mater. Res.*, Vol.22, pp. 302-313

Van Vliet, K.J., Prchlik, L. & Smith, J.F. (2004). Direct measurement of indentation frame compliance. *J. Mater. Res.*, Vol.19, pp. 325-331

Wang, J.G., Choi, B.W., Nieh, T.G. & Liu, C.T. (2000). Crystallization and nanoindentation behavior of a bulk Zr–Al–Ti–Cu–Ni amorphous alloy. *J. Mater. Res.*, Vol.15, pp. 798-807

Zhou, J., Komvopoulos, K. & Minor, A.M. (2006). Nanoscale plastic deformation and fracture of polymers studied by in situ nanoindentation in a transmission electron microscope. *Appl. Phys. Lett.*, Vol.88, pp. 181908 (1-3)